D0213756

Realizability Theory for Continuous Linear Systems

Realizability Theory for Continuous Linear Systems

A. H. Zemanian

Department of Electrical Engineering
State University of New York at Stony Brook
Stony Brook, New York

DOVER PUBLICATIONS, INC.
New York

Copyright

Published in Canada by General Publishing Company, Ltd., 30 Lesmill Road, Don Mills, Toronto, Ontario.
Published in the United Kingdom by Constable and Company, Ltd., 3 The Lanchesters, 162–164 Fulham Palace Road, London W6 9ER.

Bibliographical Note

This Dover edition, first published in 1995, is an unabridged, slightly corrected republication of the work first published by Academic Press, New York, 1972, as Volume 97 in the series "Mathematics in Science and Engineering."

Library of Congress Cataloging-in-Publication Data

Zemanian, A. H. (Armen H.)
Realizability theory for continuous linear systems / A.H. Zemanian.
p. cm.
"An unabridged, slightly corrected republication of the work first published by Academic Press, New York, 1972, as volume 97 in the series 'Mathematics in science and engineering' "—T.p. verso.
Includes bibliographical references (p. –) and index.
ISBN 0-486-68823-2 (pbk.)
1. System analysis. 2. Linear systems. 3. Operator theory. 4. Functional analysis. I. Title.
QA402.Z45 1995
003′.74—dc20 95-22735
CIP

Manufactured in the United States of America
Dover Publications, Inc., 31 East 2nd Street, Mineola, N.Y. 11501

To the Memory of My Parents

PARSEGH AND FILOR ZEMANIAN

6.3. Analyticity and the Exchange Formula 119
6.4. Inversion and Uniqueness 120
6.5. A Causality Criterion 121

CHAPTER 7. **The Scattering Formulism**

7.1. Introduction 124
7.2. Preliminary Considerations Concerning L_p-Type Distributions 125
7.3. Scatter-Passivity 128
7.4. Bounded* Scattering Transforms 131
7.5. The Realizability of Bounded* Scattering Transforms 134
7.6. Bounded*-Real Scattering Transforms 137
7.7. Lossless Hilbert Ports 139
7.8. The Lossless Hilbert n-Port 143

CHAPTER 8. **The Admittance Formulism**

8.1. Introduction 149
8.2. Passivity 150
8.3. Linearity and Semipassivity Imply Continuity 153
8.4. The Fourier Transformation on $\mathscr{S}(H)$ 155
8.5. Local Mappings 157
8.6. Positive Sesquilinear Forms on $\mathscr{D} \times \mathscr{D}$ 160
8.7. Positive Sesquilinear Forms on $\mathscr{D}(H) \times \mathscr{D}(H)$ 162
8.8. Certain Semipassive Mappings of $\mathscr{D}(H)$ into $\mathscr{E}(H)$ 168
8.9. An Extension of the Bochner–Schwartz Theorem 174
8.10. Representations for Certain Causal Semipassive Mappings 175
8.11. A Representation for Positive* Transforms 178
8.12. Positive* Admittance Transforms 183
8.13. Positive* Real Admittance Transforms 186
8.14. A Connection between Passivity and Semipassivity 187
8.15. A Connection between the Admittance and Scattering Formulisms 189
8.16. The Admittance Transform of a Lossless Hilbert Port 192

APPENDIX A. **Linear Spaces** 194

APPENDIX B. **Topological Spaces** 198

APPENDIX C. **Topological Linear Spaces** 201

APPENDIX D. **Continuous Linear Mappings** 206

APPENDIX E. **Inductive-Limit Spaces** 211

APPENDIX F. **Bilinear Mappings and Tensor Products** 213

APPENDIX G. **The Bochner Integral** 216

References 222

Index of Symbols 225
Index 227

Preface

"Realizability theory" is a part of mathematical systems theory and is concerned with the following ideas. Any physical system defines a relation between the stimuli imposed on the system and the corresponding responses. Moreover, any such system is always causal and may possess other properties such as time-invariance and passivity. Two questions: How are the physical properties of the system reflected in various analytic descriptions of the relation? Conversely, given an analytic description of a relation, does there exist a corresponding physical system possessing certain specified properties? If the latter is true, the analytic description is said to be realizable. Considerations of this sort arise in a number of physical sciences. For example, see McMillan (1952), Newcomb (1966), or Wohlers (1969) for electrical networks, Toll (1956) or Wu (1954) for scattering phenomena, and Gross (1953), Love (1956), or Meixner (1954) for viscoelasticity.

This book is an exposition of realizability theory as applied to the operators generated by physical systems as mappings of stimuli into responses. This constitutes the so-called "black box" approach since we do not concern ourselves with the internal structure of the system at hand. Physical characteristics such as linearity, causality, time-invariance, and passivity are defined as mathematical restrictions on a given operator. Then, the two questions are answered by obtaining a description of the operator in the form of a kernel or convolution representation and establishing a variety of necessary and sufficient conditions for that representation to possess the indicated properties. Thus, the present work is an abstraction of classical realizability theory in the following way. A given representation is realized not by a physical system but rather by an operator possessing mathematically defined properties, such as causality and passivity, which have physical significance. We may state this in another way. Our primary concern is the study of physical properties and their mathematical characterizations and not the design of particular systems.

Two properties we shall always impose on any operator under consideration are linearity and continuity. They are quite commonly (but by no means always) possessed by physical systems. Of course, continuity only has a meaning with respect to the topologies of the domain and range spaces of the operator. We can in general take into account a wider class of continuous linear operators by choosing a smaller domain space with a stronger topology and a larger range space with a weaker topology. With this as our motivation, we choose the basic testing-function space of distribution theory as the domain for our operators and the space of distributions as the range space. The imposition of other properties upon the operator will in general allow us to extend the operator onto wider domains in a continuous fashion. For example, time-invariance implies that the operator has a convolution representation and can therefore be extended onto the space of all distributions with compact supports. This distributional setting also provides the following facility. It allows us to obtain certain results, such as Schwartz's kernel theorem, which simply do not hold under any formulation that permits the use of only ordinary functions. Thus, distribution theory provides a natural language for the realizability theory of continuous linear systems.

Still another facet of this book should be mentioned. Almost all the realizability theories for electrical systems deal with signals that take their instantaneous values in n-dimensional Euclidean space. However, there are many systems whose signals have instantaneous values in a Hilbert or Banach space. Section 4.2 gives an example of this. For this reason, we assume that the domain and range spaces for the operator at hand consist of Banach-space-valued distributions. Many of the results of earlier realizability theories readily carry over to this more general setting, other results go over but with difficulty, and some do not generalize at all. Moreover, the theory of Banach-space-valued distributions is somewhat more complicated than that of scalar distributions; Chapter 3 presents an exposition of it. Still other analytical tools we shall need as a consequence of our use of Banach-space-valued distributions are the elementary calculus of functions taking their values in locally convex spaces, which is given in Chapter 1, and Hackenbroch's theory for the integration of Banach-space-valued functions with respect to operator-valued measures, a subject we discuss in Chapter 2.

The systems theory in this book occurs in Chapters 4, 5, 7, and 8. Chapter 4 is a development of Schwartz's kernel theorem in the present context and ends with a kernel representation for our continuous linear operators. Causality appears as a support condition on the kernel. How time-invariance converts a kernel operator into a convolution operator is indicated in Chapter 5. We digress in Chapter 6 to develop those properties of the Laplace transformation that will be needed in our subsequent frequency-domain discussions. Passivity is a very strong assumption; it is from this that we get the richest realizability theory. Chapter 7 imposes a passivity condition that is appropriate for scattering phenomena, whereas a passivity condition that is suitable for an admittance formulism is exploited in Chapter 8.

It is assumed that the reader is familiar with the material found in the customary undergraduate courses on advanced calculus, Lebesgue integration, and functions of a complex variable. Furthermore, a variety of standard results concerning topological linear spaces and the Bochner integral will be used. In order to make this book accessible to readers who may be unfamiliar with either of these topics, a survey of them is given in the appendixes. Although no proofs are presented, enough definitions and discussions are given to make what is presented there understandable, it is hoped, to someone with no knowledge of either subject. Almost every result concerning the aforementioned two topics that is used in this book can be found in the appendixes, and a reference to the particular appendix where it occurs is usually given. For the few remaining results of this nature that are employed, we provide references to the literature.

The problems usually ask the reader either to supply the proofs of certain assertions that were made but not proved in the text or to extend the theory in various ways. On occasion, we employ a result that was stated only in a previous problem. For this reason, it is advisable for the reader to pay some attention to the problems.

All theorems, corollaries, lemmas, examples, and figures are triple-numbered; the first two numbers coincide with the corresponding section numbers. On the other hand, equations are single-numbered starting with (1) in each section.

Acknowledgments

This book was conceived while I was a Research Fellow at the Mathematical Institute of the University of Edinburgh during the 1968–1969 academic year. My tenure in that post was supported by a grant to Professor A. Erdelyi from the Science Research Council of Great Britain. It is a pleasure to express my gratitude for that support. The subsequent development of this book was assisted by Grants GP-7577, GP-18060, and GP-27958 from the Applied Mathematics Division of the National Science Foundation under the administration of Dr. B. R. Agins. I also wish to express my gratitude to R. K. Bose, V. Dolezal, and W. Hackenbroch for various suggestions that have been incorporated into the text. Finally and once again, to my wife, thanks.

Chapter 1

Vector-Valued Functions

1.1. INTRODUCTION

The purpose of this first chapter is to present a number of results concerning the calculus of functions that map n-dimensional real Euclidean space into a locally convex space. In addition, the theory of analytic functions from the complex plane into a Banach space is also introduced. The discussion is quite similar to the elementary calculus and the theory of complex-valued analytic functions. Complications arise at various points, however, because of the more involved structure of locally convex and Banach spaces. Nevertheless, there are no surprises. The primary results of Riemann integration, differentiation, and functions of a complex variable carry over to the present context. The reader may, if he wishes, go directly to Chapter 2 after a perusal of the next section and refer back to the present chapter as the occasion demands.

As was mentioned in the Preface, certain facts concerning topological linear spaces will be used. A compilation of them can be found in the appendixes of this book.

1.2. NOTATIONS AND TERMINOLOGY

The present section is devoted to a description of some of the notations and terminology that are used throughout this book; almost all of them adhere to customary usage. The Index of Symbols at the end of the book lists our frequently used symbols and indicates the pages on which they are defined.

Let Φ be a set. The notation $\{\phi \in \Phi: P(\phi)\}$, or simply $\{\phi: P(\phi)\}$ if Φ is understood, denotes the set of all $\phi \in \Phi$ for which the proposition $P(\phi)$ concerning ϕ is true. $\{\phi_i\}_{i \in I}$ is a collection of indexed elements where the index i traverses the set I. A sequence is denoted by $\{\phi_k\}_{k=1}^{\infty}$ or $\{\phi_k: k = 1, 2, \ldots\}$, a finite collection by $\{\phi_k\}_{k=1}^{n}$ or $\{\phi_1, \ldots, \phi_n\}$, and a set containing a single element ϕ by $\{\phi\}$. We also use the abbreviated notation $\{\phi_k\}$ when it is clear what type of set we are dealing with.

Let Ω and Λ be subsets of Φ. The notation $\phi, \psi, \theta, \ldots \in \Omega$ means that all the elements $\phi, \psi, \theta, \ldots$ are members of Ω; $\phi, \psi, \theta, \ldots \notin \Omega$ is the corresponding negation. If Φ is a complex linear space (see Appendix A), $\Omega + \Lambda$ (or $\Omega - \Lambda$) denotes the set of elements of the form $\psi + \theta$ (or respectively $\psi - \theta$) where $\psi \in \Omega$ and $\theta \in \Lambda$. Similarly, if ϕ is a fixed member of Φ and α is a complex number, $\Omega + \phi$ (or $\alpha\Omega$) is the set of all elements of the form $\psi + \phi$ (or respectively $\alpha\psi$), where $\psi \in \Omega$. On the other hand, $\Omega\backslash\Lambda$ is the set of all elements in Ω that are not in Λ. Thus, $\Phi\backslash\Lambda$ is the complement of Λ in Φ. We also have the customary union $\cup$, intersection $\cap$, and inclusion symbols $\subset$ or $\supset$. Furthermore, $\bigcup_{k=1}^{\infty} \Omega_k$ and $\bigcap_{k=1}^{\infty} \Omega_k$ denote unions and intersections, respectively, over sequences $\{\Omega_k\}$ of sets in Φ. A similar notation is used for finite collections and indexed collections of sets, for example, $\bigcup_{k=1}^{n} \Omega_k$ and $\bigcap_{i \in I} \Omega_i$.

If Φ is a topological space (Appendix B), $\mathring{\Omega}$ is the interior of Ω, and $\overline{\Omega}$ the closure of Ω. However, for any complex number α, $\bar{\alpha}$ is the complex conjugate of α.

Now, let Φ_k, where $k = 1, \ldots, n$, be sets. The *Cartesian product* $\Phi_1 \times \cdots \times \Phi_n$ is the set of all ordered n-tuples $\{\phi_1, \ldots, \phi_n\}$, where $\phi_k \in \Phi_k$ and $k = 1, \ldots, n$.

A rule f that assigns one or more elements ψ in a set Ψ to each element ϕ in another set Φ is called a *relation*. Thus, f determines a subset of $\Phi \times \Psi$ called the *graph of f* and consisting of all ordered pairs $\{\phi, \psi\}$, where ψ is an element assigned to ϕ by f. We also use the notation $f: \Phi \to \Psi$ as well as $f: \phi \mapsto \psi$ to denote the rule, the sets for which the rule is defined, and the typical elements related by f. The *domain of f* is the set of all elements ϕ on which f is defined, in this case, Φ; ϕ is called the *independent variable*. The *range of f* is the subset of Ψ consisting of all elements ψ that are assigned by f to members of Φ; ψ is called the *dependent variable*.

A *function* f is a relation $f\colon \Phi \to \Psi$ that assigns precisely one element in Ψ to each member of Φ; in other words, f is a function if and only if the equations $\psi = f\phi$ and $\theta = f\phi$ imply that $\psi = \theta$. Synonymous with the word "function" are the terms *operator*, *mapping*, and *transformation*. We say that $f\colon \phi \mapsto \psi$ *maps*, *carries*, or *transforms* ϕ *into* Ψ and that f is a *function on* (or *from*) Φ *into* Ψ or a *mapping of* Φ *into* Ψ. We also say that f is Ψ-valued; when Ψ is the real line or the complex plane, the phrase "Ψ-valued" is replaced by *real-valued* or *complex-valued*, respectively. If Ω is a subset of Φ, the symbol $f(\Omega)$ denotes the set $\{\psi \in \Psi\colon \psi = f\phi,\ \phi \in \Omega\}$. The function g that is defined only on Ω but coincides with f on Ω is called the *restriction of* f *to* Ω. On the other hand, f is called an *extension of* g.

A function f is said to be *one-to-one* or *injective* and is also called an *injection* if the equations $f\phi = \psi$ and $f\chi = \psi$ imply that $\phi = \chi$. In this case, we have the function $f^{-1}\colon \psi \mapsto \phi$, which maps the range of f into Φ; f^{-1} is called the *inverse of* f. A function $f\colon \Phi \to \Psi$ is said to be *onto* or *surjective* and is also called a *surjection* if the range of f coincides with Ψ. Thus, f^{-1} is a surjection onto Φ whenever f is injective. If $f\colon \Phi \to \Psi$ is both injective and surjective, it is said to be *bijective* or a *bijection*.

We denote the elements in the range of f by the alternative notations $\psi = f\phi = f(\phi) = \langle f, \phi\rangle$. On occasion, it will be convenient to violate this symbolism by using $f(\phi)$ to denote the function f rather than its range value, as was commonly done in classical mathematics. Whenever we do so, it will be clear from the context what is meant. On still other occasions when the symbol for f is rather complicated, we may use the dot notation $f(\cdot) = \langle f, \cdot\rangle$ in order to indicate where the independent variable should appear. For example, when $t \in R$, we may denote the function $t \mapsto \cos(\sin t)$ by $\cos(\sin \cdot)$.

When $\mathscr{V}$ and $\mathscr{W}$ are topological linear spaces (see Appendix D and especially Sections D8 and D11), the symbol $[\mathscr{V}; \mathscr{W}]$ denotes the linear space of all continuous linear mappings of $\mathscr{V}$ into $\mathscr{W}$. Unless the opposite is explicitly stated, it is always understood that $[\mathscr{V}; \mathscr{W}]$ is equipped with its bounded topology. When it has the pointwise topology, it is denoted by $[\mathscr{V}; \mathscr{W}]^{\sigma}$. Thus, if A and B are Banach spaces, $[A; B]$ possesses the uniform operator topology, whereas $[A; B]^{\sigma}$ has the strong operator topology.

R^n and C^n denote, respectively, the real and complex n-dimensional Euclidean spaces. Thus, an arbitrary point $t \in R^n$ (or $t \in C^n$) is an ordered n-tuple $t = \{t_k\}_{k=1}^{n}$ of real (respectively complex) numbers t_k whose magnitude is

$$|t| = \left[\sum_{k=1}^{n} |t_k|^2\right]^{1/2}.$$

We set $R = R^1$ and $C = C^1$. R_+ denotes the positive half-line $\{t \in R\colon 0 < t < \infty\}$. $R_e{}^n$ is the set of all ordered n-tuples each of whose components is a real number or ∞; for example, $\{2, \infty, -1\}$ is such a triplet. As before, we set $R_e = R_e^1$. We do not allow $-\infty$ as a possible component of any $t \in R_e^n$. Thus, if $a \in R_e^n$, $-a$ is an n-tuple whose components are either real numbers or $-\infty$. If $x \in R_e$, the symbol $[x]$ will denote the n-tuple in R_e^n each of whose components is x; however, the n-tuple $[0]$ is denoted simply by 0. The number n of components in $[x]$ will be implied by the context in which the symbol $[x]$ is used. An integer $k \in R^n$ is an n-tuple all of whose components are integers. A *compact set in* R^n is a closed bounded set in R^n. Given any set Ω in R^n, the *diameter of* Ω is denoted by diam Ω and defined by

$$\operatorname{diam} \Omega = \sup\{|t - x| : t, x \in \Omega\}.$$

If $x = \{x_k\}_{k=1}^n$ and $t = \{t_k\}_{k=1}^n$ are members of R^n or R_e^n, the notations $x \leq t$ and $x < t$ mean that $x_k \leq t_k$ and respectively $x_k < t_k$ for $k = 1, \ldots, n$. If $t \geq 0$ (or if $t > 0$), t is said to be *nonnegative* (respectively *positive*). An *interval in* R^n is the Cartesian product of n intervals in R. As special cases, we have

$$(x, y) = \{t \in R^n : x < t < y\}, \qquad -x, y \in R_e^n, \tag{1}$$

$$(x, y] = \{t \in R^n : x < t \leq y\}, \qquad -x \in R_e^n, \quad y \in R^n, \tag{2}$$

$$[x, y) = \{t \in R^n : x \leq t < y\}, \qquad x \in R^n, \quad y \in R_e^n, \tag{3}$$

$$[x, y] = \{t \in R^n : x \leq t \leq y\}, \qquad x, y \in R^n. \tag{4}$$

Equation (1) is an open interval, whereas (4) is a closed interval. The symbol (x, y) is the same as that for the inner product in a Hilbert space, but they can be distinguished by the way they are used. The *volume of any interval* I in R^n with endpoints x and y, where $x \leq y$, is denoted by vol I and defined by

$$\operatorname{vol} I \triangleq \prod_{k=1}^{n} (y_k - x_k).$$

Let us mention a few other customary symbols. If f is a continuous function on R^n or C^n, its *support* is the closure of the set of points t for which $f(t) \neq 0$ and is denoted by supp f. (The support of a distribution on R^n, which is defined subsequently, is also denoted by supp f.) The symbols sup and inf denote, respectively, the operations of taking the supremum and infimum of a subset of R, whereas max and min denote, respectively, the taking of the maximum and minimum of a finite subset of R. Also, $\overline{\lim}$ and $\underline{\lim}$ symbolize the limit superior and limit inferior, respectively.

Two unusual symbols are the following. Whenever we wish to emphasize that a particular equation is a definition, we replace the equality sign by $\triangleq$. The notation $\diamond$ denotes the end of a proof or example.

Throughout this book, A and B will denote complex Banach spaces, whereas H will be a complex Hilbert space.

1.3. CONTINUOUS FUNCTIONS

In the rest of this chapter, $\mathscr{V}$ always denotes a separated sequentially complete locally convex space, and Γ a generating family of seminorms for the topology of $\mathscr{V}$. Also J is a subset of R^n equipped with the topology induced by R^n.

Let f be a mapping on J into $\mathscr{V}$. The continuity of f is defined in Appendix B4. Since J is a metric space, the continuity of f is equivalent to its sequential continuity (Appendix B8). The mapping f is said to be *uniformly continuous on* J if, for every neighborhood Ω of 0 in $\mathscr{V}$, there exists an $r \in R_+$ such that the conditions t, $x \in J$ and $|t - x| < r$ imply that $f(t) - f(x) \in \Omega$. This condition is equivalent to the following. Given any $\varepsilon \in R_+$ and any $\gamma \in \Gamma$, there exists an $r \in R_+$ such that the conditions t, $x \in J$ and $|t - x| < r$ imply that $\gamma[f(t) - f(x)] < \varepsilon$. The mapping f is said to be bounded on J if $f(J)$ is a bounded set in $\mathscr{V}$ (see Appendix C8).

Theorem 1.3-1. *Let J be a compact subset of R^n, and let f be a continuous mapping of J into $\mathscr{V}$. Then f is uniformly continuous and bounded on J.*

PROOF. We first establish the uniform continuity of f. Since $\mathscr{V}$ is locally convex, any neighborhood of 0 in $\mathscr{V}$ contains a balanced convex neighborhood Ω of 0 in $\mathscr{V}$. Set $\Lambda = \frac{1}{2}\Omega$. By the convexity of Ω, we have that $\Lambda + \Lambda \subset \Omega$. The continuity of f implies that, for each $t \in J$, there exists a neighborhood $N(t, r) \triangleq \{x \in J: |x - t| < r\}$ such that

$$f(x) \in \Lambda + f(t) \tag{1}$$

whenever $x \in N(t, r)$. Now, the collection of neighborhoods $\{N(t, r/2)\}_{t \in J}$ covers J. But J is compact, and therefore a finite subset of this collection also covers J. That is,

$$J \subset N(t_1, r_1/2) \cup \cdots \cup N(t_m, r_m/2).$$

Next, set $\eta = \min_k r_k/2$, and consider any two points $t, x \in J$ such that $|t - x| < \eta$. There exists some $N(t_k, r_k/2)$ containing t. Hence,

$$f(t) - f(t_k) \in \Lambda. \tag{2}$$

Moreover,

$$|x - t_k| \leq |x - t| + |t - t_k| < \eta + \tfrac{1}{2}r_k \leq r_k.$$

Consequently, $x \in N(t_k, r_k)$. By (1), $f(x) - f(t_k) \in \Lambda$. Therefore, since Λ is balanced,

$$f(t_k) - f(x) \in \Lambda. \tag{3}$$

We can combine (2) and (3) to write

$$f(t) - f(x) = f(t) - f(t_k) + f(t_k) - f(x) \in \Lambda + \Lambda \subset \Omega.$$

This proves the uniform continuity of f on J.

To show its boundedness, let γ be any member of Γ. Since both γ and f are continuous functions, so too is the mapping $t \mapsto \gamma[f(t)]$ of J into R. Hence, $\gamma[f(J)]$ is bounded in R. Since γ is arbitrary, this implies that $f(J)$ is bounded in $\mathscr{V}$. ◇

1.4. INTEGRATION

Integrals of continuous vector-valued functions on compact intervals can be constructed in the same way as are Riemann integrals of continuous complex-valued functions.

Let $P = \{P_k\}_{k=1}^n \in R^n$, $Q = \{Q_k\}_{k=1}^n \in R^n$, and $P \le Q$. Then, $[P, Q]$ is a compact interval in R^n. For each k, partition the one-dimensional compact interval $[P_k, Q_k]$ into m_k subintervals with the nondecreasing set of endpoints $P_k = t_{k,0}, t_{k,1}, \ldots, t_{k,m_k} = Q_k$. Upon selecting a particular endpoint t_{k,μ_k} for each k, we get a point

$$t_\mu = \{t_{1,\mu_1}, \ldots, t_{n,\mu_n}\} \in [P, Q],$$

where $\mu \triangleq \{\mu_1, \ldots, \mu_n\}$ and $0 \le \mu \le m \triangleq \{m_1, \ldots, m_n\}$. Let I_μ denote the n-dimensional subinterval $[t_{\mu-[1]}, t_\mu]$, where $[1] \le \mu \le m$ and $[1] = \{1, 1, \ldots, 1\}$. Set

$$\operatorname{vol} I_\mu \triangleq \prod_{k=1}^{n} (t_{k,\mu_k} - t_{k,\mu_k-1}).$$

We let π denote the collection of subintervals $\{I_\mu\}_{[1]\le\mu\le m}$ and call π a *rectangular partition of* $[P, Q]$. Finally, we set

$$|\pi| \triangleq \max_{[1]\le\mu\le m} |t_\mu - t_{\mu-[1]}|.$$

Given two rectangular partitions π_1 and π_2 of $[P, Q]$, we say that π_2 *is a refinement of* π_1 if every subinterval in π_1 is the union of a set of subintervals in π_2.

Let f be a continuous function on $[P, Q]$ into $\mathscr{V}$, where, as usual, $[P, Q]$ has the topology induced by R^n. Corresponding to f and any given rectangular partition of $[P, Q]$, we can set up the *Riemann sum*

$$S(f, \pi) \triangleq \sum_{[1]\le\mu\le m} f(t_\mu) \operatorname{vol} I_\mu. \tag{1}$$

Next, let $\{\pi_k\}_{k=1}^{\infty}$ be a sequence of rectangular partitions of $[P, Q]$ such that $|\pi_k| \to 0$. Then, $\{S(f, \pi_k)\}_{k=1}^{\infty}$ is a Cauchy sequence in $\mathscr{V}$. Indeed, by Theorem 1.3-1, f is uniformly continuous on $[P, Q]$. Hence, given any $\gamma \in \Gamma$ and $\varepsilon \in R_+$, there exists an $\eta \in R_+$ such that

$$\gamma[f(t) - f(x)] < \frac{\varepsilon}{\text{vol}[P, Q]}$$

whenever $|t - x| < \eta$ and $t, x \in [P, Q]$. Now, for any π_k and π_m such that $|\pi_k| < \eta$ and $|\pi_m| < \eta$, we may write

$$S(f, \pi_k) - S(f, \pi_m) = S(g, \pi_{k,m}). \tag{2}$$

Here, $\pi_{k,m}$ is a partition of $[P, Q]$ that is a refinement of both π_k and π_m, and g is a function on $[P, Q]$ into $\mathscr{V}$ such that, on each subinterval of $\pi_{k,m}$, g is a constant function of the form $g(t) = f(t_\nu) - f(t_\xi)$, where t_ν and t_ξ are fixed points satisfying $|t_\nu - t_\xi| < \eta$. Hence, by (1) and (2),

$$S(f, \pi_k) - S(f, \pi_m) = \sum_\mu g(t_\mu) \text{ vol } I_\mu,$$

where the summation on μ corresponds to the rectangular partition $\pi_{k,m}$. Thus,

$$\gamma[S(f, \pi_k) - S(f, \pi_m)] \le \sum_\mu \gamma[g(t_\mu)] \text{ vol } I_\mu$$

$$\le \frac{\varepsilon}{\text{vol}[P, Q]} \sum_\mu \text{vol } I_\mu = \varepsilon.$$

So, truly, $\{S(f, \pi_k)\}_{k=1}^{\infty}$ is a Cauchy sequence in $\mathscr{V}$.

Since $\mathscr{V}$ is sequentially complete and separated, this sequence has a unique limit in $\mathscr{V}$, which we denote alternatively by

$$\int_{[P,Q]} f(t)\, dt = \int_P^Q f(t)\, dt = \int_P^Q f\, dt$$

and refer to as the *(Riemann) integral of f on* $[P, Q]$.

It is also a fact that the integral of f is independent of the choice of the sequence $\{\pi_k\}$ of rectangular partitions. This can be shown by making use of the next lemma.

Lemma 1.4-1. *Let $\{\phi_k\}_{k=1}^{\infty}$ tend to ϕ in $\mathscr{V}$ and $\{\psi_k\}_{k=1}^{\infty}$ tend to ψ in $\mathscr{V}$. Assume that, for every $\gamma \in \Gamma$, $\gamma(\phi_k - \psi_k) \to 0$ as k and m tend to infinity independently. Then $\phi = \psi$.*

PROOF. If $\phi \neq \psi$, then by the separatedness of $\mathscr{V}$, there exists a $\gamma \in \Gamma$ such the $\gamma(\phi - \psi) > 0$. But then,

$$\begin{aligned}\gamma(\phi_k - \psi_m) &= \gamma(\phi_k - \phi + \phi - \psi + \psi - \psi_m)\\ &\geq \gamma(\phi - \psi) - \gamma(\phi_k - \phi) - \gamma(\psi - \psi_m),\end{aligned}$$

and the last two terms on the right-hand side tend to zero as $k \to \infty$ and $m \to \infty$. Hence $\underline{\lim}\, \gamma(\phi_k - \psi_k) \geq \gamma(\phi - \psi) > 0$, which contradicts the hypothesis. ◇

Now, let $\{\pi_k\}$ and $\{\pi_k'\}$ be two sequences of rectangular partitions of $[P, Q]$ such that $|\pi_k| \to 0$ and $|\pi_k'| \to 0$. By virtue of Lemma 1.4-1, to prove that $S(f, \pi_k)$ and $S(f, \pi_k')$ tend to the same limit, we need merely show that, for each $\gamma \in \Gamma$, $\gamma[S(f, \pi_k) - S(f, \pi_m')] \to 0$ as k and m tend to infinity independently. But this can be done by setting $S(f, \pi_k) - S(f, \pi_m') = S(g, \pi_{k,m})$ as in (2) and arguing exactly as before.

We summarize the results obtained as far as follows.

Theorem 1.4-1. *Let f be a continuous function that maps the compact interval $[P, Q] \subset R^n$, where $P \leq Q$, into $\mathscr{V}$. Then, for any sequence $\{\pi_k\}$ of rectangular partitions of $[P, Q]$ such that $|\pi_k| \to 0$, the Riemann sums $S(f, \pi_k)$ tend in $\mathscr{V}$ to a unique limit $\int_P^Q f(t)\, dt$, called the integral of f on $[P, Q]$. Moreover, the limit $\int_P^Q f(t)\, dt$ is independent of the choice of $\{\pi_k\}$.*

When $Q \leq P$, we use the definition

$$\int_P^Q f(t)\, dt \triangleq (-1)^n \int_Q^P f(t)\, dt.$$

We now list a number of facts, which are all easily established from the definition of $\int_P^Q f(t)\, dt$. In the following, both $\mathscr{V}$ and $\mathscr{W}$ are separated sequentially complete locally convex spaces, and f and g are continuous functions mapping the compact interval $[P, Q] \subset R^n$, where $P \leq Q$, into $\mathscr{V}$.

I. $\int_P^Q f(t)\, dt$ is the zero member of $\mathscr{V}$ if any component of P is equal to the corresponding component of Q.

II. Let $F \in [\mathscr{V}; \mathscr{W}]$. (For example, we may have $F \in \mathscr{V}' = [\mathscr{V}; C]$.) Then,

$$F \int_P^Q f(t)\, dt = \int_P^Q Ff(t)\, dt.$$

III. For any continuous seminorm γ on $\mathscr{V}$,

$$\gamma\left[\int_P^Q f(t)\, dt\right] \leq \int_P^Q \gamma[f(t)]\, dt.$$

IV. Let $\mathscr{V} = [A; B]$, where A and B are complex Banach spaces. Then, for any $a \in A$,

$$\int_P^Q f(t)\,dt\,a = \int_P^Q f(t)a\,dt.$$

V. Let $\mathscr{V} = [H; H]$, where H is a complex Hilbert space with the inner product $(\cdot, \cdot)$. For any $a, b \in H$,

$$\left(\int_P^Q f(t)\,dt\,a, b\right) = \int_P^Q (f(t)a, b)\,dt.$$

VI. Integration is a linear process. That is, for $\alpha, \beta \in C$,

$$\int_P^Q [\alpha f(t) + \beta g(t)]\,dt = \alpha \int_P^Q f(t)\,dt + \beta \int_P^Q g(t)\,dt.$$

VII. Let F be a continuous mapping of $[P, Q]$ into $[\mathscr{V}; \mathscr{W}]$. Let h denote the function $t \mapsto F(t)f(t)$, where $t \in [P, Q]$. Then, for any sequence $\{\pi_k\}_{k=1}^{\infty}$ of rectangular partitions of $[P, Q]$ such that $|\pi_k| \to 0$,

$$\int_P^Q F(t)f(t)\,dt \triangleq \lim_{k \to \infty} S(h, \pi_k)$$

exists as a limit in $\mathscr{W}$ and is independent of the choice of $\{\pi_k\}$.

VIII. Assume that $\{f_j\}_{j=1}^{\infty}$ is a sequence of continuous $\mathscr{V}$-valued functions that converges *uniformly* on $[P, Q]$ to the $\mathscr{V}$-valued function f. (By this, we mean that, given any $\gamma \in \Gamma$ and $\varepsilon \in R_+$, there exists an integer k such that $\gamma[f(t) - f_j(t)] < \varepsilon$ for all $j > k$ and all $t \in [P, Q]$.) Then, f is also continuous and

$$\int_P^Q f_j(t)\,dt \to \int_P^Q f(t)\,dt.$$

Problem 1.4-1. Prove the preceding eight assertions.

1.5. REPEATED INTEGRATION AND IMPROPER INTEGRALS

Under the assumptions of Theorem 1.4-1, the integral

$$\int_P^Q f(t)\,dt \tag{1}$$

can be written as the repeated integral

$$\int_{P_1}^{Q_1} dt_1 \cdots \int_{P_n}^{Q_n} f(t)\,dt_n = \int_{P_1}^{Q_1} \cdots \int_{P_n}^{Q_n}. \tag{2}$$

To see this, we first rewrite $S(f, \pi)$ as follows:

$$S(f, \pi) = \sum_{\mu_1=1}^{m_1} (t_{1,\mu_1} - t_{1,\mu_1-1}) \cdots \sum_{\mu_n=1}^{m_n} (t_{n,\mu_n} - t_{n,\mu_n-1}) f(t_\mu)$$
$$= \sum_{\mu_1} \cdots \sum_{\mu_n} \tag{3}$$

Set

$$|\pi^k| = \max_{1 \le \mu_k \le m_k} (t_{k,\mu_k} - t_{k,\mu_k-1}).$$

As $|\pi^n| \to 0$, the innermost summation in (3) tends to

$$\int_{P_n}^{Q_n} f(t)\, dt_n \tag{4}$$

because, with all the components of t fixed except for t_n, $t_n \mapsto f(t)$ is a continuous mapping of $[P_n, Q_n]$ into $\mathscr{V}$.

Now, (4) is a continuous function on $[P_1, Q_1] \times \cdots \times [P_{n-1}, Q_{n-1}]$ into $\mathscr{V}$. Indeed, let Δt denote an increment in t such that t_n remains unchanged. Then, by note III of the preceding section,

$$\gamma \int_{P_n}^{Q_n} [f(t + \Delta t) - f(t)]\, dt_n \le \int_{P_n}^{Q_n} \gamma[f(t + \Delta t) - f(t)]\, dt_n.$$

By the uniform continuity of f, the right-hand side tends to zero as $|\Delta t| \to 0$, which verifies our assertion. As a consequence, the two innermost summations in (3) tend to

$$\int_{P_{n-1}}^{Q_{n-1}} dt_{n-1} \int_{P_n}^{Q_n} f(t)\, dt_n$$

as first $|\pi^n| \to 0$ and then $|\pi^{n-1}| \to 0$.

Continuing in this way, we see that (3) tends to (2) as all the $|\pi^k|$ are taken to zero one at a time in the order of decreasing k.

To show that (1) is equal to (2), choose $\gamma \in \Gamma$ and $\varepsilon \in R_+$ arbitrarily. We may write

$$\gamma\left[\int_P^Q f(t)\, dt - \int_{P_1}^{Q_1} \cdots \int_{P_n}^{Q_n}\right]$$
$$\le \gamma\left[\int_P^Q f(t)\, dt - \sum_{\mu_1} \cdots \sum_{\mu_n}\right] + \gamma\left[\sum_{\mu_1} \cdots \sum_{\mu_n} - \sum_{\mu_1} \cdots \sum_{\mu_{n-1}} \int_{P_n}^{Q_n}\right]$$
$$+ \cdots + \gamma\left[\sum_{\mu_1} \int_{P_2}^{Q_2} \cdots \int_{P_n}^{Q_n} - \int_{P_1}^{Q_1} \cdots \int_{P_n}^{Q_n}\right]. \tag{5}$$

Now, by what we have already shown in this section and by Theorem 1.4-1, we can choose a sufficiently fine rectangular partition π of $[P, Q]$ such that every term on the right-hand side is less than $\varepsilon/(n+1)$. [Indeed, for one such term, say, the pth term, we can choose a π_p such that this term is less than $\varepsilon/(n+1)$ for every refinement of π_p. Then, a π can be found that is a refinement of every π_p.] Hence, the left-hand side of (5) is less than ε, which proves the equality between (1) and (2).

The order of integration in (2) can be changed in any fashion. This follows from the fact that we can change the order of summation in (3) and then take repeated limits as above to get another repeated integral equal to (1).

The next theorem summarizes all this.

Theorem 1.5-1. *Let f be a continuous function that maps the compact interval $[P, Q] \subset R^n$, where $P \leq Q$, into $\mathscr{V}$. Then,*

$$\int_P^Q f(t)\,dt = \int_{P_1}^{Q_1} dt_1 \cdots \int_{P_n}^{Q_n} f(t)\,dt_n. \tag{6}$$

Moreover, the order of integration in the right-hand side can be changed in any fashion.

The rest of this section is devoted to a discussion of certain improper integrals. Let $Q^\infty \in R_e^n$ be such that one or more of its components are ∞. We shall say that $Q \in R^n$ tends to Q^∞ and shall write $Q \to Q^\infty$ if each component of Q tends to the corresponding component of Q^∞. Now, let $P \in R^n$ be such that $P \leq Q^\infty$ and assume that f is a $\mathscr{V}$-valued continuous function on $[P, Q]$ for every $Q \in R^n$ such that $P \leq Q \leq Q^\infty$. Assume in addition that

$$\lim_{Q \to Q^\infty} \int_P^Q f(t)\,dt \tag{7}$$

exists in $\mathscr{V}$ and is independent of the fashion in which $Q \to Q^\infty$. [Thus, for example, the components of Q could tend to those of Q^∞ simultaneously or one at a time and in any order without changing (7).] The *improper integral*

$$\int_P^{Q^\infty} f(t)\,dt \tag{8}$$

is defined to be the limit (7).

In quite the same way, we can define the improper integral

$$\int_{P^{-\infty}}^{Q} f(t)\,dt, \tag{9}$$

where $P^{-\infty}$ is an n-tuple with $-\infty$ for some or all of its components and real numbers for the remaining components. When all the components of $P^{-\infty}$ are $-\infty$ and those of Q^{∞} are ∞, we write

$$\int_{P^{-\infty}}^{Q^{\infty}} f(t)\,dt = \int_{R^n} f(t)\,dt = \int_{R^n} f\,dt. \tag{10}$$

It is readily seen that all the assertions of notes I–VI of Section 1.4 continue to hold for (8)–(10). Now, however, the right-hand side of the inequality in note III may become ∞.

Problem 1.5-1. Prove the two statements of the preceding paragraph.

1.6. DIFFERENTIATION

Let $\Delta t_k \in R$. In the following, $\Delta t|_k$ will denote a member of R^n all of whose components are zero except for the kth, which is instead Δt_k. Thus, for any $t = \{t_1, \ldots, t_n\} \in R^n$,

$$t + \Delta t|_k = \{t_1, \ldots, t_{k-1}, t_k + \Delta t_k, t_{k+1}, \ldots, t_n\}.$$

Let f be a $\mathscr{V}$-valued function on some subset of R^n. (We remind the reader that throughout this chapter, we are assuming that $\mathscr{V}$ is a separated sequentially complete locally convex space with Γ as a generating family of seminorms for the topology of $\mathscr{V}$.) We say that f is *differentiable at the point t with respect to t_k* if f is defined on some neighborhood of t and if the quantity

$$\frac{f(t + \Delta t|_k) - f(t)}{\Delta t_k}$$

converges to a limit in $\mathscr{V}$ as $|\Delta t_k| \to 0$. That limit is called the *derivative of f with respect to t_k evaluated at t* and is denoted by $\partial_k f(t)$ as well as by $\partial_{t_k} f(t)$. Let Ω be a subset of R^n. f is said to be *differentiable on Ω with respect to t_k* if f is differentiable at every $t \in \Omega$; in this case, its *derivative* $\partial_k f = \partial_{t_k} f$ is also a $\mathscr{V}$-valued function on Ω. If $\partial_k f$ is in turn differentiable on Ω with respect to, say, t_j, the *second derivative $\partial_j \partial_k f$ of f* is defined as the derivative of $\partial_k f$ on Ω with respect to t_j. Continuing in this way, we obtain the higher derivatives of f.

Integration and differentiation can be related as follows.

Theorem 1.6-1. *Let $P = \{P_k\}_{k=1}^n \in R^n$ and $Q = \{Q_k\}_{k=1}^n \in R^n$ be such that $P_k = Q_k$ for all components except the jth component, for which we have* *instead* $P_j < Q_j$. *Let* f *be a* $\mathscr{V}$*-valued function having a continuous derivative* $\partial_j f$ *on* $[P, Q]$. *Then,*

$$\int_{P_j}^{Q_j} \partial_j f(t)\, dt_j = f(Q) - f(P) \tag{1}$$

PROOF. The left-hand side of (1) exists because of the continuity of $\partial_j f(t)$ as a function of t_j. Now, let $\mathscr{V}'$ be the dual of $\mathscr{V}$ and let $F \in \mathscr{V}'$ be arbitrary. It follows from the definition of $\partial_j f$ that $F\,\partial_j f(t) = \partial_j Ff(t)$. So, by note II of Section 1.4 and the fundamental theorem of integral calculus,

$$\begin{aligned} F \int_{P_j}^{Q_j} \partial_j f(t)\, dt_j &= \int_{P_j}^{Q_j} F\, \partial_j f(t)\, dt_j \\ &= \int_{P_j}^{Q_j} \partial_j Ff(t)\, dt_j = Ff(Q) - Ff(P). \end{aligned}$$

But, the weak topology of $\mathscr{V}$ separates $\mathscr{V}$ (Appendix D7), and thus (1) is obtained. ◇

In regard to a change in the order of differentiation in a second derivative, we have the following result.

Theorem 1.6-2. *Let* f *be a* $\mathscr{V}$*-valued function on a subset of* R^2. *If* $\partial_1 f$, $\partial_2 f$, *and* $\partial_2 \partial_1 f$ *exist and are continuous on a neighborhood* Ω *of a fixed point* $t \in R^2$, *then* $\partial_1 \partial_2 f$ *exists at* t *and*

$$\partial_1 \partial_2 f(t) = \partial_2 \partial_1 f(t). \tag{2}$$

Note. We could apply an argument such as that of the preceding proof. But this would only establish that $\partial_1 \partial_2 f(t)$ exists as a limit in the weak topology of $\mathscr{V}$. To conclude that $\partial_1 \partial_2 f(t)$ exists in the initial topology of $\mathscr{V}$, we proceed as follows.

PROOF. Set $t = \{x_0, y_0\}$. In the following, $h \in R$ and $l \in R$ are so restricted that $\{x_0 + h, y_0 + l\}$ remains in a compact interval Λ contained in Ω and containing $\{x_0, y_0\}$ in its interior $\mathring{\Lambda}$. We want to show that

$$\lim_{h \to 0} (1/h)[\partial_2 f(x_0 + h, y_0) - \partial_2 f(x_0, y_0)] \tag{3}$$

exists and has the value $\partial_2 \partial_1 f(x_0, y_0)$.

Now, (3) may be written as

$$\lim_{h \to 0} \lim_{l \to 0} (1/hl)[f(x_0 + h, y_0 + l) - f(x_0 + h, y_0) - f(x_0, y_0 + l) + f(x_0, y_0)].$$

We may invoke Theorem 1.6-1 to rewrite this expression as

$$\lim_{h\to 0}\lim_{l\to 0}(1/hl)\int_0^h[\partial_1 f(x_0+z,\, y_0+l)-\partial_1 f(x_0+z,\, y_0)]\,dz$$

$$=\lim_{h\to 0}\lim_{l\to 0}(1/hl)\int_0^h dz\int_0^l \partial_2\,\partial_1 f(x_0+z,\, y_0+\zeta)\,d\zeta.$$

But, for any $\gamma\in\Gamma$, we may employ note III of Section 1.4 to obtain

$$\gamma\left[(1/hl)\int_0^h dz\int_0^l \partial_2\,\partial_1 f(x_0+z,\, y_0+\zeta)\,d\zeta-\partial_2\,\partial_1 f(x_0,\, y_0)\right]$$

$$=\gamma\left\{(1/hl)\int_0^h dz\int_0^l[\partial_2\,\partial_1 f(x_0+z,\, y_0+\zeta)-\partial_2\,\partial_1 f(x_0,\, y_0)]\,d\zeta\right\}$$

$$\leq (1/|hl|)\left|\int_0^h dz\int_0^l \gamma[\partial_2\,\partial_1 f(x_0+z,\, y_0+\zeta)-\partial_2\,\partial_1 f(x_0,\, y_0)]\,d\zeta\right|$$

$$\leq \sup_{\substack{|z|\leq|h|\\|\zeta|\leq|l|}}\gamma[\partial_2\,\partial_1 f(x_0+z,\, y_0+\zeta)-\partial_2\,\partial_1 f(x_0,\, y_0)].$$

In view of the continuity of $\partial_2\,\partial_1 f$ at $\{x_0, y_0\}$, the last expression tends to zero as first l and then h tend to zero. This proves that (3) exists as a limit in $\mathscr{V}$ and is equal to $\partial_2\,\partial_1 f(x_0, y_0)$. ◇

Let $k=\{k_1,\ldots,k_n\}$ be a nonnegative integer in R^n. D^k denotes the differential operator

$$D^k \triangleq \partial_1^{k_1}\cdots\partial_n^{k_n}. \tag{4}$$

We also use the notation $D^k f=f^{(k)}$ as well as $D^k f(t)=D_t^k f(t)=f^{(k)}(t)$. Throughout this book, we adhere to the following conventions. Whenever k is used as in (4), the notation $|k|$ will denote the sum $k_1+\cdots+k_n$, rather than the magnitude of k as a member of R^n. Moreover, we shall refer to k as the *order of* D^k. (This is in contrast to the usual practice of referring to $|k|$ as the order of D^k.) Whenever we say that the derivatives of a function f up to the order k are continuous on a set Ω, we mean that the functions f, $\partial_k f$, $\partial_j\,\partial_k f,\ldots, D^k f$, obtained by successively applying to f the various operators ∂_k to obtain $D^k f$, exist and are continuous on Ω. If Ω is an open set, it follows from Theorem 1.6-2 that any change in the order of differentiation in $D^k f$ will not alter $D^k f$ on Ω. The same is true when Ω is a compact interval, because the values of $D^k f$ on the boundary on Ω are continuous extensions of its values on the interior of Ω. We shall say that a $\mathscr{V}$-valued function defined on a subset of R^n is *smooth* if the function possesses continuous derivatives of all orders at all points of its domain.

Theorem 1.6-3. *Let $\mathscr{W}$ be a separated locally convex space. Also, let F and f be, respectively, $[\mathscr{V}; \mathscr{W}]$-valued and $\mathscr{V}$-valued functions having the continuous derivatives $\partial_k F$ and $\partial_k f$ on a neighborhood of the point $t \in R^n$. Then, at t,*

$$\partial_k(Ff) = (\partial_k F)f + F\,\partial_k f. \tag{5}$$

[Here, Ff denotes the function $x \mapsto F(x)f(x)$.] Moreover, if F and f have continuous derivatives up to order k on a neighborhood of a point t, then, at t, we have

$$D^k(Ff) = \sum_{0 \leq p \leq k} \binom{k}{p}(D^{k-p}F)(D^p f) \tag{6}$$

where

$$\binom{k}{p} \triangleq \binom{k_1}{p_1} \cdots \binom{k_n}{p_n}.$$

Note. The last quantity is the *n-dimensional binomial coefficient*. For $n = 1$, we have

$$\binom{m}{q} \triangleq \frac{m!}{q!(m-q)!},$$

where m and q are nonnegative integers in R with $q < m$. Also, (6) is called *Leibniz's rule for the differentiation of a product*.

PROOF. In this proof, Δt_k tends to zero inside an interval so small that $\Delta t_k \mapsto F(t + \Delta t|_k)$ and $\Delta t_k \mapsto f(t + \Delta t|_k)$ are continuous and therefore bounded functions on that interval (Theorem 1.3-1). The left-hand side of (5) is the limit of $\mathscr{W}$ (if it exists) of

$$\frac{F(t + \Delta t|_k) - F(t)}{\Delta t_k} f(t + \Delta t|_k) + F(t)\frac{f(t + \Delta t|_k) - f(t)}{\Delta t_k}. \tag{7}$$

Since $[\mathscr{V}; \mathscr{W}]$ possesses the bounded topology, the assumption that F has a derivative at t means that

$$G(\Delta t_k) \triangleq \frac{F(t + \Delta t|_k) - F(t)}{\Delta t_k}, \qquad \Delta t_k \neq 0$$

tends in $[\mathscr{V}; \mathscr{W}]$ to $G(0) \triangleq \partial_k F(t)$ uniformly on the bounded sets in $\mathscr{V}$. By the assumption on f,

$$g(\Delta t_k) \triangleq f(t + \Delta t|_k), \qquad \Delta t_k \neq 0$$

tends in $\mathscr{V}$ to $g(0) \triangleq f(t)$. Moreover, $\{g(\Delta t_k)\}$, where Δt_k traverses its permissible values, is a bounded set in $\mathscr{V}$. Now,

$$G(\Delta t_k)g(\Delta t_k) - G(0)g(0) = [G(\Delta t_k) - G(0)]g(\Delta t_k) + G(0)[g(\Delta t_k) - g(0)].$$

Thus, both terms on the right-hand side tend to 0 in $\mathscr{W}$, which shows that the first term in (7) tends to $(\partial_k F)(t)f(t)$. The second term in (7) clearly tends to $F(t)\, \partial_k f(t)$ in $\mathscr{W}$. Thus, (5) is established.

Equation (6) is established by repeated application of (5). ◇

As a last consideration, we take up differentiation under the integral sign.

Theorem 1.6-4. *Let* $[P, Q]$ *be a compact interval in* R^n *and* Ξ *an open set in* R; t *and* x *will denote variables in* $[P, Q]$ *and* Ξ, *respectively. Assume that* f *is a continuous* $\mathscr{V}$*-valued function on* $[P, Q] \times \Xi$ *and that* $\partial_x f$ *exists and is continuous on* $[P, Q] \times \Xi$. *Set*

$$g(x) = \int_P^Q f(t, x)\, dt, \qquad x \in \Xi. \tag{8}$$

Then, $\partial_x g$ *exists at each point of* Ξ *and*

$$\partial_x g(x) = \int_P^Q \partial_x f(t, x)\, dt, \qquad x \in \Xi. \tag{9}$$

PROOF. Fix $x \in \Xi$ and restrict $\Delta x \in R$ so that $x + \Delta x \in \Xi$. For $\Delta x \neq 0$, consider

$$\begin{aligned} h_{\Delta x}(x) &\triangleq \frac{g(x + \Delta x) - g(x)}{\Delta x} - \int_P^Q \partial_x f(t, x)\, dt \\ &= \int_P^Q dt\, \frac{1}{\Delta x} \int_x^{x+\Delta x} [\partial_\xi f(t, \xi) - \partial_x f(t, x)]\, d\xi. \end{aligned}$$

By Theorem 1.3-1, $\partial_x f$ is uniformly continuous on $[P, Q] \times \Xi$. Therefore, given any $\varepsilon > 0$ and any $\gamma \in \Gamma$, there exists an $\eta \in R_+$ such that

$$\gamma[\partial_\xi f(t, \xi) - \partial_x f(t, x)] < \frac{\varepsilon}{\operatorname{vol}[P, Q]} \tag{10}$$

for every $t \in [P, Q]$ and all ξ such that $|\xi - x| < \eta$. Appealing to note III of Section 1.4 and using (10), we may write

$$\begin{aligned} \gamma[h_{\Delta x}(x)] &\leq \int_P^Q dt\, \frac{1}{|\Delta x|} \left| \int_x^{x+\Delta x} \gamma[\partial_\xi f(t, \xi) - \partial_x f(t, x)]\, d\xi \right| \\ &< \varepsilon \end{aligned}$$

for all $|\Delta x| < \eta$. This proves the theorem. ◇

Problem 1.6-1. Let $\mathscr{W}$, as well as $\mathscr{V}$, be a separated sequentially complete locally convex space. Also, let F and f be, respectively, $[\mathscr{V}; \mathscr{W}]$-valued and

$\mathscr{V}$-valued functions such that $F, f, \partial_k F$, and $\partial_k f$ are continuous on a compact interval $[P, Q] \subset R^n$. Establish the formula for *integration by parts*:

$$\int_P^Q (\partial_k F) f \, dt_k = F(Q)f(Q) - F(P)f(P) - \int_P^Q F(\partial_k f) \, dt_k. \tag{11}$$

Next, assume that F and f are smooth on R^n and that at least one of them has a compact support. Show that, for any nonnegative integer $k \in R^n$,

$$\int_{R^n} (D^k F) f \, dt = (-1)^{|k|} \int_{R^n} F D^k f \, dt. \tag{12}$$

Problem 1.6-2. Let $[P, Q]$ and Ξ be as in Theorem 1.6-4 and let f be a continuous $\mathscr{V}$-valued function on $[P, Q] \times \Xi$. Define g by (8). Show that g is a continuous function on Ξ.

Problem 1.6-3. Let $\{f_m\}_{m=1}^{\infty}$ be a sequence of continuous $\mathscr{V}$-valued functions on an open set $\Omega \subset R^n$. With ∂_k denoting a differentiation, assume that $\partial_k f_m$ exists and is continuous on Ω for every m. Also, assume that $\{f_m\}$ converges uniformly on Ω to a function f, whereas $\{\partial_k f_m\}$ converges uniformly on Ω to a function g. Then, $\partial_k f$ exists on Ω, and $\partial_k f = g$.

1.7. BANACH-SPACE-VALUED ANALYTIC FUNCTIONS

We turn to a discussion of analytic functions that take their values in a complex Banach space A. Let Ω be an open set in C and let f be a mapping of Ω into A. The mapping f is said to be *differentiable* (or *weakly differentiable*) *at a given point* $\zeta \in \Omega$ if, as $\Delta\zeta \to 0$ in C,

$$\frac{f(\zeta + \Delta\zeta) - f(\zeta)}{\Delta\zeta} \tag{1}$$

tends to a limit $Df(\zeta) \in A$ in the topology (or respectively the weak topology) of A and if this limit is independent of the path along which $\Delta\zeta$ tends to zero. $Df(\zeta)$ is called the *derivative* (or respectively the *weak derivative*) *of f at* ζ. Alternative notations that we use for $Df(\zeta)$ are $D_\zeta f(\zeta)$ and $f^{(1)}(\zeta)$. The differentiability of f at ζ immediately implies its continuity at ζ. If f possesses a derivative (or weak derivative) at every point of Ω, then f is said to be *analytic* (or *weakly analytic*) *on* Ω. At times, we say that f is *analytic at a point* ζ; by this, we mean that f is analytic at every point of some open neighborhood of ζ.

If B is another complex Banach space and if F is an $[A; B]$-valued function on Ω, then, as $\Delta\zeta \to 0$,

$$\frac{F(\zeta + \Delta\zeta) - F(\zeta)}{\Delta\zeta} \tag{2}$$

may converge to a limit $DF(\zeta) \triangleq D_\zeta F(\zeta) \triangleq F^{(1)}(\zeta)$ in $[A; B]$ with respect to any one of three topologies in $[A; B]$, namely the uniform operator topology, the strong operator topology, and the weak operator topology (see Appendix D11), in such a way that the limit is independent of the path along which $\Delta\zeta$ tends to zero. In this case, we say that F possesses respectively either a *derivative* a *strong derivative*, or a *weak derivative at* ζ. When this is so at every point of Ω, we say that F is respectively either *analytic*, *strongly analytic*, or *weakly analytic on* Ω.

Clearly, the analyticity of f on Ω implies its weak analyticity, whereas the analyticity of F on Ω implies its strong analyticity, which in turn implies its weak analyticity. It is quite a useful fact that the reverse implications are also true, as is stated later, by Theorem 1.7-1. To show this, we first prove the following.

Lemma 1.7-1. *Let g be a complex-valued analytic function on an open set $\Omega \subset C$ and let Ξ be a closed disc $\{\zeta \in C : |\zeta - \zeta_0| \leq r,\ \zeta_0 \in C,\ r > 0\}$ contained in Ω. Then, there exists a finite number $M = M(f, \Xi)$ depending on f and Ξ such that, for every ζ, $\zeta + \alpha$, and $\zeta + \beta$ in the interior $\mathring{\Xi}$ of Ξ, we have $|Q(\zeta, \alpha, \beta,)| \leq M$, where*

$$Q(\zeta, \alpha, \beta) \triangleq \frac{1}{\alpha - \beta}\left[\frac{f(\zeta + \alpha) - f(\zeta)}{\alpha} - \frac{f(\zeta + \beta) - f(\zeta)}{\beta}\right].$$

PROOF. Let S denote the boundary of Ξ. We can choose a circle P in Ω which encircles Ξ such that

$$d \triangleq \{|\tau - \zeta| : \tau \in P, \quad \zeta \in S \cup \Xi\,|\} > 0.$$

Then, upon fixing ζ, $\zeta + \alpha$, and $\zeta + \beta$ as points in $\mathring{\Xi}$, we may use Cauchy's integral formula to write

$$Q(\zeta, \alpha, \beta) = \frac{1}{2\pi i}\int_P \frac{f(\tau)}{(\tau - \zeta)(\tau - \zeta - \alpha)(\tau - \zeta - \beta)}\, d\tau.$$

Since $|\tau - \zeta|$, $|\tau - \zeta - \alpha|$, and $|\tau - \zeta - \beta|$ are no less than d,

$$|Q(\zeta, \alpha, \beta)| \leq d^{-3} \sup_{\tau \in P} |f(\tau)| \triangleq M. \quad \diamond$$

Theorem 1.7-1. *An A-valued function on an open set $\Omega \subset C$ is analytic on Ω if and only if it is weakly analytic on Ω. Similarly, an $[A; B]$-valued function is analytic on Ω if and only if it is strongly analytic on Ω, and this is the case if and only if it is weakly analytic on Ω.*

PROOF. We prove only the second sentence, the proof of the first one being quite similar. In view of our previous remarks, we need merely show that the weak analyticity of the $[A; B]$-valued function F on Ω implies its analyticity on Ω.

Let $a \in A$ and $b' \in B'$, where B' is the dual of B. Set

$$R(\zeta, \alpha, \beta) \triangleq \frac{1}{\alpha - \beta}\left[\frac{F(\zeta + \alpha) - F(\zeta)}{\alpha} - \frac{F(\zeta + \beta) - F(\zeta)}{\beta}\right].$$

Choose Ξ as in Lemma 1.7-1 and let ζ, $\zeta + \alpha$, and $\zeta + \beta$ be points in $\mathring{\Xi}$. By Lemma 1.7-1, there exists a constant M depending on a, b', F, and Ξ but independent of ζ, α, and β such that

$$|b'R(\zeta, \alpha, \beta)a| \leq M.$$

By Appendix D12,

$$\|R(\zeta, \alpha, \beta)a\|_B \leq M_1,$$

where M_1 depends on a, F, and Ξ but is independent of ζ, α, and β. An application of the principle of uniform boundedness (Appendix D12) now shows that

$$\|R(\zeta, \alpha, \beta)\|_{[A, B]} \leq M_2,$$

where M_2 depends on F and Ξ but is also independent of ζ, α, and β.

Now, let $\varepsilon \in R_+$ be arbitrarily chosen. For all permissible α and β such that $|\alpha|$ and $|\beta|$ are both less than $\varepsilon/2M_2$, we have that $|\alpha - \beta| < \varepsilon/M_2$, so that

$$\left\|\frac{F(\zeta + \alpha) - F(\zeta)}{\alpha} - \frac{F(\zeta + \beta) - F(\zeta)}{\beta}\right\| < \varepsilon.$$

By virtue of the completeness of $[A; B]$, this implies that, as $\alpha \to 0$,

$$[F(\zeta + \alpha) - F(\zeta)]/\alpha$$

tends to a limit in the uniform operator topology of $[A; B]$. Hence, F has a derivative at ζ. Since ζ can be selected as any point of Ω by appropriately choosing Ξ, F is analytic on Ω. ◇

As a consequence of Theorem 1.7-1, many of the results for complex-valued analytic functions can be carried directly over to A-valued or $[A; B]$-valued analytic functions. For example, let F be an $[A; B]$-valued analytic function on an open set Ω and let $a \in A$ and $b' \in B'$. Then, $b'F(\cdot)a$ is a complex-valued analytic function on Ω, so that $D^k b'F(\zeta)a$ exists for every k and every $\zeta \in \Omega$. This implies that $D^k F$ is weakly analytic and therefore analytic on Ω for every k, which shows that F is smooth on Ω.

We finally observe that an adaptation of the proof of Theorem 1.6-3 establishes Liebniz's rule [see (6) of Section 1.6] for the differentiation of Ff, where F is an $[A; B]$-valued analytic function and f is an A-valued analytic function.

1.8. CONTOUR INTEGRATION

Theorem 1.7-1 also allows us to extend contour integration to Banach-space-valued functions. We do so in this section for the integral of an A-valued function f on a contour P in C. Since $[A; B]$ is also a Banach space, our results immediately extend to an $[A; B]$-valued function F on P.

The integral $\int_P f(\zeta)\, d\zeta$ of a continuous A-valued function f on a contour P in C is defined and shown to exist exactly as in the case of a complex-valued function (Copson, 1962, pp. 52–59). We also have, as in the scalar case, the estimate

$$\left\| \int_P f(\zeta)\, d\zeta \right\|_A \leq \sup_{\zeta \in P} \left\| f(\zeta) \right\|_A \quad \text{(length of } P\text{)}. \tag{1}$$

As an immediate consequence of the definition of $\int_P f(\zeta)\, d\zeta$ as a limit of certain Riemann sums, we can state that, for any $a \in A$, $a' \in A'$, and $b' \in B'$,

$$a' \int_P f(\zeta)\, d\zeta = \int_P a' f(\zeta)\, d\zeta \tag{2}$$

and

$$b' \int_P F(\zeta)\, d\zeta\, a = \int_P b' F(\zeta) a\, d\zeta. \tag{3}$$

Contour integration is a linear process as in note VI of Section 1.4. Moreover,

$$\int_{-P} f(\zeta)\, d\zeta = -\int_P f(\zeta)\, d\zeta,$$

where $-P$ is the contour obtained by reversing the orientation of P.

Theorem 1.8-1 (Cauchy's theorem). *Let Ω be a simply connected open set in C, and let P be a closed contour in Ω. If f is analytic on Ω, then*

$$\int_P f(\zeta)\, d\zeta = 0. \tag{4}$$

PROOF. When P is a closed contour in Ω, the right-hand side of (2) is zero according to Cauchy's theorem for complex-valued functions. Equation (4) now follows from the fact that the weak topology of A separates A (Appendix D7). ◇

Arguments like that of the preceding proof can be used to extend other standard results from the theory of complex-valued analytic functions.

Theorem 1.8-2 (Cauchy's integral formula). *Let* Ω, P, *and* f *be as in Theorem* 1.8-1 *and let* $\zeta \in C$ *be a point inside* P. *Then, for each nonnegative integer* $k \in R$,

$$f^{(k)}(\zeta) = \frac{k!}{2\pi i} \int_P \frac{f(\tau)}{(\tau - \zeta)^{k+1}} \, d\tau. \tag{5}$$

Theorem 1.8-3. *Let* $\{f_n\}_{n=1}^{\infty}$ *be a sequence of* A*-valued analytic functions on an open set* $\Omega \subset C$ *such that* $f_n \to f$ *in* A *uniformly on each compact subset of* Ω. *Then,* f *is also an* A*-valued analytic function on* Ω *and for each nonnegative integer* k $f_n^{(k)} \to f^{(k)}$ *in* A *uniformly on each compact subset of* Ω.

[For this theorem, an argument like the proof of Theorem 1.8-1 would only prove that $f_n^{(k)}(\zeta) \to f^{(k)}(\zeta)$ in the weak topology of A. However, an estimation of $f_n^{(k)}(\zeta) - f^{(k)}(\zeta)$ using (5) leads to our stronger conclusion.]

Theorem 1.8-4. *Let* $\{\zeta, \tau\} \mapsto f(\zeta, \tau)$ *be a continuous* A*-valued function on* $\Omega \times P$, *where* $\zeta \in \Omega$, $\tau \in P$, Ω *is an open set in* C, *and* P *is a contour in* C. *Assume that* $f(\cdot, \tau)$ *is an analytic function on* Ω *for each* $\tau \in P$. *Set*

$$G(\zeta) \triangleq \int_P f(\zeta, \tau) \, d\tau. \tag{6}$$

Then, G *is an* A*-valued analytic function on* Ω, *and*

$$G^{(k)}(\zeta) = \int_P D_\zeta^k f(\zeta, \tau) \, d\tau, \qquad k = 1, 2, \ldots .$$

In the next theorem, P is an oriented path in C extending to infinity such that any finite portion P_n obtained by tracing P from one of its points to another is a contour. Let $\{P_n\}_{n=1}^{\infty}$ be a collection of such finite portions of P with the properties that $P_n \subset P_{n+1}$ and $\bigcup_n P_n = P$. We define

$$\int_P f(\tau) \, d\tau \triangleq \lim_{n \to \infty} \int_{P_n} f(\tau) \, d\tau$$

if the limit exists. Upon combining Theorems 1.8-3 and 1.8-4, we obtain the following.

Theorem 1.8-5. *Assume that* f *satisfies the hypothesis of Theorem* 1.8-4 *on* $\Omega \times P_n$ *for each* n. *Suppose that, as* $n \to \infty$, $\int_{P_n} f(\zeta, \tau) \, d\tau$ *converges uniformly with respect to all* ζ *in each compact subset of* Ω. *Define* G *by* (6). *Then, the conclusion of Theorem* 1.8-4 *holds once again.*

Problem 1.8-1. Prove Theorems 1.8-2–1.8-4.

Problem 1.8-2. Let P be a contour starting at $w \in C$ and ending at $z \in C$. Assume that the $[A; B]$-valued function F and the A-valued function f are analytic on P. Show that

$$\int_w^z f^{(1)}(\zeta)\, d\zeta = F(z) - F(w) \tag{7}$$

and

$$\int_w^z F^{(1)}(\zeta) f(\zeta)\, d\zeta = F(z) f(z) - F(w) f(w) - \int_w^z F(\zeta) f^{(1)}(\zeta)\, d\zeta. \tag{8}$$

Chapter 2

Integration with Vector-Valued Functions and Operator-Valued Measures

2.1. INTRODUCTION

An essential tool needed in our subsequent development of an admittance formulism for time-invariant passive systems is a certain theory of integration due to Hackenbroch (1968), wherein functions taking their values in a given Banach space A are integrated with respect to measures that take their values in a space of operators mapping A into another Banach space B. Chapter 2 is devoted to a presentation of this theory.

2.2. OPERATOR-VALUED MEASURES

Throughout this chapter, T is an arbitrary nonvoid set, $\mathfrak{C}$ is a σ-algebra of subsets of T, $\pi = \{E_k\}_{k=1}^{r}$ is an arbitrary partition of T, and $\mathcal{Q}$ is the collection of all partitions π. (See Appendix G, Sections, G1, G2, and G7.) As always, A and B are complex Banach spaces, and H is a complex Hilbert space with the inner product $(\cdot, \cdot)$.

The concept of the total variation $|\mu|(T) = \text{Var}\, \mu$ of μ on T, where μ is a complex measure, can be extended to any mapping P of $\mathfrak{C}$ into $[A; B]$.

$$\text{Total variation of } P \text{ on } T \triangleq \text{Var}\, P$$

$$\triangleq \sup\left\{\sum_k \|P(E_k)\|_{[A;\, B]} \colon \pi \in \mathcal{Q}\right\}.$$

Similarly, the semivariation is defined as follows:

$$\text{Semivariation of } P \text{ on } T \triangleq \text{SVar}\, P$$

$$\triangleq \sup\left\{\left\|\sum_k P(E_k) a_k\right\|_B \colon \pi \in \mathcal{Q}, \quad a_k \in A, \quad \|a_k\| \leq 1\right\}.$$

These concepts are commonly used in the theory of operator-valued measures (Dinculeanu, 1967). On the other hand, Hackenbroch bases his theory of integration on still another concept of variation, namely, the following:

$$\text{Scalar semivariation of } P \text{ on } T$$

$$\triangleq \text{SSVar}\, P$$

$$\triangleq \sup\left\{\left\|\sum_k P(E_k) \alpha_k\right\|_{[A;\, B]} \colon \pi \in \mathcal{Q}, \quad \alpha_k \in C, \quad |\alpha_k| \leq 1\right\}.$$

This definition of SSVar P continues to have a sense when P is replaced by a mapping N on $\mathfrak{C}$ that takes its values in an arbitrary Banach space A, and not merely in the space $[A; B]$ of operators.

The integration of certain complex-valued functions with respect to an A-valued function N on $\mathfrak{C}$ or an $[A; B]$-valued function P on $\mathfrak{C}$ can be defined as an extension of the integration of the so-called simple functions so long as N and P are additive and have finite scalar semivariations. The process is similar to the usual construction of the Riemann integral of a continuous function. Let us be explicit.

As is indicated in Appendix G8, a *simple function f from T into A* is any mapping of the form $f = \sum_k a_k \chi_{E_k}$, where $a_k \in A$, $\{E_k\} \in \mathcal{Q}$, and χ_{E_k} denotes the characteristic function of E_k. We let $\mathcal{G}_0(A) \triangleq \mathcal{G}_0(T, \mathfrak{C}; A)$ denote the linear space of all simple functions from T into A. Moreover, $G(A)$ is taken to be the Banach space of all bounded A-valued functions f from T into A with the norm $\|\cdot\|_{G(A)}$, where

$$\|f\|_{G(A)} \triangleq \sup_{t \in T} \|f(t)\|_A.$$

Hence, $\mathscr{G}_0(A) \subset G(A)$. Finally, $\mathscr{G}(A)$ will denote the closure of $\mathscr{G}_0(A)$ in $G(A)$. When $A = C$, we simplify our notation by setting $\mathscr{G}_0(C) = \mathscr{G}_0$, $\mathscr{G}(C) = \mathscr{G}$, and $G(C) = G$. $\mathscr{G}(A)$ is in general a proper subspace of $G(A)$.

A mapping N of $\mathfrak{C}$ into A is said to be *additive* if, for any pair of disjoint sets $E, F \in \mathfrak{C}$, we have $N(E \cup F) = N(E) + N(F)$. In this case, we can define the *integral of any complex-valued simple function* $f = \sum_k \alpha_k \chi_{E_k}$, where $\alpha_k \in C$, *with respect to* N by means of the expression

$$\int_T dN_t \; f(t) \triangleq \sum_k N(E_k)\alpha_k .$$

The subscript T on the integral sign is at times dropped when the set T on which the integration is taking place is evident. The subscript t on N signifies that the points in T are denoted by t.

Here are some immediate results of our definitions. The function

$$f \mapsto \int dN_t \; f(t) \tag{1}$$

is a linear mapping of $\mathscr{G}_0$ into A, and, for all $f \in \mathscr{G}_0$,

$$\left\| \int dN_t \; f(t) \right\|_A \leq \sup_{t \in T} |f(t)| \text{ SSVar } N. \tag{2}$$

Now, assume that SSVar $N < \infty$. In accordance with Appendix D, Sections D2 and D5, this implies that the mapping (1) has a unique extension that is a continuous linear mapping of $\mathscr{G}$ into A. We use the same symbolism for the extended mapping. The inequality (2) continues to hold for all $f \in \mathscr{G}$. Assume still further that N is the mapping P of $\mathfrak{C}$ into $[A; B]$. For any $a \in A$ and $b' \in B'$, where B' is the dual of B, and for all $f \in \mathscr{G}$, we have

$$\left[\int dP_t \; f(t) \right] a = \int d(P_t a) \; f(t) \tag{3}$$

and

$$b' \left[\int dP_t \; f(t) \right] a = \int d(b' P_t a) \; f(t). \tag{4}$$

Here, $P(\cdot)a$ is a B-valued additive function on $\mathfrak{C}$, and

$$\text{SSVar } P(\cdot)a \leq \|a\| \text{ SSVar } P.$$

Also, $b'P(\cdot)a$ is a complex-valued additive function on $\mathfrak{C}$, and

$$\text{SSVar } b'P(\cdot)a \leq \|b'\| \, \|a\| \text{ SSVar } P.$$

In regard to the functions on $\mathfrak{C}$, our attention will be primarily confined to those functions that take their values in an operator space $[A; B]$ and possess the following σ-additivity property.

Definition 2.2-1. A function P on $\mathfrak{C}$ into $[A; B]$ is said to be *σ-additive in the strong operator topology of* $[A; B]$ if it is additive and, given any sequence $\{E_k\}_{k=1}^{\infty} \subset \mathfrak{C}$ such that $E_k \cap E_j$ is the void set whenever $k \neq j$, we have that

$$P\left(\bigcup_{k=1}^{\infty} E_k\right)a = \sum_{k=1}^{\infty} P(E_k)a$$

for every $a \in A$. When this is the case, P is called an *operator-valued measure* or more explicitly an $[A; B]$*-valued measure*. If, in addition, $A = B = H$ and the range of P is contained in the space $[H; H]_+$ of positive operators (see Appendix D15), P is called a PO *measure* or a *positive-operator-valued measure*.

Note that P is an $[A; B]$-valued measure if and only if P is additive as a function on $\mathfrak{C}$ into $[A; B]$ and, for every increasing sequence $\{E_k\}_{k=1}^{\infty}$ in $\mathfrak{C}$ (i.e., $E_k \subset E_{k+1}$ for all k) and for all $a \in A$,

$$P\left(\bigcup_k E_k\right)a = \lim_{k \to \infty} P(E_k)a.$$

We now investigate in some detail the properties of operator-valued measures and PO measures. Let $\{E_k\}_{k=1}^{\infty}$ be an increasing sequence in $\mathfrak{C}$ with $\bigcup_k E_k = T$. Then, for any PO measure P,

$$\|P(E_k)\| = \sup_{\|a\|=1} \big(P(E_k)a, a\big) \leq \sup_{\|a\|=1} \big(P(T)a, a\big) = \|P(T)\|. \tag{5}$$

Moreover, since $(P(E_k)a, a) \to (P(T)a, a)$ for each $a \in H$, it follows that

$$\|P(E_k)\| \to \|P(T)\|, \qquad k \to \infty. \tag{6}$$

Lemma 2.2-1. *If $M \in [H; H]_+$ and $M_k \in [H; H]_+$, where $k = 1, 2, \ldots,$ and if, as $k \to \infty$, $(M_k a, a)$ increases monotonically to the limit (Ma, a) for every $a \in H$, then $M_k \to M$ in the strong operator topology of $[H; H]$.*

PROOF. If $Q \in [H; H]_+$ and $a \in H$, we may employ Schwarz's inequality (Appendix A6) to write

$$\|Qa\|^4 = |(Qa, Qa)|^2 \leq (Qa, a)(QQa, Qa).$$

But $(QQa, Qa) \leq \|Q\| \, \|Qa\|^2$, and so

$$\|Qa\|^2 \leq (Qa, a)\|Q\|. \tag{7}$$

We may replace Q by $M - M_k$ because $M - M_k \in [H; H]_+$ according to the hypothesis. Moreover, $((M - M_k)a, a) \to 0$ as $k \to \infty$. On the other hand, upon invoking the polarization equation (Appendix A7) and making two applications of the principle of uniform boundedness (Appendix D12), we see that $\|M - M_k\|$ is bounded for all k. This proves the lemma. ◇

Theorem 2.2-1. *Let P be an additive mapping of $\mathfrak{C}$ into $[H; H]_+$. The following three assertions are equivalent.*

(i) *P is a PO measure on $\mathfrak{C}$.*
(ii) *For all $a, b \in H$, $(P(\cdot)a, b)$ is a complex measure on $\mathfrak{C}$.*
(iii) *For every $a \in H$, $(P(\cdot)a, a)$ is a positive finite measure on $\mathfrak{C}$.*

Note. As is indicated in Appendix G4, complex and positive measures are by definition σ-additive.

PROOF. That (i) implies (ii) and (ii) implies (iii) is clear. To show that (iii) implies (i), we need merely prove that P is σ-additive in the strong operator topology of $[H; H]$. Let $\{E_k\}_{k=1}^{\infty}$ be an increasing sequence in $\mathfrak{C}$, and set $E = \bigcup_k E_k$. Hence, $E \in \mathfrak{C}$. Consequently, $P(E_k) \in [H; H]_+$ and $P(E) \in [H; H]_+$. Also, $(P(E_k)a, a)$ increases monotonically to $(P(E)a, a)$ because of the positivity and σ-additivity of the measure $(P(\cdot)a, a)$. Hence, by Lemma 2.2-1, $P(E_k) \to P(E)$ in the strong operator topology of $[H; H]$. $\diamond$

Theorem 2.2-2. *If $P\colon \mathfrak{C} \to [A; B]$ is an operator-valued measure, then*

$$\text{SSVar } P \leq 4 \sup_{E \in \mathfrak{C}} \|P(E)\|_{[A;B]} < \infty. \tag{8}$$

PROOF. Let F be an arbitrary member of the dual of $[A; B]$. Then, $FP \triangleq FP(\cdot)$ is a complex measure on $\mathfrak{C}$. Let $FP = L + iM$, where L is the real part of FP and M the imaginary part. For any $\pi = \{E_k\}_{k=1}^{r} \in \mathscr{Q}$, we may write

$$\begin{aligned}\sum_k |L(E_k)| &= \sum{}^{+} L(E_k) - \sum{}^{-} L(E_k)\\ &= L(\textstyle\bigcup^{+} E_k) - L(\bigcup^{-} E_k),\end{aligned}$$

where $\sum^{+}$ and $\bigcup^{+}$ ($\sum^{-}$ and $\bigcup^{-}$) are taken over those k for which $L(E_k) \geq 0$ [respectively $L(E_k) < 0$]. Therefore,

$$\begin{aligned}\sup_{\pi \in \mathscr{Q}} \sum |L(E_k)| &= \sup_{\pi \in \mathscr{Q}} [L(\textstyle\bigcup^{+} E_k) - L(\bigcup^{-} E_k)]\\ &\leq 2 \sup_{E \in \mathfrak{C}} |L(E)|.\end{aligned}$$

A similar inequality holds for M, and therefore

$$\begin{aligned}\text{SSVar } FP \leq \text{Var } FP &\leq \sup_{\pi \in \mathscr{Q}} \sum |L(E_k)| + \sup_{\pi \in \mathscr{Q}} \sum |M(E_k)|\\ &\leq 2 \sup_{E \in \mathfrak{C}} |L(E)| + 2 \sup_{E \in \mathfrak{C}} |M(E)|\\ &\leq 4 \sup_{E \in \mathfrak{C}} |FP(E)|\\ &\leq 4 \|F\| \sup_{E \in \mathfrak{C}} \|P(E)\|.\end{aligned}$$

So,

$$\sup_{\|F\|=1} \text{SSVar } FP \leq 4 \sup_{E \in \mathfrak{C}} \|P(E)\|.$$

But

$$\sup_{\|F\|=1} \text{SSVar } FP = \sup\{|F \textstyle\sum P(E_k)\alpha_k| : \|F\| = 1, \quad \{E_k\} \in \mathcal{Q}, \quad \alpha_k \in C, \quad |\alpha_k| \leq 1\}. \tag{9}$$

According to Appendix D13,

$$\sup_{\|F\|=1} \left|F \textstyle\sum P(E_k)\alpha_k\right| = \left\| \textstyle\sum P(E_k)\alpha_k \right\|_{[A;B]},$$

and therefore the right-hand side of (9) is equal to SSVar P. This establishes the first inequality of (8).

To obtain the second inequality, let $b' \in B'$ and $a \in A$. According to Appendix G7, the complex measure $b'Pa = b'P(\cdot)a$ satisfies

$$\sup_{E \in \mathfrak{C}} |b'P(E)a| \leq \text{Var } b'Pa < \infty.$$

Two applications of the principle of uniform boundedness (Appendix D12) complete the proof. ◇

Theorem 2.2-3. *If* $P\colon \mathfrak{C} \rightsquigarrow [H; H]_+$ *is a PO measure, then*

$$\text{SSVar } P = \|P(T)\|_{[H;H]}. \tag{10}$$

PROOF. Choose arbitrarily a partition $\pi = \{E_k\}_{k=1}^r \in \mathcal{Q}$, two members a and b in H, and the complex numbers α_k such that $|\alpha_k| \leq 1$, where $k = 1, \ldots, r$. We may write

$$\left|\left(\sum_{k=1}^r \alpha_k P(E_k)a, b\right)\right| \leq \sum |(P(E_k)a, b)|. \tag{11}$$

By the Schwarz inequality (Appendix A6), the right-hand side is bounded by

$$\sum (P(E_k)a, a)^{1/2}(P(E_k)b, b)^{1/2}.$$

But, by the Schwarz inequality for sums, the last expression is bounded by

$$\begin{aligned}[\textstyle\sum (P(E_k)a, a)]^{1/2}[\textstyle\sum (P(E_k)b, b)]^{1/2} &= (P(T)a, a)^{1/2}(P(T)b, b)^{1/2} \\ &\leq \|P(T)\| \; \|a\| \; \|b\|.\end{aligned}$$

Take a supremum of the left-hand side of (11) over all $\pi \in \mathcal{Q}$, all α_k such that $|\alpha_k| \leq 1$, and all a, $b \in H$ such that $\|a\| = \|b\| = 1$. This yields SSVar $P \leq \|P(T)\|$. On the other hand, upon choosing $r = 1$, $\alpha_1 = 1$, and $E_1 = T$ in the definition of SSVar P, we see that SSVar $P \geq \|P(T)\|$. ◇

Problem 2.2-1. A fact we will use later on is the following. Let $T = R^n$ and let $\mathfrak{C}$ be the collection of Borel subsets of T. For $\eta = \{\eta_k\}_{k=1}^n \in R^n$ and $t = \{t_k\}_{k=1}^n \in R^n$, we set $\eta t = \sum_k \eta_k t_k$. Then, for fixed η, the function $t \mapsto e^{i\eta t}$ is a member of $\mathscr{G}$. Show this.

Problem 2.2-2. Let P be a mapping of $\mathfrak{C}$ into $[A; B]$. Show that

$$\|P(T)\| \leq \text{SSVar}\, P \leq \text{SVar}\, P \leq \text{Var}\, P \leq \infty.$$

Problem 2.2-3. With P being an additive mapping of $\mathfrak{C}$ into $[A; B]$, assume that SVar $P < \infty$. Show that an integral $\int dP_t f(t)$ can be defined for any $f \in \mathscr{G}(A)$ as an extension of the integrals (defined in a natural way) of the functions in $\mathscr{G}_0(A)$. Also, show that

$$\left\| \int dP_t f(t) \right\|_B \leq \sup_{t \in T} \|f(t)\|_A \, \text{SVar}\, P$$

and, for any $b' \in B'$,

$$b' \int dP_t f(t) = \int d(b'P_t) f(t),$$

where $b'P(\cdot)$ is an A'-valued additive function on $\mathfrak{C}$ and

$$\text{SVar}\, b'P(\cdot) \leq \|b'\|_{B'}\, \text{SVar}\, P.$$

[Thus, upon assuming that SVar $P < \infty$, we obtain hereby a means of integrating A-valued functions with respect to $[A; B]$-valued measures. However, Hackenbroch's theory encompasses greater generality so far as the measures P are concerned; it takes into account certain measures P for which SSVar $P < \infty$ but SVar $P = \infty$. For an example of such a measure, see Hackenbroch (1968, pp. 332–333.)]

2.3. σ-FINITE OPERATOR-VALUED MEASURES

Let $\{T_k\}_{k=1}^\infty$ be an increasing sequence of sets with $T_k \in \mathfrak{C}$ and $\bigcup_k T_k = T$. Set

$$\mathfrak{C}_k \triangleq \{E \in \mathfrak{C}: \quad E \subset T_k\}, \qquad k = 1, 2, \ldots$$

and

$$\mathfrak{C}_\infty \triangleq \bigcup_{k=1}^\infty \mathfrak{C}_k.$$

It follows that $\mathfrak{C}_k$ is a σ-algebra of subsets in T_k.

Definition 2.3-1. Let P be a mapping of $\mathfrak{C}_\infty$ into $[H: H]_+$ such that the restriction of P to each $\mathfrak{C}_k$ is a PO measure on $\mathfrak{C}_k$. Then, P is called a *σ-finite PO measure on* $\mathfrak{C}_\infty$.

As an example of a σ-finite PO measure, we may take $T = R^n$, $T_k = \{t \in R^n\colon |t| \leq k\}$, $\mathfrak{C}_k$ equal to the collection of Borel subsets of T_k, and P equal to the Lebesgue measure on $\mathfrak{C}_\infty$. P is finite on each bounded Borel subset of R^n but is not defined on all the Borel subsets of R^n.

Throughout this section, P will always be a σ-finite PO measure, and h will be a member of $\mathscr{G}$. It follows that $|h(\cdot)| \in \mathscr{G}$ also. By virtue of Theorem 2.2-3,

$$\int_{T_k} dP_t h(t) \tag{1}$$

exists as a member of $[H; H]$ for each k in accordance with the preceding section.

Definition 2.3-2. h is said to be *integrable with respect to P* if, as $k \to \infty$, (1) converges in the strong operator topology of $[H; H]$; the limit is denoted by

$$\int_T dP_t h(t) = \int dP_t h(t). \tag{2}$$

This defines a linear mapping $h \mapsto \int dP_t h(t)$ of $\mathscr{G}$ into $[H; H]$. Upon invoking (4) of the preceding section, we get, for all $a, b \in H$,

$$\begin{aligned}\left(\int dP_t h(t)a, b\right) &= \left(\lim_{k\to\infty} \int_{T_k} dP_t h(t)a, b\right)\\ &= \lim_{k\to\infty} \int_{T_k} d(P_t a, b)h(t)\\ &\triangleq \int d(P_t a, b)h(t).\end{aligned} \tag{3}$$

Here, $(P(\cdot)a, b)$ is an additive complex-valued function on $\mathfrak{C}_\infty$ whose restriction to each $\mathfrak{C}_k$ is a complex measure. Similarly,

$$\left[\int dP_t h(t)\right]a = \int d(P_t a)h(t). \tag{4}$$

Theorem 2.3-1. *Let g be a nonnegative function in $\mathscr{G}$. g is integrable with respect to P if and only if the sequence $\{F_k\}_{k=1}^\infty$, where*

$$F_k \triangleq \int_{T_k} dP_t g(t) \in [H; H],$$

is bounded in the strong operator topology.

Note. By the principle on uniform boundedness, the boundedness of $\{F_k\}$ in the strong operator topology is equivalent to the boundedness of $\{F_k\}$ in the uniform operator topology.

PROOF. Since convergent sequences are bounded, the "only if" part of the theorem is clear. Conversely, assume that $\{F_k\}$ is bounded in the strong operator topology. Clearly, $(F_k a, a) \geq 0$ for every $a \in H$, and therefore $F_k \in [H; H]_+$. Similarly, $F_m - F_k \in [H; H]_+$ whenever $m > k$. Also, by the principle of uniform boundedness again, $\|F_k\| \leq M$, where M is a constant not depending on k. Moreover, upon appealing to (7) of the preceding section, we may write, for each $a \in H$,

$$\|(F_m - F_k)a\|^2 \leq 2M((F_m - F_k)a, a).$$

The right-hand side tends to zero as $k \to \infty$ because $(F_k a, a)$ is an increasing bounded sequence. Therefore, $\{F_k\}$ is a Cauchy sequence in the strong operator topology and hence must converge because of the sequential completeness of $[H, H]$ under that topology (Appendix D11). ◇

Theorem 2.3-2. *If $g \in \mathscr{G}$ is a nonnegative function and is integrable with respect to P and if $h \in \mathscr{G}$ is such that $|h(t)| \leq Mg(t)$ for all $t \in T$, where M is a constant, then h is integrable with respect to P. Moreover,*

$$\left\| \int dP_t\, h(t) \right\| \leq 4M \left\| \int dP_t\, g(t) \right\|. \tag{5}$$

PROOF. We may decompose h into

$$h = h_1 - h_2 + ih_3 - ih_4, \tag{6}$$

where, for each $j = 1, 2, 3, 4$, h_j is a nonnegative function in $\mathscr{G}$ and $h_j(t) \leq Mg(t)$. Hence, for each k and j,

$$\begin{aligned}\left\| \int_{T_k} dP_t\, h_j(t) \right\| &= \sup_{\|a\|=1} \int_{T_k} d(P_t a, a) h_j(t) \\ &\leq M \sup_{\|a\|=1} \int_T d(P_t a, a) g(t) = M \left\| \int_T dP_t\, g(t) \right\|. \end{aligned} \tag{7}$$

(See Appendix D15). This inequality coupled with Theorem 2.3-1 shows that the h_j, and therefore h as well, are integrable with respect to P. The estimate (5) follows from (7) and Appendix D11. ◇

The last theorem implies that h is integrable with respect to P whenever $|h(\cdot)|$ is.

Theorem 2.3-3. *Let $g \in \mathscr{G}$ be a nonnegative function that is integrable with respect to P. Define a mapping Q on $\mathfrak{C}$ as follows:*

$$Q(E) \triangleq \int_E dP_t g(t) \triangleq \int_T dP_t \chi_E(t) g(t), \qquad E \in \mathfrak{C}, \tag{8}$$

where χ_E is the characteristic function for E. Then, Q is a PO measure on $\mathfrak{C}$.

PROOF. Observe that $\chi_E g$ satisfies the hypothesis of Theorem 2.3-2, so that the right-hand side of (8) exists as a member of $[H; H]$. In fact, $Q(E) \in [H; H]_+$, since

$$\big(Q(E)a, a\big) = \int_T d(P_t a, a)\, \chi_E(t) g(t) \geq 0$$

for every $a \in H$. It is straightforward to show that Q is additive on $\mathfrak{C}$.

Next, we shall show that, for any $a \in H$, $(Q(\cdot)a, a)$ is a positive finite measure on $\mathfrak{C}$. To do this, we have only to show that $(Q(\cdot)a, a)$ is σ-additive (Appendix G4). For any $J \in \mathfrak{C}$, $\chi_J g$ is integrable with respect to P according to Theorem 2.3-2. Therefore,

$$\big(Q(J)a, a\big) = \lim_{m \to \infty} \int_{T_m} d(P_t a, a) \chi_J(t) g(t), \qquad J \in \mathfrak{C}. \tag{9}$$

Now, let $\{E_k\}$ be an increasing sequence in $\mathfrak{C}$ with $\bigcup_k E_k = E$. Hence, $E \in \mathfrak{C}$ also. Note that

$$\int_{T_m} d(P_t a, a)\, \chi_{E_k}(t) g(t)$$

is an increasing function of both m and k. By virtue of (9), given any $\varepsilon > 0$, we can choose an m such that

$$0 \leq \int_T d(P_t a, a) \chi_E(t) g(t) - \int_{T_m} d(P_t a, a) \chi_E(t) g(t) < \tfrac{1}{2}\varepsilon.$$

Then, we can choose a k such that

$$\begin{aligned} 0 &\leq \int_{T_m} d(P_t a, a) \chi_E(t) g(t) - \int_{T_m} d(P_t a, a) \chi_{E_k}(t) g(t) \\ &\leq \big(P(T_m \cap E \backslash E_k)a, a\big) \sup_{t \in T} |g(t)| < \tfrac{1}{2}\varepsilon \end{aligned}$$

because of the σ-additivity of $(P(\cdot)a, a)$ on $\mathfrak{C}_m$. Hence,

$$\begin{aligned} 0 &\leq \big(Q(E)a, a\big) - \big(Q(E_k)a, a\big) \\ &\leq \int_T d(P_t a, a) \chi_E(t) g(t) - \int_{T_m} d(P_t a, a) \chi_{E_k}(t) g(t) < \varepsilon. \end{aligned}$$

So, truly, $(Q(\cdot)a, a)$ is a positive finite measure on $\mathfrak{C}$. By virtue of Theorem 2.2-1, Q is a PO measure on $\mathfrak{C}$. ◇

The preceding theorem implies that the integrability of h with respect to P and the value of $\int dP_t\, h(t)$ do not depend on the choice of the sequence $\{T_k\}$. Indeed, let $\{E_k\}_{k=1}^{\infty}$ be another increasing sequence in $\mathfrak{C}_\infty$ such that $\bigcup_k E_k = T$. Let $g \in \mathscr{G}$ be nonnegative and integrable with respect to P. Then, by the σ-additivity of the PO measure Q defined by (8), as $k \to \infty$, $Q(E_k) \to Q(T)$ or equivalently

$$\int_{E_k} dP_t\, g(t) \to \int_T dP_t\, g(t)$$

in the strong operator topology. Upon decomposing h in accordance with (6), we see that the same is true when g is replaced by h.

Theorem 2.3-4. *Let P, g, and h be as in Theorem* 2.3-2. *Let Q be the PO measure defined by* (8). *Set $E_g = \{t \in T : g(t) \neq 0\}$. Then,*

$$\int dP_t\, h(t) = \int_{E_g} dQ_t \frac{h(t)}{g(t)} \triangleq \int_T dQ_t\, \chi(t) \frac{h(t)}{g(t)}, \tag{10}$$

where χ is the characteristic function for E_g and it is understood that $\chi(t)h(t)/g(t) = 0$ for $t \notin E_g$.

PROOF. We have from the theory of measurable real-valued functions that $E_g \in \mathfrak{C}$. Moreover, it is straightforward to show that the condition

$$|\chi(t)h(t)/g(t)| \leq M, \tag{11}$$

where M is a constant, implies that $\chi h/g \in \mathscr{G}$. [See Rudin (1966, pp. 8–16).] Now, the left-hand side of (10) exists by virtue of Theorem 2.3-2. In view of (11) and the fact that any constant function is integrable with respect to Q, we see that the right-hand side of (10) also exists by Theorem 2.3-2 again.

Upon appealing to (3), we may write, for any $a, b \in H$,

$$\left(\int dP_t\, h(t)a, b\right) = \int d(P_t a, b)h(t). \tag{12}$$

By a standard result for scalar integrals (Dunford and Schwartz, 1966, p. 180) the right-hand side of (12) is equal to

$$\int_{E_g} dQ_{a,b,t} \frac{h(t)}{g(t)}, \tag{13}$$

where, for any $E \in \mathfrak{C}$,

$$Q_{a,b}(E) = \int_E d(P_t a, b)g(t) = (Q(E)a, b). \tag{14}$$

Upon combining (12)–(14) and noting that a and b are arbitrary, we obtain the equality in (10). ◇

Problem 2.3-1. Verify the first two sentences in the proof of Theorem 2.3-4.

2.4. TENSOR PRODUCTS AND VECTOR-VALUED FUNCTIONS

The objective of this section is to relate the tensor product $\mathscr{G} \otimes A$ of $\mathscr{G}$ and A and its completion $\mathscr{G} \hat{\otimes} A$ under the π-topology to certain subspaces of $\mathscr{G}(A)$. These concepts are discussed in Appendix F, and we shall freely use various definitions and results that are discussed there. In particular, given any $g \in \mathscr{G}$ and $a \in A$, the tensor product $g \otimes a$ is defined in Appendix F3, where it is also pointed out that the mapping $\{g, a\} \mapsto g \otimes a$ is bilinear. Furthermore, the tensor product $\mathscr{G} \otimes A$ of $\mathscr{G}$ and A is defined as the span of all such $g \otimes a$. Each $u \in \mathscr{G} \otimes A$ has a nonunique representation of the form $u = \sum_{k=1}^{r} g_k \otimes a_k$, where $g_k \in \mathscr{G}$, $a_k \in A$, and r is finite. The function ρ is defined on any $u \in \mathscr{G} \otimes A$ by

$$\rho(u) = \inf\left\{\sum_{k=1}^{r} \|g_k\|_G \, \|a_k\|_A \colon \quad u = \sum g_k \otimes a_k\right\}, \tag{1}$$

where the infimum is taken over all representations of u. ρ is a norm. Moreover, the topology generated by ρ is called the π-topology (Appendix F5). The completion of $\mathscr{G} \otimes A$ under the π-topology is denoted by $\mathscr{G} \hat{\otimes} A$; thus, $\mathscr{G} \hat{\otimes} A$ is a Banach space. Any $w \in \mathscr{G} \hat{\otimes} A$ has a nonunique representation of the form $w = \sum_{k=1}^{\infty} g_k \otimes a_k$, where $g_k \in \mathscr{G}$, $a_k \in A$, and

$$\sum_{k=1}^{\infty} \|g_k\|_G \, \|a_k\|_A \tag{2}$$

is finite. Two series $\sum g_k \otimes a_k$ and $\sum h_k \otimes b_k$ represent the same $w \in \mathscr{G} \hat{\otimes} A$ if and only if

$$\rho\left(\sum_{k=1}^{r} g_k \otimes a_k - \sum_{k=1}^{s} h_k \otimes b_k\right) \to 0 \tag{3}$$

as r and s tend to infinity independently. The value that the norm ρ assigns to any $w \in \mathscr{G} \hat{\otimes} A$ is the infimum of the values (2) taken over all representations for w (see Appendix F7).

We define a mapping I of $\mathscr{G} \hat{\otimes} A$ into $\mathscr{G}(A)$ as follows. Given any $w \in \mathscr{G} \hat{\otimes} A$, choose any representation $w = \sum_{k=1}^{\infty} g_k \otimes a_k$. Then, define Iw by

$$Iw = \sum_{k=1}^{\infty} g_k a_k. \tag{4}$$

The right-hand side of (4) is a member of $\mathscr{G}(A)$ because the series converges under the norm of $\mathscr{G}(A)$ by virtue of the finiteness of (2). (Clearly, this convergence is absolute, a fact we shall make use of later on.) We have to show that this definition of Iw is independent of the choice of the representation for w.

Let $\sum h_k \otimes b_k$, where $h_k \in \mathscr{G}$ and $b_k \in A$, be another representation for w. By virtue of (3), we have, for any e in the dual of $\mathscr{G} \otimes A$,

$$\sum_{k=1}^{r} e(g_k \otimes a_k) - \sum_{k=1}^{s} e(h_k \otimes b_k) \to 0, \qquad r, s \to \infty.$$

According to Appendix F, Sections F4 and F6, there exists a bijection $e \mapsto j$ from the dual of $\mathscr{G} \otimes A$ onto the space $B(\mathscr{G}, A)$ of continuous bilinear forms on $\mathscr{G} \times A$ for which

$$e(g_k \otimes a_k) = j(g_k, a_k).$$

Therefore,

$$\sum_{k=1}^{\infty} j(g_k, a_k) = \sum_{k=1}^{\infty} j(h_k, b_k) \tag{5}$$

for every $j \in \mathscr{B}(\mathscr{G}, A)$. In particular, choose j such that $j(g, a) = g(t_0)q(a)$ where t_0 is an arbitrarily fixed point of T and $q \in A'$, A' being the dual of A. The finiteness of (2) implies that $\sum g_k(t_0)a_k$ and $\sum h_k(t_0)b_k$ both converge in A. Consequently, (5) implies that

$$q(\sum g_k(t_0)a_k) = q(\sum h_k(t_0)b_k)$$

for all $q \in A'$. But the weak topology of A separates A (Appendix D7), and therefore

$$\sum g_k(t_0)a_k = \sum h_k(t_0)b_k$$

for every $t_0 \in T$. So, truly, Iw does not depend on the representation used for w.

Theorem 2.4-1. *The mapping $I: \mathscr{G} \hat{\otimes} A \to \mathscr{G}(A)$ defined by (4) is linear, continuous, and injective.*

PROOF. The linearity of I follows easily from the fact that any representation $\sum g_k \otimes a_k$ of $w \in \mathscr{G} \hat{\otimes} A$ and its image $\sum g_k a_k$ in $\mathscr{G}(A)$ converge absolutely under the norms ρ and $\|\cdot\|_{G(A)}$ respectively and therefore can be rearranged (Appendix C12).

To show the continuity of I, let $w \in \mathscr{G} \hat{\otimes} A$ and let $\sum g_k \otimes a_k$ be any one of its representations. Set $f = Iw = \sum g_k a_k$. Then,

$$\|f\|_{G(A)} \leq \sum \|g_k\|_G \|a_k\|_A.$$

Since this holds for every representation of w, we have that $\|f\|_{G(A)} \leq \rho(w)$, which implies the continuity of I.

We now set about proving the injectivity of I. We have to show that, if $\{g_k\}_{k=1}^{\infty} \subset \mathscr{G}$ and $\{a_k\}_{k=1}^{\infty} \subset A$ are such that $\sum \|g_k\| \, \|a_k\| < \infty$ and $f \triangleq \sum g_k a_k$ is the zero member of $\mathscr{G}(A)$, then $w \triangleq \sum g_k \otimes a_k$ is the zero member of $\mathscr{G} \hat{\otimes} A$. That $w = 0 \in \mathscr{G} \hat{\otimes} A$ means that, given any $\varepsilon > 0$, there exists an equivalent representation $w = \sum h_k \otimes b_k$, where $h_k \in \mathscr{G}$, $b_k \in A$, and $\sum \|h_k\| \, \|b_k\| < \varepsilon$. We shall prove that such an equivalent representation exists.

Let $\{E_i\}_{i=1}^{r}$ be an arbitrary partition of T, let t_i be a point in E_i, and let χ_i be the characteristic function for E_i. Therefore, $\sum_i \chi_i(t) = 1$ for every $t \in T$. Since $f(t_i) = \sum_k g_k(t_i) a_k = 0$, we may write

$$\sum_{i=1}^{r} \chi_i \otimes \sum_{k=1}^{\infty} g_k(t_i) a_k = 0.$$

Now, $\mathscr{G} \otimes A$ is equipped with the π-topology, and therefore the mapping $\{g, a\} \mapsto g \otimes a$ is continuous (Appendix F5). Therefore, since $\sum_k g_k(t_i) a_k$ converges in A,

$$\sum_{i=1}^{r} \sum_{k=1}^{\infty} g_k(t_i) \chi_i \otimes a_k = 0.$$

Upon interchanging the summations on i and k and subtracting the resulting representation of $0 \in \mathscr{G} \hat{\otimes} A$ from $w = \sum_k g_k \otimes a_k \in \mathscr{G} \hat{\otimes} A$, we obtain

$$w = \sum_{k=1}^{\infty} \left\{ \sum_{i=1}^{r} [g_k - g_k(t_i)] \chi_i \right\} \otimes a_k. \tag{6}$$

Now,

$$\left\| \sum_{i=1}^{r} g_k(t_i) \chi_i \right\|_G \leq \|g_k\|_G$$

and (2) is finite. Therefore, given any $\varepsilon > 0$, there exists an integer s not depending on the choices of the partition $\{E_i\}$ or the points $t_i \in E_i$ such that

$$\sum_{k=s+1}^{\infty} \left\| \sum_{i=1}^{r} [g_k - g_k(t_i)] \chi_i \right\|_G \|a_k\|_A \leq 2 \sum_{k=s+1}^{\infty} \|g_k\| \, \|a_k\| < \tfrac{1}{2}\varepsilon. \tag{7}$$

We now state a lemma but postpone its proof.

Lemma 2.4-1. *For any given $g \in \mathscr{G}$ and $\varepsilon > 0$, there exists a partition $\{E_i\}_{i=1}^{r}$ of T such that*

$$\left\| g - \sum_i g(t_i) \chi_i \right\|_G < \varepsilon \tag{8}$$

whatever be the choices of $t_i \in E_i$.

According to this lemma, we can choose a partition $\{E_i\}_{i=1}^r$ such that, for each $k = 1, \ldots, s$,

$$\left\| \sum_i [g_k - g_k(t_i)]\chi_i \right\|_G < \tfrac{1}{2}\varepsilon \left[\sum_{k=1}^{s} \|a_k\| \right]^{-1}. \tag{9}$$

Hence, for that partition, (7) and (9) imply that (6) is the equivalent representation for w that we have been seeking. This completes the proof of Theorem 2.4-1. ◇

Proof of Lemma 2.4-1. By the definition of $\mathscr{G}$, there exists a partition $\{E_i\}_{i=1}^r$ of T and a corresponding simple function

$$\sum_{i=1}^{r} c_i \chi_i, \qquad c_i \in C, \quad i = 1, \ldots, r,$$

such that, for every i and every $t \in E_i$,

$$|g(t) - c_i| < \tfrac{1}{2}\varepsilon.$$

Therefore,

$$\left| g(t) - \sum_i c_i \chi_i(t) \right| < \tfrac{1}{2}\varepsilon$$

for every $t \in T$. The last two inequalities may be combined to yield (8). ◇

We now define some additional notations and terminology that we will be using. The image of $\mathscr{G} \hat{\otimes} A$ under the mapping I is denoted by $\mathscr{G} \hat{\odot} A$. Thus, $\mathscr{G} \hat{\odot} A$ is the set of all $f \in \mathscr{G}(A)$ having representations of the form (4) such that (2) is finite.

Furthermore, the image of $\mathscr{G} \otimes A$ under I is denoted by $\mathscr{G} \odot A$. Hence, $\mathscr{G} \odot A$ is the set of all $f \in \mathscr{G}(A)$ having representations of the form $f = \sum_{k=1}^{r} g_k a_k$. Thus, for any $f \in \mathscr{G} \odot A$, $f(T)$ is contained in a finite-dimensional subspace of A. For this reason, the members of $\mathscr{G} \odot A$ will be called *finite-dimensionally-ranging functions.*

Recall that $\mathscr{G}_0(A)$ is the space of all A-valued simple functions on T. Clearly,

$$\mathscr{G}_0(A) \subset G \odot A \subset \mathscr{G} \hat{\odot} A. \tag{10}$$

Corollary 2.4-1a. *Define the functional* $\|\cdot\|_1$ *on each* $f \in \mathscr{G} \hat{\odot} A$ *by*

$$\|f\|_1 \triangleq \inf\left\{ \sum_k \|g_k\|_G \|a_k\|_A : \ f = \sum g_k a_k \right\}. \tag{11}$$

Then, $\|\cdot\|_1$ *is a norm on* $\mathscr{G} \hat{\odot} A$. *Moreover,* $\mathscr{G} \hat{\odot} A$ *equipped with the topology generated by* $\|\cdot\|_1$ *is a Banach space, and the mapping I defined by* (4) *is an isomorphism from* $\mathscr{G} \hat{\otimes} A$ *onto* $\mathscr{G} \hat{\odot} A$.

PROOF. By Theorem 2.4-1 and the definition of $\mathscr{G} \hat{\odot} A$, I is linear, continuous, and bijective. Therefore, its inverse I^{-1} is also linear and continuous (Appendix D14). Also, for $w \in \mathscr{G} \hat{\otimes} A$ and $f = Iw$ and for the norm ρ defined by (1), we have that $\|f\|_1 = \rho(w)$. Thus, I is an isomorphism, $\|\cdot\|_1$ is a norm on $\mathscr{G} \hat{\odot} A$, and $\mathscr{G} \hat{\odot} A$ must be complete since $\mathscr{G} \hat{\otimes} A$ is complete. ◇

Henceforth, it will be understood that $\mathscr{G} \hat{\odot} A$ possesses the topology generated by $\|\cdot\|_1$. $\mathscr{G} \hat{\odot} A$ is the completion of $\mathscr{G} \odot A$ with respect to the norm $\|\cdot\|_1$.

2.5. INTEGRATION OF VECTOR-VALUED FUNCTIONS

We are now ready to define the integral of an A-valued function with respect to an $[A; B]$-valued measure. In this section, P will be an $[A; B]$-valued measure on $\mathfrak{C}$. Consequently, SSVar $P < \infty$ according to Theorem 2.2-2, and $\int dP_t\, g(t)$ exists as a member of $[A; B]$ for each $g \in \mathscr{G}$.

Definition 2.5-1. Let $f \in \mathscr{G} \odot A$ and choose any representation $f = \sum_{k=1}^{r} g_k a_k$, where $g_k \in \mathscr{G}$ and $a_k \in A$. We define the *integral of f with respect to P* by

$$\int dP_t\, f(t) \triangleq \sum_{k=1}^{r} \left[\int dP_t\, g_k(t) \right] a_k. \tag{1}$$

To justify this definition, we have to show that the right-hand side of (1) does not depend on the choice of the representation for f. Let $f = \sum_{i=1}^{s} h_i b_i$, where $h_i \in \mathscr{G}$ and $b_i \in A$, be another representation. Now, we can find l linearly independent elements $e_1, \ldots, e_l \in A$ such that, for each k and i,

$$a_k = \sum_{j=1}^{l} \alpha_{kj} e_j, \qquad b_i = \sum_{k=1}^{l} \beta_{ij} e_j,$$

where $\alpha_{kj}, \beta_{ij} \in C$. (See Appendix A3.) Upon substituting these sums into the two representations of f and invoking the linear independence of the e_j, we obtain

$$\sum_{k=1}^{r} g_k \alpha_{kj} = \sum_{i=1}^{s} h_i \beta_{ij}.$$

Hence,

$$\sum_{k=1}^{r}\left[\int dP_t\, g_k(t)\right]a_k = \sum_{j=1}^{l}\left[\int dP_t \sum_{k=1}^{r} g_k(t)\alpha_{kj}\right]e_j$$
$$= \sum_{j=1}^{l}\left[\int dP_t \sum_{i=1}^{s} h_i(t)\beta_{ij}\right]e_j$$
$$= \sum_{i=1}^{s}\left[\int dP_t\, h_i(t)\right]b_i.$$

This is what we wished to show.

Clearly, then, $f \mapsto \int dP_t f(t)$ is a linear mapping of $\mathscr{G} \odot A$ into B. Still more is true. It is a continuous mapping because of the inequality

$$\left\|\int dP_t\, f(t)\right\|_B \leq \|f\|_1 \text{ SSVar } P, \tag{2}$$

which is established as follows. We know that, for any $g \in \mathscr{G}$,

$$\left\|\int dP_t\, g(t)\right\|_{[A;\,B]} \leq \|g\|_G \text{ SSVar } P.$$

Consequently, from (1), we have

$$\left\|\int dP_t\, f(t)\right\|_B \leq \sum_{k=1}^{r}\left\|\int dP_t\, g_k(t)\right\| \|a_k\|$$
$$\leq \sum \|g_k\| \; \|a_k\| \text{ SSVar } P.$$

Since this holds for every representation of f, (2) follows.

We can now conclude that the mapping $f \mapsto \int dP_t f(t)$ possesses a unique extension that is a continuous linear mapping of $\mathscr{G} \hat{\odot} A$ into B. This is how we define the integral $\int dP_t f(t)$ on any $f \in \mathscr{G} \hat{\odot} A$. But, since every such f is the limit under the norm $\|\cdot\|_1$ of a series $\sum_{k=1}^{\infty} g_k a_k$, an equivalent definition is the following.

Definition 2.5-2. Let $f \in \mathscr{G} \hat{\odot} A$ and choose any representation $f = \sum_{k=1}^{\infty} g_k a_k$, where $g_k \in \mathscr{G}$ and $a_k \in A$. The *integral of f with respect to P* is defined by

$$\int dP_t\, f(t) \triangleq \sum_{k=1}^{\infty}\left[\int dP_t\, g_k(t)\right]a_k. \tag{3}$$

It follows that this definition is independent of the choice of the representation for f, that the inequality (2) continues to hold for all $f \in \mathscr{G} \hat{\odot} A$, and that $f \mapsto \int dP_t f(t)$ is a continuous linear mapping of $\mathscr{G} \hat{\odot} A$ into B.

In the next two theorems, T and X are two nonvoid sets, $\mathfrak{C}$ and $\mathfrak{C}'$ are σ-algebras of subsets of T and X, respectively, and μ is either a complex-valued measure or a positive measure on $\mathbf{C}'$ (Appendix G4). Also, $\mathfrak{C} \times \mathfrak{C}'$ denotes the product σ-algebra in $T \times T$ (Appendix G3). $L_1(\mu; A) = L_1(X, \mathfrak{C}'; \mu; A)$ is the linear space of Bochner-integrable A-valued functions on X with respect to the measure μ; as is explained in Appendix G13, $L_1(\mu; A)$ is really a space of equivalence classes of functions, but we speak of its members as being individual functions. Finally, $\mathscr{G}(T \times X, \mathfrak{C} \times \mathfrak{C}'; C)$ is the Banach space of all complex-valued bounded functions g on $T \times X$ that are the limits of sequences of simple functions under the norm $\|\cdot\|_G$, where

$$\|g\|_G \triangleq \sup\{|g(t, x)| : t \in T, \quad x \in X\}. \tag{4}$$

We now state a representation theorem for any $f \in \mathscr{G} \hat{\odot} A$.

Theorem 2.5-1. *Let $F \in L_1(X, \mathfrak{C}'; \mu; A)$ and $l \in \mathscr{G}(T \times X, \mathfrak{C} \times \mathfrak{C}'; C)$. Then,*

$$f(t) \triangleq \int_X d\mu_x\, l(t, x)F(x) \tag{5}$$

exists as a Bochner integral for each $t \in T$ and defines a function $f \in \mathscr{G} \hat{\odot} A$, where $\mathscr{G} = \mathscr{G}(T, \mathfrak{C}; C)$. Moreover,

$$\|f\|_1 \le \|F\|_{L_1} \|l\|_G, \tag{6}$$

where

$$\|F\|_{L_1} \triangleq \int_X d|\mu|_x \, \|F(x)\|_A$$

and $|\mu|$ denotes the total-variation measure of μ (Appendix G7).

Conversely, every $f \in \mathscr{G} \hat{\odot} A$ has a representation of the form (5).

PROOF. We prove the last statement first. If $f \in \mathscr{G} \hat{\odot} A$, then $f = \sum_{k=1}^{\infty} g_k a_k$, where $g_k \in \mathscr{G}$, $a_k \in A$, and $\sum \|g_k\| \, \|a_k\| < \infty$. Let $X = \{k\}_{k=1}^{\infty}$, let $\mathfrak{C}'$ be the set of all subsets of X, and let $\mu(E)$ be the cardinality of $E \subset \mathfrak{C}'$. [This μ is called the *counting measure*; see Rudin (1966, p. 17).] We can choose an l such that $l(t, k) = g_k(t)$ and an F such that $F(k) = a_k$. This immediately yields the representation (5) for f.

Conversely, the existence of (5) as a Bochner integral is asserted by Appendix G14. To prove the rest of Theorem 2.5-1, let $L_1^{\,0}(\mu; A) \triangleq L_1^{\,0}(X, \mathfrak{C}'; \mu; A)$ be the space of all simple functions in $L_1(\mu; A)$. Thus, $L_1^{\,0}(\mu; A)$ is the space of simple functions $F = \sum_{k=1}^{r} a_k \chi_{E_k}$ such that, if $a_k \neq \{0\}$, then $\mu(E_k)$ is finite. We define a mapping

$$J: \quad G(T \times X, \mathfrak{C} \times \mathfrak{C}'; C) \times L_1^{\,0}(\mu; A) \to \mathscr{G}(T, \mathfrak{C}; C) \odot A$$

by

$$[J(l, F)](t) \triangleq \int_X d\mu_x\, l(t, x)F(x) = \sum_{k=1}^{r} \int_{E_k} d\mu_x\, l(t, x)a_k,$$

where again $F = \sum_{k=1}^{r} a_k \chi_{E_k}$. [That

$$\int_{E_k} d\mu_x\, l(t, x) \in \mathscr{G}(T, \mathfrak{C}; C)$$

can be seen by taking a sequence of simple functions on $T \times X$ that converges to l and then using the estimate in Appendix G13.] We next observe that

$$\begin{aligned}\|J(l, F)\|_{G(A)} &\triangleq \sup_{t \in T} \|[J(l, F)](t)\|_A \leq \|J(l, F)\|_1 \\ &\leq \sum_{k=1}^{r} \sup_{t \in T} \left| \int_{E_k} d\mu_x\, l(t, x) \right| \|a_k\| \\ &\leq \sum_{k=1}^{r} |\mu|(E_k)\, \|a_k\| \sup_{\substack{t \in T \\ x \in X}} |l(t, x)| = \|F\|_{L_1} \|l\|_G. \qquad (7)\end{aligned}$$

This shows that the linear mapping $F \mapsto J(l, F)$ is continuous from $L_1{}^0(\mu; A)$, supplied with the topology induced by $L_1(\mu; A)$, into $\mathscr{G} \hat{\odot} A$. By virtue of the density of $L_1{}^0(\mu; A)$ in $L_1(\mu; A)$ (Appendix G16), that mapping has a unique extension that is a continuous linear mapping of $L_1(\mu; A)$ into $\mathscr{G} \hat{\odot} A$. We can conclude that, for any $F \in L_1(\mu; A)$, the f given by (5) is a member of $\mathscr{G} \hat{\odot} A$ and that the inequality (6) is a consequence of (7). ◇

Our next objective is to develop a Fubini-type theorem.

Theorem 2.5-2. *Let $F \in L_1(X, \mathfrak{C}'; \mu; A)$ and $l \in \mathscr{G}(T \times X, \mathfrak{C} \times \mathfrak{C}'; C)$. Then,*

$$\int_T dP_t \int_X d\mu_x\, l(t, x)F(x) = \int_X d\mu_x \left[\int_T dP_t\, l(t, x) \right] F(x), \qquad (8)$$

where the outer integral on the left-hand side exists in the sense of Definition 2.5-2 and the outer integral on the right-hand side is a Bochner integral.

PROOF. That the left-hand side of (8) exists in the sense of Definition 2.5-2 follows from Theorem 2.5-1.

On the other hand, for each $x \in X$, $l(\cdot, x) \in \mathscr{G} = \mathscr{G}(T, \mathfrak{C}; C)$ according to Appendix G14. So, $\int dP_t\, l(t, x)$ exists in accordance with Section 2.2. We can choose a sequence $\{l_n\}_{n=1}^{\infty}$ of simple functions in $\mathscr{G}(T \times X, \mathfrak{C} \times \mathfrak{C}'; C)$ such that $\|l - l_n\| \to 0$. Therefore, for each $x \in X$,

$$\int_T dP_t\, l_n(t, x) \to \int_T dP_t\, l(t, x) \qquad (9)$$

in $[A; B]$. But, for each n, the left-hand side of (9) is a simple $[A; B]$-valued function of x. Therefore, the right-hand side is a measurable function of x (Appendix G9).

Since $F \in L_1(\mu; A)$, we can choose a sequence $\{F_n\}$ such that $F_n \in L_1{}^0(\mu; A)$ and $F_n \to F$ in $L_1(\mu; A)$ as well as almost everywhere on X (Appendix G16). Then, for each n, $\int_T dP_t\, l_n(t, x)F_n(x)$ is a simple B-valued function of x, and, as $n \to \infty$, it converges almost everywhere on X to

$$\int_T dP_t\, l(t, x)F(x). \tag{10}$$

Therefore, (10) is a measurable function of x. Moreover, $\|F(\cdot)\|_A \in L_1(\mu; R)$, and

$$\left\| \int_T dP_t\, l(t, x)F(x) \right\|_B \le \|F(x)\|_A \sup_{t,x} |l(t, x)| \text{ SSVar } P. \tag{11}$$

Consequently, by Appendix G15, the right-hand side of (8) truly exists as a Bochner integral.

Next, we observe that

$$\left\| \int_X d\mu_x \left[\int_T dP_t\, l(t, x) \right] F(x) \right\|_B \le \|F\|_{L_1} \sup_{t,x} |l(t, x)| \text{ SSVar } P \tag{12}$$

according to the inequality (2) of Section 2.2 and Appendix G11. On the other hand, for f defined by (5), we have that $f \in \mathscr{G} \hat{\odot} A$. So, by (2) and (6),

$$\begin{aligned} \left\| \int_T dP_t\, f(t) \right\|_B &\le \|f\|_1 \text{ SSVar } P \\ &\le \|F\|_{L_1} \sup_{t,x} |l(t, x)| \text{ SSVar } P. \end{aligned} \tag{13}$$

The inequalities (12) and (13) and the density of $L_1{}^0(\mu; A)$ in $L_1(\mu; A)$ imply that we need merely establish the equality in (8) for every $F \in L_1{}^0(\mu; A)$.

So, let $F = \sum_{k=1}^r a_k \chi_{E_k} \in L_1{}^0(\mu; A)$ and let $b' \in B'$ be arbitrary. We may write

$$b' \int d\mu_x \left[\int dP_t\, l(t, x) \right] F(x) = \sum_{k=1}^r \int_{E_k} d\mu_x \int d(b'P_t a_k)\, l(t, x) \tag{14}$$

according to Appendix G17 and (4) of Section 2.2. But, when $a_k \ne 0$, μ and $b'P(\cdot)a_k$ have finite total variations on E_k and T, respectively (Theorem 2.2-1 and Appendix G7), and therefore we may apply the scalar Fubini theorem to the right-hand side of (14) to interchange the integrations. Then, upon extracting b' and the a_k, we obtain

$$b' \int dP_t \int d\mu_x\, l(t, x)F(x).$$

Since $b' \in B'$ is arbitrary, (8) has been established for every $F \in L_1{}^0(\mu; A)$. This completes the proof. ◇

Problem 2.5-1. Show that Definition 2.5-1 is consistent with the definition of $\int dP_t f(t)$ indicated in Problem 2.2-3, for the case where SVar $P < \infty$ and $f \in \mathscr{G} \odot A$. [The last two conditions imply that SSVar $P < \infty$, according to Problem 2.2-2, and that $f \in \mathscr{G}(A)$.]

2.6. SESQUILINEAR FORMS GENERATED BY PO MEASURES

We end this chapter with a discussion of certain positive sesquilinear forms generated by PO measures. In the first part of this section, P is restricted to being a PO measure on $\mathfrak{C}$. Hence, P maps $\mathfrak{C}$ into $[H; H]_+$, where H is a complex Hilbert space with the inner product $(\cdot, \cdot)$. Moreover, SSVar $P = \|P(T)\| < \infty$ according to Theorem 2.2-3. As usual, $\bar{\alpha}$ denotes the complex conjugate of any number, function, or measure α.

Lemma 2.6-1. *Let f and v be any two members of $\mathscr{G} \odot H$. Define a function $\mathfrak{B}_P$ on $\mathscr{G} \odot H \times \mathscr{G} \odot H$ by choosing any two representations*

$$f = \sum_{k=1}^{r} g_k a_k, \qquad v = \sum_{j=1}^{s} h_j b_j,$$

where $g_k, h_j \in \mathscr{G}$ and $a_k, b_j \in H$, and then setting

$$\mathfrak{B}_P(f, v) \triangleq \sum_{k=1}^{r} \sum_{j=1}^{s} \left(\left[\int dP_t\, g_k(t)\overline{h_j(t)} \right] a_k, b_j \right). \tag{1}$$

Then, $\mathfrak{B}_P$ is a positive sesquilinear form on the space $\mathscr{G} \odot H \times \mathscr{G} \odot H$, and

$$|\mathfrak{B}_P(f, v)| \le \|P(T)\| \, \|f\|_1 \|v\|_1. \tag{2}$$

PROOF. An argument like the one following Definition 2.5-1 shows that the right-hand side of (1) is independent of the choices of the representations for f and v. Moreover, $\mathfrak{B}_P$ is clearly a sesquilinear form on $\mathscr{G} \odot H \times \mathscr{G} \odot H$.

To prove the inequality (2), we write

$$\begin{aligned} |\mathfrak{B}_P(f, v)| &\le \sum_{k=1}^{r} \sum_{j=1}^{s} \left\| \int dP_t\, g_k(t)\overline{h_j(t)} \right\| \|a_k\| \, \|b_j\| \\ &\le \|P(T)\| \sum_k \|g_k\| \, \|a_k\| \sum_j \|h_j\| \, \|b_j\|. \end{aligned} \tag{3}$$

Since the left-hand side does not depend on the representations for f and v, we may take the infimum over all such representations to get (2).

Finally, we show that $\mathfrak{B}_P$ is a positive form. Let $g_{k,0}$ be a simple function that approximates g_k, and set

$$f_0 = \sum_{k=1}^{r} g_{k,0}\, a_k. \tag{4}$$

Then,

$$\mathfrak{B}_P(f,f) - \mathfrak{B}_P(f_0, f_0) = \sum_{k=1}^{r} \sum_{j=1}^{r} \left(\int dP_t[g_k(t)\overline{g_j(t)} - g_{k,0}(t)\overline{g_{j,0}(t)}]a_k, a_j \right).$$

Through an estimate similar to (3), we see that the right-hand side can be made arbitrarily small by choosing the $g_{k,0}$ appropriately. Moreover, functions of the form (4) are themselves simple functions. Thus, to complete the proof, we need merely establish the positivity of $\mathfrak{B}_P$ on functions of the form $f_0 = \sum_{i=1}^{n} a_i \chi_{E_i}$, where the E_i comprise a partition of T and are therefore pairwise disjoint. Whence,

$$\int dP_t\, \chi_{E_i}(t)\chi_{E_m}(t) = P(E_i \cap E_m) = \begin{cases} P(E_i), & i = m \\ 0, & i \neq m, \end{cases}$$

and therefore

$$\mathfrak{B}_P(f_0, f_0) = \sum_{i=1}^{n} (P(E_i)a_i, a_i) \geq 0.$$

Lemma 2.6-1 has been completely established. ◇

In view of Lemma 2.6-1, we can extend $\mathfrak{B}_P$ continuously onto the Cartesian product $\mathscr{G} \hat{\odot} H \times \mathscr{G} \hat{\odot} H$ supplied with the product topology (Appendix D5). The resulting mapping, which we also denote by $\mathfrak{B}_P$, will be a positive sesquilinear form on $\mathscr{G} \hat{\odot} H \times \mathscr{G} \hat{\odot} H$ that satisfies the inequality (2) for all $f, v \in \mathscr{G} \hat{\odot} H$. We use this result to define still another kind of integral.

Definition 2.6-1. For any $f, v \in \mathscr{G} \hat{\odot} H$, we set

$$\int d(P_t f(t), v(t)) \triangleq \mathfrak{B}_P(f, v). \tag{5}$$

The next theorem presents an explicit formula for (5) that can be used when representations for f and v are given in accordance with Theorem 2.5-1. The following notation is used. T, X, and Y are three nonvoid sets, $\mathfrak{C}$, $\mathfrak{C}'$, and $\mathfrak{C}''$ are σ-algebras of subsets of T, X, and Y, respectively, and μ and ν are complex-valued measures or positive measures on $\mathfrak{C}'$ and $\mathfrak{C}''$, respectively. We know from Theorem 2.5-1 that $f \in \mathscr{G} \hat{\odot} H$ if and only if it has the representation

$$f(t) = \int_X d\mu_x\, l(t, x)F(x), \tag{6}$$

where $F \in L_1(X, \mathfrak{C}'; \mu; H)$ and $l \in \mathscr{G}(T \times X, \mathfrak{C} \times \mathfrak{C}'; C)$. Similarly, $g \in \mathscr{G} \hat{\odot} H$ if and only if

$$v(t) = \int_Y d\nu_y\, m(t, y) V(y), \tag{7}$$

where $V \in L_1(Y, \mathfrak{C}''; \nu; H)$ and $m \in \mathscr{G}(T \times Y, \mathfrak{C} \times \mathfrak{C}''; C)$.

Theorem 2.6-1. *Let f and v have the representations* (6) *and* (7). *Then,*

$$\begin{aligned} \mathfrak{B}_P(f, v) &\triangleq \int d(P_t f(t), v(t)) \\ &= \int d\mu_x \int d\bar{\nu}_y \left(\left[\int dP_t\, l(t, x)\overline{m(t, y)} \right] F(x), V(y) \right). \end{aligned} \tag{8}$$

PROOF. We first note that the mapping

$$\{t, x, y\} \mapsto l(t, x)\overline{m(t, y)}$$

is a member of $\mathscr{G}(T \times X \times Y, \mathfrak{C} \times \mathfrak{C}' \times \mathfrak{C}''; C)$. This fact and a straightforward estimate using (2) of Section 2.2 show that the right-hand side of (8) exists as a scalar integral.

For the integrable simple functions

$$F = \sum_{k=1}^{r} a_k \chi_{E_k} \in L_1^0(X, \mathfrak{C}'; \mu; H)$$

and

$$V = \sum_{j=1}^{s} b_j \chi_{I_j} \in L_1^0(Y, \mathfrak{C}''; \nu; H),$$

the right-hand side of (8) is equal to

$$\begin{aligned} &\sum_{k=1}^{r} \sum_{j=1}^{s} \int_{E_k} d\mu_x \int_{I_j} d\bar{\nu}_y \left(\left[\int dP_t\, l(t, x)\overline{m(t, y)} \right] a_k, b_j \right) \\ &\qquad = \sum \sum \int_{E_k} d\mu_x \int_{I_j} d\bar{\nu}_y \int d(P_t a_k, b_j) l(t, x)\overline{m(t, y)}. \end{aligned}$$

Since μ, $\bar{\nu}$, and $(P(\cdot)a_k, b_j)$ have finite total variations on E_k, I_j, and T, respectively, whenever $a_k \neq 0$ and $b_j \neq 0$, we may apply the scalar Fubini theorem and then extract the a_k and b_j to obtain

$$\sum \sum \left(\left[\int dP_t \int_{E_k} d\mu_x\, l(t, x) \int_{I_j} d\bar{\nu}_y\, \overline{m(t, y)} \right] a_k, b_j \right) = \mathfrak{B}_P(f, v).$$

Thus, (8) is true for any integrable simple functions F and V.

To show that (8) remains true for all $F \in L_1(X, \mathfrak{C}'; \mu; H)$ and

$$V \in L_1(Y, \mathfrak{C}''; \nu; H),$$

we need merely show that both of its sides depend continuously on F and V with respect to the L_1 norms (Appendix D5). For the left-hand side, this follows from

$$\begin{aligned}|\mathfrak{B}_P(f, v)| &\le \|P(T)\| \; \|f\|_1 \; \|v\|_1 \\ &\le \|P(T)\| \; \|F\|_{L_1} \sup_{t,x} |l(t, x)| \; \|V\|_{L_1} \sup_{t,y} |m(t, y)|. \qquad (9)\end{aligned}$$

It is easily seen that the right-hand side of (8) is also bounded by the right-hand side of (9). ◇

In the remainder of this chapter, $P: \mathfrak{C}_\infty \to [H; H]_+$ is a σ-finite PO measure. Also, we let $g \in \mathscr{G}$, $g(t) > 0$ for all $t \in T$, and g be integrable with respect to P (Definition 2.3-2). Moreover, we set

$$Q(E) \triangleq \int_E dP_t \, g(t), \qquad E \in \mathfrak{C}.$$

By Theorem 2.3-3, Q is a PO measure on $\mathfrak{C}$. Finally, we set

$$\mathscr{G}_g(H) \triangleq \mathscr{G}_g(T, \mathfrak{C}; H) \triangleq \{f \in \mathscr{G}(H): f/g \in \mathscr{G} \mathbin{\hat{\odot}} H\}. \qquad (10)$$

Lemma 2.6-2. $\mathscr{G}_g(H) \subset \mathscr{G}_{\sqrt{g}}(H) \subset \mathscr{G} \mathbin{\hat{\odot}} H$.

PROOF. We will use the obvious fact that, if $u \in \mathscr{G}$ and $q \in \mathscr{G} \mathbin{\hat{\odot}} H$, then $uq \in \mathscr{G} \mathbin{\hat{\odot}} H$. Since $g \in \mathscr{G}$, $\sqrt{g} \in \mathscr{G}$ also. By definition, for any $f \in \mathscr{G}_g(H)$, we have that $f/g \in \mathscr{G} \mathbin{\hat{\odot}} H$. Therefore, $\sqrt{g}f/g = f/\sqrt{g} \in \mathscr{G} \mathbin{\hat{\odot}} H$, which implies that $\mathscr{G}_g(H) \subset \mathscr{G}_{\sqrt{g}}(H)$. Another multiplication by $\sqrt{g}$ shows that $\mathscr{G}_{\sqrt{g}}(H) \subset \mathscr{G} \mathbin{\hat{\odot}} H$. ◇

Lemma 2.6-3. *If $f \in \mathscr{G}_g(H)$, then f has the representation* (6) *where, in addition, the function $\{t, x\} \mapsto l(t, x)/g(t)$ is a member of $\mathscr{G}(T \times X, \mathfrak{C} \times \mathfrak{C}'; C)$.*

This lemma follows directly from Theorem 2.5-1.

Definition 2.6-2. For any $f \in \mathscr{G}_g(H)$, we set

$$\int dP_t \, f(t) \triangleq \int dQ_t \, \frac{f(t)}{g(t)}.$$

Note that the right-hand side has a sense according to (10) and Definition 2.5-2. Clearly, $f \mapsto \int dP_t \, f(t)$ is a linear mapping of $\mathscr{G}_g(H)$ into H. Moreover, this definition of $\int dP_t f(t)$ is independent of the choice of g. To see this, first note that, for every $f \in \mathscr{G}_g(H) \cap \mathscr{G} \odot H$, we may choose a representation, $f = \sum_{k=1}^{r} g_k a_k$, where $a_k \in H$, $g_k \in \mathscr{G}$, and $|g_k(t)| < c_k g(t)$ for all $t \in T$, the c_k being constants. [Indeed, since $f/g \in \mathscr{G} \odot H$, we can write $f/g = \sum_{k=1}^{r} q_k a_k$ and choose $c_k \triangleq \sup_t |q_k(t)|$.] Therefore,

$$\int dP_t\, f(t) = \sum_{k=1}^{r} \left[\int dP_t\, g_k(t)\right] a_k .$$

According to Theorem 2.3-2, the right-hand side does not depend on the choice of g. Next, it can be seen as before that every $f \in \mathscr{G}_g(H)$ has a representation of the form $f = \sum_{k=1}^{\infty} g_k a_k$, where $a_k \in H$, $g_k \in \mathscr{G}$, $|g_k(t)| < c_k g(t)$ for all $t \in T$, the c_k are constants, and $\sum_{k=1}^{\infty} c_k \|a_k\| < \infty$. Set $f_r = \sum_{k=1}^{r} g_k a_k$. As $r \to \infty$,

$$\int dP_t\ f_r(t) \to \int dP_t\, f(t) \tag{11}$$

in H because

$$\left\| \int dP_t[f(t) - f_r(t)] \right\| = \left\| \int dQ_t \frac{1}{g(t)} \sum_{k=r+1}^{\infty} g_k(t) a_k \right\|$$

$$\leq Q(T) \sum_{k=r+1}^{\infty} c_k \|a_k\| \to 0.$$

Since the left-hand side of (11) does not depend on the choice of g, neither does the right-hand side.

Definition 2.6-3. For any $f,\, v \in \mathscr{G}_g(H)$, we set

$$\int d(P_t f(t), v(t)) \triangleq \int d\left(Q_t \frac{f(t)}{[g(t)]^{1/2}}, \quad \frac{v(t)}{[g(t)]^{1/2}}\right). \tag{12}$$

In view of Definition 2.6-1 and Lemma 2.6-2, the right-hand side has a sense and defines a sesquilinear mapping on $\mathscr{G}_g(H) \times \mathscr{G}_g(H)$. Moreover, (12) is independent of the choice of g, as can be seen through an argument similar to the one used for Definition 2.6-2.

Theorem 2.6-2. *Let $f \in \mathscr{G}_g(H)$ and $v \in \mathscr{G}_g(H)$ have the representations* (6) *and* (7). *Then,* (8) *of Section* 2.5 *and* (8) *of this section still hold true in the present situation, where P is a σ-finite PO measure.*

PROOF. By Definition 2.6-2,

$$\int dP_t \int d\mu_x\, l(t, x) F(x) = \int dQ_t \int d\mu_x \frac{l(t, x)}{g(t)} F(x).$$

By virtue of Lemma 2.6-3, Theorem 2.5-2, and Theorem 2.3-4, the right-hand side is equal to

$$\int d\mu_x \int dQ_t \frac{l(t, x)}{g(t)} F(x) = \int d\mu_x \int dP_t \, l(t, x)F(x),$$

and this justifies (8) of Section 2.5. A similar manipulation establishes the other equation. ◇

Problem 2.6-1. Show that the definition of $\mathfrak{B}_P$ given by (1) does not depend on the choices of the representations for f and v.

Problem 2.6-2. Show that (12) does not depend on the choice of g.

Problem 2.6-3. Prove the other part of Theorem 2.6-2.

Problem 2.6-4. Let P be a σ-finite measure and let $g \in \mathscr{G}$ be a positive function [i.e., $g(t) > 0$ for all t] that is integrable with respect to P. Furthermore, let f, $v \in \mathscr{G}_g(H)$ and let $\{E_k\}_{k=1}^{\infty}$ be an increasing sequence in $\mathfrak{C}$ with $\bigcup E_k = T$. Show that

$$\lim_k \int_{E_k} dP_t \, f(t) = \int_T dP_t \, f(t) \tag{13}$$

and

$$\lim_k \int_{E_k} d(P_t f(t), v(t)) = \int_T d(P_t f(t), v(t)). \tag{14}$$

Chapter 3

Banach-Space-Valued Testing Functions and Distributions

3.1. INTRODUCTION

As was mentioned in the Preface, the natural framework for a realizability theory of continuous linear systems is distribution theory. Since the signals in the systems of concern to us take their values in Banach spaces, the properties of Banach-space-valued distributions are essential to our purposes. The present chapter is devoted to a discussion of such distributions; they constitute a special case of the vector-valued distributions of Schwartz (1957). We start with a description of the primary testing-function space in the theory of distributions, namely $\mathscr{D}^m(A)$. As always, A and B denote complex Banach spaces.

3.2. THE BASIC TESTING-FUNCTION SPACE $\mathscr{D}^m(A)$

Let m be an n-tuple each of whose components is either a nonnegative integer in R^1 or ∞. Also, let K be a compact set in R^n. $\mathscr{D}_K{}^m(A)$ denotes the linear space of all functions ϕ from R^n into A such that $\operatorname{supp} \phi \subset K$ and, for every integer $k \in R^n$ with $0 \le k \le m$, $\phi^{(k)}$ is continuous. We assign to $\mathscr{D}_K{}^m(A)$ the topology generated by the collection $\{\gamma_k : 0 \le k \le m\}$ of seminorms, where

$$\gamma_k(\phi) \triangleq \sup_{t \in K} \|\phi^{(k)}(t)\|_A . \tag{1}$$

Since γ_0 is a norm, $\mathscr{D}_K{}^m(A)$ is separated. Moreover, it is metrizable because the collection $\{\gamma_k\}$ is countable (Appendix C7).

Lemma 3.2-1. *$\mathscr{D}_K{}^m(A)$ is complete and therefore a Fréchet space.*

PROOF. Since $\mathscr{D}_K{}^m(A)$ is metrizable, we need only establish its sequential completeness (see Appendix C11). Let $\{\phi_i\}_{i=1}^{\infty}$ be a Cauchy sequence in $\mathscr{D}_K{}^m(A)$. In view of (1) and the completeness of A, we have that, for every k as restricted above, there exists an A-valued function ψ_k on R^n for which $\phi_i^{(k)} \to \psi_k$ uniformly on R^n. By note VIII of Section 1.4, ψ_k is continuous. Also, by Problem 1.6-3, $\psi_0^{(k)} = \psi_k$. Clearly, $\operatorname{supp} \psi_0 \subset K$. Hence, ψ_0 is the limit in $\mathscr{D}_K{}^m(A)$ of $\{\phi_i\}$. ◇

Note that, if every component of m is finite, $\mathscr{D}_K{}^m(A)$ is a Banach space because its topology is the same as that generated by the single norm ρ, where

$$\rho(\phi) \triangleq \max_{0 \le k \le m} \gamma_k(\phi).$$

When all the components of m are ∞ (i.e., when $m = [\infty]$), we denote $\mathscr{D}_K{}^m(A)$ by $\mathscr{D}_K(A)$. Moreover, we set $\mathscr{D}_K{}^m(C) = \mathscr{D}_K{}^m$ and $\mathscr{D}_K(C) = \mathscr{D}_K$.

Now, let $\{K_j\}_{j=1}^{\infty}$ be a sequence of compact sets in R^n such that $K_1 \subset K_2 \subset K_3 \subset \cdots$, $\bigcup_j K_j = R^n$, and every compact set $J \subset R^n$ is contained in some K_j. We define $\mathscr{D}^m(A) = \mathscr{D}^m_{R^n}(A)$ as the inductive-limit space generated by the $\mathscr{D}^m_{K_j}(A)$. That is,

$$\mathscr{D}^m(A) = \mathscr{D}^m_{R^n}(A) = \bigcup_{k=1}^{\infty} \mathscr{D}^m_{K_j}(A),$$

and this space possesses the inductive-limit topology (see Appendix E1). As before, we set $\mathscr{D}^{[\infty]}(A) = \mathscr{D}(A)$, $\mathscr{D}^m(C) = \mathscr{D}^m$, and $\mathscr{D}(C) = \mathscr{D}$.

This definition of $\mathscr{D}^m(A)$ does not depend on the choice of $\{K_j\}$. Indeed, for any other sequence $\{H_i\}_{i=1}^{\infty}$ of compact sets with the required properties, $\bigcup_j \mathscr{D}^m_{K_j}(A)$ and $\bigcup_i \mathscr{D}^m_{H_i}(A)$ are identical as linear spaces because every $\mathscr{D}^m_{K_j}$

is contained in some $\mathscr{D}^m_{H_i}$ and conversely. To show that the topologies are the same, let Λ be a convex neighborhood of 0 in $\bigcup_i \mathscr{D}^m_{H_i}(A)$. Given any K_j, we can find an H_i containing K_j. By definition of the inductive-limit topology, $\Lambda \cap \mathscr{D}^m_{H_i}(A)$ is a neighborhood of 0 in $\mathscr{D}^m_{H_i}(A)$. Moreover, $\mathscr{D}^m_{K_j}(A)$ is a subspace of $\mathscr{D}^m_{H_i}(A)$, and its topology is the same as that induced on it by $\mathscr{D}^m_{H_i}(A)$ because both topologies are generated by the same seminorms γ_k. Hence, $\Lambda \cap \mathscr{D}^m_{K_j}$ is a neighborhood of 0 in $\mathscr{D}^m_{K_j}(A)$. Consequently, Λ is a convex neighborhood of 0 in $\bigcup_j \mathscr{D}^m_{K_j}(A)$. Similarly, every convex neighborhood of 0 in $\bigcup_j \mathscr{D}^m_{K_j}(A)$ is a convex neighborhood of 0 in $\bigcup_i \mathscr{D}^m_{H_i}(A)$. Consequently, the two inductive limit topologies are identical

$\mathscr{D}^m(A)$ is clearly a strict inductive-limit space (Appendix E3). Moreover, it possesses the closure property defined in Appendix E4 because each $\mathscr{D}^m_{K_j}(A)$ is complete. As a consequence, the following assertions hold. $\mathscr{D}^m(A)$ is a complete separated locally convex space. A linear mapping f of $\mathscr{D}^m(A)$ into another locally convex space $\mathscr{W}$ is continuous if and only if its restriction to each $\mathscr{D}^m_{K_j}(A)$ is either sequentially continuous or bounded. A set is bounded in $\mathscr{D}^m(A)$ if and only if it is contained and bounded in some $\mathscr{D}^m_{K_j}(A)$. Similarly, a sequence $\{\phi_j\}$ converges in $\mathscr{D}^m(A)$ if and only if it is contained and converges in some $\mathscr{D}^m_{K_j}(A)$. Thus, a linear mapping on $\mathscr{D}^m(A)$ is continuous if and only if it is sequentially continuous. (See Appendix E.)

Lemma 3.2-2. *Let J and K be two compact intervals in R^n such that J contains a neighborhood of K. Then, given any $\phi \in \mathscr{D}_K{}^m(A)$, there exists a sequence $\{\phi_j\}_{j=1}^\infty \subset \mathscr{D}_J(A)$ such that $\phi_j \to \phi$ in $\mathscr{D}_J{}^m(A)$.*

PROOF. Set

$$\zeta(t) = \begin{cases} \exp[1/(|t|^2 - 1)], & |t| < 1 \\ 0, & |t| \geq 1 \end{cases}$$

and

$$\eta_p(t) = \zeta(pt)\left[\int \zeta(pt)\, dt\right]^{-1}, \qquad p = 1, 2, \ldots.$$

(In this proof, all integrations are over R^n.) It follows that η_p is a smooth nonnegative function, diam supp $\eta_p = 2/p$, and $\int \eta_p(t)\, dt = 1$. Next, set

$$\phi_p(t) = \int \phi(x)\eta_p(t - x)\, dx.$$

For all sufficiently large p, supp $\phi_p \subset J$. Moreover, we may differentiate under the integral sign (Theorem 1.6-4) and integrate by parts (Problem 1.6-1) to obtain, for any fixed $k \in R^n$ such that $0 \leq k \leq m$,

$$\phi_p^{(k)}(t) = \int \phi^{(k)}(x)\eta_p(t - x)\, dx.$$

Hence,

$$\|\phi^{(k)}(t) - \phi_p^{(k)}(t)\|_A = \left\| \int [\phi^{(k)}(t) - \phi^{(k)}(x)]\eta_p(t-x)\,dx \right\|_A$$
$$\leq \sup_{|t-x|<p^{-1}} \|\phi^{(k)}(t) - \phi^{(k)}(x)\|_A .$$

By the uniform continuity of $\phi^{(k)}$ on R^n (Theorem 1.3-1), the right-hand side tends to zero uniformly for all t as $p \to \infty$. Thus, $\{\phi_p\}$ with a sufficient number of initial terms deleted is the sequence we seek. ◇

Lemma 3.2-3. *$\mathscr{D}(A)$ is a dense subspace of $\mathscr{D}^m(A)$.*

PROOF. Let $\phi \in \mathscr{D}^m(A)$ be given and choose arbitrarily a neighborhood Λ of ϕ in $\mathscr{D}^m(A)$. Also, choose the compact intervals J and K such that supp $\phi \subset \mathring{K} \subset K \subset \mathring{J}$. Then, $\Lambda \cap \mathscr{D}_J{}^m(A)$ is a neighborhood of ϕ in $\mathscr{D}_J{}^m(A)$ and, by the preceding lemma, we can find a $\psi \in \mathscr{D}_J(A)$ that is contained in $\Lambda \cap \mathscr{D}_J^m(A)$. Hence, every neighborhood Λ of ϕ in $\mathscr{D}^m(A)$ contains a $\psi \in \mathscr{D}(A)$. The last statement is equivalent to our lemma. ◇

3.3. DISTRIBUTIONS

An $[A; B]$*-valued distribution* f *on* R^n is by definition any continuous linear mapping of $\mathscr{D}(A)$ into B; i.e., $f \in [\mathscr{D}(A); B]$. It will be shown in Section 3.5 that f can be identified with a unique continuous linear mapping of $\mathscr{D}$ into $[A; B]$. This is the reason for calling f "$[A; B]$-valued."

The canonical injection of $\mathscr{D}(A)$ into $\mathscr{D}^m(A)$ is clearly continuous for every m. Consequently, the restriction of any $f \in [\mathscr{D}^m(A); B]$ to $\mathscr{D}(A)$ is a member of $[\mathscr{D}(A); B]$, and this restriction uniquely determines f because of the density of $\mathscr{D}(A)$ in $\mathscr{D}^m(A)$. Thus,

$$[\mathscr{D}^m(A); B] \subset [\mathscr{D}(A); B].$$

In the same way, we have that

$$[\mathscr{D}^m(A); B] \subset [\mathscr{D}^p(A); B]$$

whenever $m \leq p$.

Upon setting $B = C$, we obtain the dual $[\mathscr{D}^m(A); C]$ of $\mathscr{D}^m(A)$. Similarly, with $A = C$, we get a space $[\mathscr{D}^m; B]$ of *B-valued distributions.* Note that $[C; B]$ can be identified with B so that $[\mathscr{D}^m; [C; B]]$ and $[\mathscr{D}^m; B]$ can be considered to be the same space. Finally, when $A = B = C$ and $m = [\infty]$, $[\mathscr{D}^m(A); B]$ becomes the customary space $[\mathscr{D}; C] = \mathscr{D}'$ of all *complex-valued distributions.*

Lemma 3.3-1. *Given an* $f \in [\mathscr{D}^m(A); B]$ *and a compact interval* $K \subset R^n$, *there exists an integer* $p \in R^n$ *with* $0 \leq p \leq m$ *and a constant* $M > 0$ *such that for all* $\phi \in \mathscr{D}_K{}^m(A)$,

$$\|\langle f, \phi\rangle\|_B \leq M\rho_p(\phi), \tag{1}$$

where

$$\rho_p(\phi) \triangleq \max_{0 \leq k \leq p} \sup_{t \in K} \|\phi^{(k)}(t)\|_A. \tag{2}$$

M and p depend in general on f and K.

This lemma follows directly from Appendix D2 and the fact that the linear mapping f is continuous on $\mathscr{D}^m(A)$ if and only if its restriction to every $\mathscr{D}_K{}^m(A)$ is continuous.

Unless something else is explicitly indicated, we always assign to $[\mathscr{D}^m(A); B]$ the *bounded topology*. This is the topology generated by the collection $\{\gamma_\Phi\}_{\Phi \in \mathfrak{S}}$ of seminorms, where $\mathfrak{S}$ denotes the set of all bounded sets in $\mathscr{D}^m(A)$ and

$$\gamma_\Phi(f) \triangleq \sup_{\phi \in \Phi} \|\langle f, \phi\rangle\|_B. \tag{3}$$

We also refer to the bounded topology as the $\mathfrak{S}$-*topology*. On occasion, a weaker topology is assigned to $[\mathscr{D}^m(A); B]$, namely the *pointwise topology*. It is generated by the collection $\{\gamma_\phi\}$ of seminorms, where ϕ traverses $\mathscr{D}^m(A)$ and

$$\gamma_\phi(f) \triangleq \|\langle f, \phi\rangle\|_B. \tag{4}$$

To indicate that the pointwise topology is being used, we employ the notation $[\mathscr{D}^m(A); B]^\sigma$. Both the bounded and pointwise topologies are separating.

Lemma 3.3-2. *Every* $f \in [\mathscr{D}^m(A); B]$ *uniquely defines a* $g \in [\mathscr{D}^m; [A; B]]$ *through the equation*

$$\langle g, \theta\rangle a \triangleq \langle f, \theta a\rangle, \qquad \theta \in \mathscr{D}^m, \quad a \in A. \tag{5}$$

This assertion remains true when $\mathscr{D}^m(A)$ *is replaced by* $\mathscr{D}_K{}^m(A)$ *and* $\mathscr{D}^m$ *by* $\mathscr{D}_K{}^m$ *for any given compact set K.*

PROOF. Having fixed upon some $\theta \in \mathscr{D}^m$, we define a mapping j_θ of A into B by $j_\theta a \triangleq \langle f, \theta a\rangle$ for all $a \in A$. It readily follows that j_θ is linear. That it is also continuous follows from Lemma 3.3-1. Indeed, for any compact set K that contains supp θ and for all $a \in A$, we have

$$\|j_\theta a\|_B = \|\langle f_\theta, \theta a\rangle\|_B \leq M\rho_p(\theta a) \leq M\|a\|_A \rho_p{}'(\theta),$$

where

$$\rho_p'(\theta) = \max_{0 \le k \le p} \sup_{t \in R^n} |\theta^{(k)}(t)|.$$

Hence,

$$\|j_\theta\|_{[A;\,B]} \le M\rho_p'(\theta). \tag{6}$$

Next, set $\langle g, \theta\rangle \triangleq j_\theta$. This uniquely defines g as a mapping from $\mathscr{D}^m$ into $[A; B]$. g is linear because, for any $a \in A$, α, $\beta \in C$, and θ, $\psi \in \mathscr{D}^m$,

$$\begin{aligned}\langle g, \alpha\theta + \beta\psi\rangle a &= \langle f, \alpha\theta a + \beta\psi a\rangle \\ &= \alpha\langle f, \theta a\rangle + \beta\langle f, \psi a\rangle \\ &= (\alpha\langle g, \theta\rangle + \beta\langle g, \psi\rangle)a.\end{aligned}$$

Moreover, (6) implies that g is continuous. The second statement of the lemma is established in the same way. ◇

For any nonnegative integer $k \in R^n$, we define the *generalized differentiation* D^k *on* $[\mathscr{D}^m(A); B]$ as follows. First, note that D^k is a continuous linear mapping of $\mathscr{D}^{m+k}(A)$ into $\mathscr{D}^m(A)$ since its restriction to every $\mathscr{D}_K^{m+k}(A)$ is a continuous linear mapping of $\mathscr{D}_K^{m+k}(A)$ into $\mathscr{D}_K{}^m(A)$. This allows us to define D^k on any $f \in [\mathscr{D}^m(A); B]$ by

$$\langle D^k f, \phi\rangle \triangleq (-1)^{|k|}\langle f, D^k\phi\rangle, \qquad \phi \in \mathscr{D}^{m+k}(A) \tag{7}$$

because the right-hand side has a sense.

Theorem 3.3-1. *D^k is a continuous linear mapping of $[\mathscr{D}^m(A); B]$ into $[\mathscr{D}^{m+k}(A); B]$, as well as of $[\mathscr{D}^m(A); B]^\sigma$ into $[\mathscr{D}^{m+k}(A); B]^\sigma$.*

PROOF. Observe that the right-hand side of (7) is a member of B, so that $D^k f$ maps $\mathscr{D}^{m+k}(A)$ into B. It follows readily that $D^k f \in [\mathscr{D}^{m+k}(A); B]$ and that D^k is a linear mapping of $[\mathscr{D}^m(A); B]$ into $[\mathscr{D}^{m+k}(A); B]$. To show that D^k is continuous, let Φ be a bounded set in $\mathscr{D}^{m+k}(A)$. As ϕ traverses Φ, $D^k\phi$ traverses a bounded set in $\mathscr{D}^m(A)$; call it Θ. So,

$$\begin{aligned}\sup_{\phi \in \Phi}\|\langle D^k f, \phi\rangle\|_B &= \sup_{\phi \in \Phi}\|\langle f, D^k\phi\rangle\|_B \\ &= \sup_{\theta \in \Theta}\|\langle f, \theta\rangle\|_B.\end{aligned}$$

So, truly, D^k is continuous from $[\mathscr{D}^m(A); B]$ into $[\mathscr{D}^{m+k}(A); B]$. By restricting Φ to sets of one element each, we obtain the same conclusion with the pointwise topologies. ◇

Another operator of importance to us is the *shifting operator* σ_τ, which is also called the *translation operator*. Let $\tau \in R^n$ be fixed. σ_τ is defined on any

$\phi \in \mathscr{D}^m(A)$ by $(\sigma_\tau \phi)(t) = \phi(t - \tau)$. It is an automorphism on $\mathscr{D}^m(A)$. On the other hand, σ_τ is defined on any $f \in [\mathscr{D}^m(A); B]$ by

$$\langle \sigma_\tau f, \phi \rangle \triangleq \langle f, \sigma_{-\tau} \phi \rangle, \qquad \phi \in \mathscr{D}^m(A). \tag{8}$$

As a consequence, σ_τ is an automorphism on $[\mathscr{D}^m(A); B]$, as well as on $[\mathscr{D}^m(A); B]^\sigma$. (Show this.)

Two distributions f, $g \in [\mathscr{D}(A); B]$ are said to be *equal on an open set* $\Omega \subset R^n$ if $\langle f, \phi \rangle = \langle g, \phi \rangle$ for every $\phi \in \mathscr{D}(A)$ such that supp $\phi \subset \Omega$. The *null set of* f is the union of all open sets on which f is equal to zero (i.e., is equal to the zero distribution). The complement of the null set is called the *support of* f and is denoted by supp f. Thus, supp f is a closed set.

A property of distributions that we occasionally use is the following. *If* $f \in [\mathscr{D}(A); B]$ *is equal to zero on every set in a collection of open sets, then it is equal to zero on the union of these sets.* The proof of this assertion is precisely the same as it is for complex-valued distributions on $\mathscr{D}$ and can be found, e.g., in the work of Zemanian (1965, Section 1.8). One consequence of this result is that supp f is the smallest closed set outside of which f is equal to zero. Another direct result is the following. $\langle f, \phi \rangle$ depends only on the values that $\phi \in \mathscr{D}(A)$ assumes on any arbitrarily small neighborhood of supp f. Indeed, if $\psi \in \mathscr{D}(A)$ is equal to ϕ on a neighborhood of supp f, then supp$(\phi - \psi)$ is contained in the null set of f, so that $\langle f, \phi - \psi \rangle = 0$.

We conclude this section with two examples of $[A; B]$-valued distributions.

Example 3.3-1. Let F be a fixed member of $[A; B]$ and let δ be the *delta function* defined on any $\phi \in \mathscr{D}^0(A)$ by $\langle \delta, \phi \rangle \triangleq \phi(0)$. We define $F\delta$ as a mapping on $\mathscr{D}^0(A)$ by

$$\langle F\delta, \phi \rangle \triangleq F\phi(0), \qquad \phi \in \mathscr{D}^0(A). \tag{9}$$

Consequently, $F\delta \in [\mathscr{D}^0(A); B]$. As is indicated in Lemma 3.3-2, $F\delta$ generates a member of $[\mathscr{D}^0; [A; B]]$, which we will also denote by $F\delta$. Thus, in accordance with (5) and (9), we write, for any $\theta \in \mathscr{D}^0$ and $a \in A$,

$$\langle F\delta, \theta \rangle a = \langle F\delta, \theta a \rangle = F[\theta(0)a] = [F\theta(0)]a$$

or

$$\langle F\delta, \theta \rangle = F\theta(0).$$

Upon applying the generalized differentiation D^k to $F\delta$, we obtain, for any $\phi \in \mathscr{D}^k(A)$,

$$\langle D^k(F\delta), \phi \rangle = (-1)^{|k|} F\phi^{(k)}(0) = F\langle D^k\delta, \phi \rangle. \tag{10}$$

We define $FD^k\delta$ as the distribution that assigns to ϕ the value indicated on the right-hand side of (10). Thus, we have $D^k(F\delta) = FD^k\delta$. ◇

Example 3.3-2. Let $h \in L_1([A; B])$; that is, the $[A; B]$-valued function h is Bochner integrable with respect to Lebesgue measure on the Borel subsets of R^n. We define a mapping f of $\mathscr{D}^0(A)$ into B by setting

$$\langle f, \phi \rangle \triangleq \int_{R^n} h(t)\phi(t)\, dt, \qquad \phi \in \mathscr{D}^0(A).$$

By Appendix G11,

$$\|\langle f, \phi \rangle\|_B \leq \|h\|_{L_1} \sup_t \|\phi(t)\|_A .$$

It follows that $h \mapsto f$ is a continuous linear mapping of $L_1([A; B])$ into $[\mathscr{D}^0(A); B]$. This mapping is injective; indeed, if h generates the zero member of $[\mathscr{D}^0(A); B]$, then

$$\int_{R^n} h(t)\theta(t)\, dt\, a = 0$$

for all $\theta \in \mathscr{D}$ and all $a \in A$, which by Appendix G12 implies that $h(t) = 0$ for almost all t. Thus, $L_1([A; B])$ can be identified as a subspace of $[\mathscr{D}^0(A); B]$. Any member of $[\mathscr{D}^0(A); B]$ that can be generated in this way from a member of $L_1([A; B])$ will be called a *regular* $[A; B]$*-valued distribution.* ◇

Problem 3.3-1. Show that the shifting operator σ_τ is an automorphism on $[\mathscr{D}^m(A); B]$.

Problem 3.3-2. Let h be an $[H; H]$-valued function on R with the following two properties. It is *strongly measurable* with respect to Lebesgue measure on the Borel sets; that is, $h(\cdot)a$ is a measurable H-valued function for every $a \in H$. It is *locally essentially bounded*; that is, for every compact set $K \subset R$, there exists a constant M_K such that $\|h(t)\| \leq M_K$ for almost all $t \in K$. Define a mapping f on any $\phi \in \mathscr{D}^0$ by

$$\langle f, \phi \rangle a \triangleq \int h(t) a \phi(t)\, dt.$$

Show that $f \in [\mathscr{D}^0; [H; H]]$.

Problem 3.3-3. Let P be a σ-finite $[A; B]$-valued measure on the bounded Borel subsets of R^n. For every $\phi \in \mathscr{D}(A)$, define f by

$$\langle f, \phi \rangle \triangleq \int dP\, \phi.$$

Show that $f \in [\mathscr{D}(A); B]$. f is called the $[A; B]$*-valued distribution generated by* P.

3.4. LOCAL STRUCTURE

The objective of this section is to show that every Banach-space-valued distribution can be represented on any compact interval as a finite-order derivative of a continuous Banach-space-valued function. We do this by means of a method employed by Sebastiao e Silva (1960). As before, K is a compact interval in R^n, and $m = \{m_i\}_{i=1}^n$ denotes an n-tuple each component of which is either a nonnegative integer or ∞.

$\mathscr{C}_K{}^m(A)$ is the linear space of all functions ϕ from K into A such that, for each integer $k \in R^n$ with $0 \leq k \leq m$, $\phi^{(k)}$ is continuous. $\mathscr{C}_K{}^m(A)$ is equipped with the topology generated by the collection $\{\gamma_k: 0 \leq k \leq m\}$ of seminorms, where again

$$\gamma_k(\phi) \triangleq \sup_{t \in K} \| \phi^{(k)}(t) \|_A .$$

$\mathscr{C}_K{}^m(A)$ is a Fréchet space, its completeness being established as was Lemma 3.2-1. When every component of m is finite, $\mathscr{C}_K{}^m(A)$ is a Banach space. By identifying each $\psi \in \mathscr{D}_K{}^m(A)$ with its restriction to K, we can and will view $\mathscr{D}_K{}^m(A)$ as a subspace of $\mathscr{C}_K{}^m(A)$.

Now, let $p = \{p_i\}_{i=1}^n$ be a nonnegative integer in R^n, and let $[L, Q]$ be a compact interval in R^n, where $L = \{L_i\}_{i=1}^n < Q = \{Q_i\}_{i=1}^n$. It is a fact in interpolation theory that, for each i, there exist $2(p_i + 1)$ polynomials on $[L_i, Q_i]$, which we denote by g_{X_i, ν_i}, where $X_i = L_i$, Q_i and $\nu_i = 0, \ldots, p_i$, such that

$$\partial_i^{\mu_i} g_{L_i, \nu_i}(L_i) = \delta_{\nu_i, \mu_i}, \qquad \partial_i^{\mu_i} g_{Q_i, \nu_i}(L_i) = 0,$$
$$\partial_i^{\mu_i} g_{L_i, \nu_i}(Q_i) = 0, \qquad \partial_i^{\mu_i} g_{Q_i, \nu_i}(Q_i) = \delta_{\nu_i, \mu_i}.$$

Here, $\mu_i = 0, \ldots, p_i$ and δ_{ν_i, μ_i} is the *Kronecker delta* (i.e., $\delta_{\nu_i, \mu_i} = 1$ for $\nu_i = \mu_i$ and $\delta_{\nu_i, \mu_i} = 0$ for $\nu_i \neq \mu_i$). Now, consider the function ϕ_b obtained from some $\phi \in \mathscr{C}_K{}^p(A)$ by means of the following formula:

$$\begin{aligned}\phi_b(t) \triangleq & \sum_1 \sum_{X_i} \partial_i^{\nu_i} \phi(t) |_{t_i = X_i} g_{X_i, \nu_i}(t_i) \\ & - \sum_2 \sum_{X_i, X_j} \partial_i^{\nu_i} \partial_j^{\nu_j} \phi(t) |_{t_i = X_i, t_j = X_j} g_{X_i, \nu_i}(t_i) g_{X_j, \nu_j}(t_j) + \cdots \\ & + (-1)^n \sum_n \sum_{X_1, \ldots, X_n} \partial_1^{\nu_1} \cdots \partial_n^{\nu_n} \phi(t) |_{t_1 = X_1, \ldots, t_n = X_n} g_{X_1, \nu_1}(t_1) \cdots g_{X_n, \nu_n}(t_n).\end{aligned}$$

A summation on the X's means the sum of all possible terms obtained by setting each X_q equal to either L_q or Q_q. Also, $\sum_1$ is a summation over

$$1 \leq i \leq n, \qquad 0 \leq \nu_i \leq p_i .$$

$\sum_2$ is a summation over

$$1 \le i \le n, \qquad i+1 \le j \le n, \qquad 0 \le \nu_i \le p_i, \qquad 0 \le \nu_j \le p_j.$$

Finally, $\sum_n$ is a summation over

$$0 \le \nu_1 \le p_1, \qquad 0 \le \nu_2 \le p_2, \ldots, \qquad 0 \le \nu_n \le p_n.$$

Lemma 3.4-1. (i) *For every integer $k \in R^n$ with $0 \le k \le p$, $\phi_b^{(k)}(t) = \phi^{(k)}(t)$ for all t on the boundary of K.*

(ii) *$\phi \mapsto \phi_b$ is a continuous linear mapping of $\mathscr{C}_K{}^p(A)$ into $\mathscr{C}_K{}^p(A)$.*

PROOF. That $\phi_b \in \mathscr{C}_K{}^p(A)$ follows from the facts that each g_{X_i, ν_i} is a polynomial in t_i alone and $\phi \in \mathscr{C}_K{}^p(A)$. The assertion (ii) follows directly from the definition of ϕ_b. Finally the assertion (i) can be verified through some straightforward (but tedious) computations. ◇

An operator F that maps a space $\mathscr{V}$ into $\mathscr{V}$ is called a *projection* if $FF\phi = F\phi$ for all $\phi \in \mathscr{V}$.

Lemma 3.4-2. *The operator π_p: $\phi \mapsto \phi - \phi_b$ is a continuous linear projection of $\mathscr{C}_K{}^p(A)$ onto $\mathscr{D}_K{}^p(A)$.*

PROOF. According to Lemma 3.4-1, $\phi - \phi_b \in \mathscr{D}_K{}^p(A)$. (It is understood here that $\phi - \phi_b$ is extended onto the exterior of K as the zero function.) That lemma also implies that π_p is a continuous linear mapping of $\mathscr{C}_K{}^p(A)$ into $\mathscr{C}_K{}^p(A)$. But $\mathscr{D}_K{}^p(A)$ is a subspace of $\mathscr{C}_K{}^p(A)$ and possesses a topology that is identical to the topology induced on it by $\mathscr{C}_K{}^p(A)$. Therefore, π_p is a continuous linear mapping of $\mathscr{C}_K{}^p(A)$ into $\mathscr{D}_K{}^p(A)$. Finally, the definition of ϕ_b also shows that, if $\phi \in \mathscr{D}_K{}^p(A)$, then ϕ_b is the zero function in $\mathscr{C}_K{}^p(A)$. Consequently, $\pi_p\phi = \phi$ for all $\phi \in \mathscr{D}_K{}^p(A)$, so that $\pi_p\pi_p\phi = \pi_p\phi$ for all $\phi \in \mathscr{C}_K{}^p(A)$. That is, π_p is a projection. ◇

Next step: We set up the function

$$J_p(t-x) = \frac{(t_1 - x_1)^{p_1+1}}{(p_1+1)!} 1_+(t_1 - x_1) \cdots \frac{(t_n - x_n)^{p_n+1}}{(p_n+1)!} 1_+(t_n - x_n).$$

Here, 1_+ denotes the *unit-step function on the real line*; that is,

$$1_+(t_i) = \begin{cases} 0, & t_i < 0 \\ \frac{1}{2}, & t_i = 0 \\ 1, & t_i > 0. \end{cases}$$

For each fixed $t \in R^n$, we denote the function $x \mapsto J_p(t - x)$ on K by $J_p(t - \cdot)$. Also, we remind the reader that the symbol $[2]$ denotes the n-tuple each of whose components is equal to 2.

Lemma 3.4-3. *Let $\phi \in \mathscr{D}_K^{p+[2]}(A)$. Then,*

$$t \mapsto J_p(t - \cdot)D^{p+[2]}\phi(t) \tag{1}$$

is a continuous mapping of R^n into $\mathscr{C}_K{}^p(A)$. Similarly, $t \mapsto J_p(t - \cdot)$ is a continuous mapping of R^n into $\mathscr{C}_K{}^p$.

PROOF. Let the integer $k \in R^n$ be such that $0 \leq k \leq p$. Also, fix t and let I be any compact interval in R^n containing t in its interior. As a function of $\{t, x\}$,

$$D_x{}^k J_p(t - x)D_t{}^{p+[2]}\phi(t)$$

is a continuous function from $I \times K$ into A, and therefore it is uniformly continuous on $I \times K$. So, as $\Delta t \to 0$ in R^n,

$$D_x{}^k J_p(t + \Delta t - x)D_t^{p+[2]}\phi(t + \Delta t) \to D_x{}^k J_m(t - x)D_t^{p+[2]}\phi(t)$$

in A uniformly with respect to all $x \in K$. This proves our assertion for the function (1). The proof for the other function is the same. ◇

Lemma 3.4-4. *For any $\phi \in \mathscr{D}_K^{p+[2]}(A)$,*

$$\phi = \phi(\cdot) = (-1)^{|p|} \int_K J_p(t - \cdot)D^{p+[2]}\phi(t)\, dt. \tag{2}$$

Note. In the right-hand side, we have the Riemann integral of the continuous $\mathscr{C}_K{}^p(A)$-valued function (1).

PROOF. We may repeatedly integrate by parts to establish that

$$(-1)^{|p|} \int_K J_p(t - x)D^{p+[2]}\phi(t)\, dt = \int_K J_0(t - x)D^{[2]}\phi(t)\, dt = \phi(x).$$

Equation (2) follows directly. ◇

We now appeal to note II of Section 1.4 and to Lemma 3.4-2 in order to apply π_p to (2) under the integral sign. Thus, for any $\phi \in \mathscr{D}_K^{p+[2]}(A)$,

$$\phi = \phi(\cdot) = \pi_p \phi(\cdot) = (-1)^{|p|} \int_K G_p(t, \cdot)D^{p+[2]}\phi(t)\, dt, \tag{3}$$

where

$$G_p(t, \cdot) = \pi_p J_p(t - \cdot).$$

The integral in (3) is a continuous $\mathscr{D}_K{}^p(A)$-valued function of $t \in K$ because it is the result of first applying the function (1) and then applying π_p, both of which are continuous mappings. Similarly, $G_p(t, \cdot)$ is a continuous $\mathscr{D}_K{}^p$-valued function of $t \in K$.

Lemma 3.4-5. *Let $f \in [\mathscr{D}^m(A); B]$, let K and J be compact intervals in R^n such that $K \subset \mathring{J}$, and let p be the integer in R^n corresponding to f and J in accordance with Lemma 3.3-1. Then, there exists a unique $l \in [\mathscr{D}_K{}^p(A); B]$ that satisfies the relation*

$$\langle l, \theta \rangle = \lim_{j \to \infty} \langle f, \theta_j \rangle \tag{4}$$

for all $\theta \in \mathscr{D}_K{}^p(A)$ and all sequences $\{\theta_j\} \subset \mathscr{D}_J{}^m(A)$ that tend to θ in $\mathscr{D}_J{}^p(A)$. Thus, $\langle l, \psi \rangle = \langle f, \psi \rangle$ for all $\psi \in \mathscr{D}_K{}^m(A)$.

PROOF. Choose any $\theta \in \mathscr{D}_K{}^p(A)$. By Lemma 3.2-2, we can choose a sequence $\{\theta_j\}_{j=1}^\infty \subset \mathscr{D}_J{}^m(A)$ such that $\theta_j \to \theta$ in $\mathscr{D}_J{}^p(A)$. We infer from the inequality in Lemma 3.3-1 that $\{\langle f, \theta_j \rangle\}$ is a Cauchy sequence in B. Its limit is taken to be the value $\langle l, \theta \rangle$, and this value also satisfies the inequality in Lemma 3.3-1. It follows that $l \in [\mathscr{D}_K{}^p(A); B]$. Moreover, this definition of l is independent of the choice of the sequence $\{\theta_j\}$, and l is uniquely determined by the values that f assigns to the $\phi \in \mathscr{D}_J{}^m(A)$. ◇

Theorem 3.4-1. *Let $f \in [\mathscr{D}^m(A); B]$ and let K be a compact interval in R^n. Then, there exists an integer $p \in R^n$ with $0 \le p \le m$ and a continuous $[A; B]$-valued function h on K such that, for all $\phi \in \mathscr{D}_K^{m+[2]}(A)$,*

$$\langle f, \phi \rangle = \int_K h(t) D^{p+[2]} \phi(t) \, dt. \tag{5}$$

In general, p and h depend on f and K.

PROOF. Let J and p be chosen as in Lemma 3.4-5. Then, l can be applied to (3), where now $\phi \in \mathscr{D}_K^{m+[2]}(A)$, and note II of Section 1.4 can be invoked to get

$$\langle f, \phi \rangle = \langle l, \phi \rangle = (-1)^{|p|} \int_K \langle l(\cdot), G_m(t, \cdot) D^{p+[2]} \phi(t) \rangle \, dt. \tag{6}$$

Let $g \in [\mathscr{D}_K{}^p; [A; B]]$ be related to l in accordance with Lemma 3.3-2. Then, (6) can be rewritten as

$$\langle f, \phi \rangle = (-1)^{|p|} \int_K \langle g(\cdot), G_m(t, \cdot) \rangle D^{p+[2]} \phi(t) \, dt.$$

Upon setting $h(t) = (-1)^{|p|} \langle g(\cdot), G_m(t, \cdot) \rangle$ we arrive at (5). h is a continuous $[A; B]$-valued function because it is the composite of the two continuous mappings

$$G_m(t, \cdot): \quad K \rightsquigarrow \mathscr{D}_K{}^p$$

and

$$g: \quad \mathscr{D}_K{}^p \rightsquigarrow [A; B].$$

That p and h depend on f and K can be shown by examples. ◇

Corollary 3.4-1a. *Let $f \in [\mathscr{D}^m; A]$ and let K be a compact interval in R^n. Then, there exists an integer $p \in R^n$ with $0 \leq p \leq m$ and a continuous A-valued function h on K such that, for all $\phi \in \mathscr{D}_K^{m+[2]}$,*

$$\langle f, \phi \rangle = \int_K h(t) D^{p+[2]} \phi(t)\, dt. \tag{7}$$

Here, too, p and h depend in general on f and K.

PROOF. By Theorem 3.4-1, (7) holds true for a continuous $[C; A]$-valued function h on K. But $[C; A]$ is identical to A. ◇

Problem 3.4-1. Supply the details of the proof of Lemma 3.4-1.

Problem 3.4-2. Show that the quantities p and h in Theorem 3.4-1 depend in general on f and K.

3.5. THE CORRESPONDENCE BETWEEN $[\mathscr{D}(A); B]$ AND $[\mathscr{D}; [A; B]]$

The natural identification between $[\mathscr{D}(A); B]$ and $[\mathscr{D}; [A; B]]$, which we alluded to at the beginning of Section 3.3, will now be established. In the following, $\mathscr{D} \odot A$ (or $\mathscr{D}_K \odot A$) denotes the linear space of all $\phi \in \mathscr{D}(A)$ [respectively $\phi \in \mathscr{D}_K(A)$] having representations of the form $\phi = \sum \theta_k a_k$, where $\theta_k \in \mathscr{D}$ (respectively $\theta_k \in \mathscr{D}_k$), $a_k \in A$, and the summation is over a finite number of terms.

Lemma 3.5-1. *Let K and N be compact intervals in R^n such that $K \subset \mathring{N}$. Given any $\phi \in \mathscr{D}_K(A)$, there exists a sequence $\{\phi_j\}_{j=1}^{\infty} \subset \mathscr{D}_N \odot A$ that converges in $\mathscr{D}_N(A)$ to ϕ. Thus, $\mathscr{D} \odot A$ is dense in $\mathscr{D}(A)$.*

PROOF. Let L be a compact interval in R^n such that $K \subset \mathring{L}$ and $L \subset \mathring{N}$. Let

$$d_1 = \inf\{|t - x| : t \in K, \quad x \in R^n \backslash L\},$$
$$d_2 = \inf\{|t - x| : t \in L, \quad x \in R^n \backslash N\}.$$

Thus, d_1 and d_2 are greater than zero. We choose a sequence $\{\mathfrak{E}_p\}_{p=1}^{\infty}$, where each $\mathfrak{E}_p$ is a collection $\{\Omega_{p,i}\}_i$ of open sets in R^n with the following three properties. Only a finite number of the $\Omega_{p,i}$ intersect any bounded set in R^n. For every p and i, diam $\Omega_{p,i} < d_1$. Finally,

$$\sup_i \operatorname{diam} \Omega_{p,i} \to 0, \qquad p \to \infty.$$

Next step: For each p, we can choose a collection $\{\psi_{p,i}\}_i \subset \mathscr{D}$ such that, for every i and t, we have $\operatorname{supp} \psi_{p,i} \subset \Omega_{p,i}$, $0 \le \psi_{p,i}(t) \le 1$, and $\sum_i \psi_{p,i}(t) = 1$. [See, for example, Zemanian (1965, Section 1.8).]

For each p, we define the mapping J_p on $\mathscr{D}_K(A)$ into $\mathscr{D}_L \odot A$ by

$$\phi \mapsto \sum_i \phi(t_i)\psi_{p,i},$$

where each t_i is chosen as some member of $\Omega_{p,i}$. It follows that, as $p \to \infty$, $J_p\phi$ tends to ϕ in $\mathscr{D}_L{}^0(A)$. Indeed,

$$\begin{aligned}\sup_t \Big\|\phi(t) - \sum_i \phi(t_i)\psi_{p,i}(t)\Big\|_A &\le \sup_t \sum_i \|\phi(t) - \phi(t_i)\|\psi_{p,i}(t)\\ &\le \sup_t \sum_i \sup_{t \in \Omega_{p,i}} \|\phi(t) - \phi(t_i)\|\psi_{p,i}(t)\\ &\le \sup_i \sup_{t \in \Omega_{p,i}} \|\phi(t) - \phi(t_i)\|,\end{aligned}$$

and the right-hand side tends to zero as $p \to \infty$ because of the uniform continuity of $\phi \in \mathscr{D}_K(A)$.

Next, let η_q, where $q = 1, 2, \ldots,$ be the function defined in the proof of Lemma 3.2-2, and let R_q be the linear mapping on $\mathscr{D}^0(A)$ defined by

$$\theta \mapsto \int_{R^n} \theta(x)\eta_q(\cdot - x)\,dx.$$

If $q^{-1} < d_2$, then $\operatorname{supp} R_q\theta \subset N$ whenever $\operatorname{supp} \theta \subset L$. Moreover, R_q is continuous from $\mathscr{D}_L{}^0(A)$ into $\mathscr{D}_N(A)$. Indeed, for any $\theta \in \mathscr{D}_L{}^0(A)$ and any nonnegative integer $k \in R^n$,

$$\begin{aligned}\sup_t \|D_t{}^k R_q \theta(t)\| &= \sup_t \left\| \int \theta(x) D_t{}^k \eta_q(t - x)\,dx \right\|\\ &\le \int_{R_n} |D^k\eta_q(x)|\,dx \sup_t \|\theta(t)\|.\end{aligned}$$

Now, for any $\phi \in \mathscr{D}_K(A)$, consider $R_q J_p \phi \in \mathscr{D}_N \odot A$. By what we have already shown, for fixed $q > d_2^{-1}$ and as $p \to \infty$, we have that $R_q J_p \phi \to R_q \phi$ in $\mathscr{D}_N(A)$. Furthermore, by the proof of Lemma 3.2-2, $R_q\phi \to \phi$ in $\mathscr{D}_N(A)$ as $q \to \infty$. The fact that $\mathscr{D}_N(A)$ has a countable base of neighborhoods of 0 (Appendix C7) can now be exploited to extract a sequence $\{R_{q_j} J_{p_j} \phi\}_{j=1}^{\infty}$ that converges in $\mathscr{D}_N(A)$ to ϕ. This establishes the first conclusion.

The second conclusion follows from the definition of $\mathscr{D}(A)$ as an inductive-limit space.

Lemma 3.5-2. *$\mathscr{D} \odot A$ is dense in $\mathscr{D}^m(A)$ whatever be m.*

PROOF. $\mathscr{D}(A)$ is a dense subspace of $\mathscr{D}^m(A)$, and the topology of $\mathscr{D}(A)$ is stronger than that induced on it by $\mathscr{D}^m(A)$. Consequently, this lemma is an immediate result of the preceding one. ◇

Theorem 3.5-1. *There is a bijection from $[\mathscr{D}(A); B]$ onto $[\mathscr{D}; [A; B]]$ defined by*

$$\langle g, \theta \rangle a = \langle f, \theta a \rangle, \tag{1}$$

where $\theta \in \mathscr{D}$, $a \in A$, $g \in [\mathscr{D}; [A; B]]$, and $f \in [\mathscr{D}(A); B]$.

Note. Because of this result, we will subsequently denote g and f by the same symbol, say f, and will replace (1) by

$$\langle f, \theta \rangle a = \langle f, \theta a \rangle. \tag{2}$$

PROOF. Lemma 3.3-2 shows that f uniquely determines g though (1). So, let us consider the converse.

By Corollary 3.4-1a, for any given $g \in [\mathscr{D}; [A; B]]$ and compact interval $K \subset R^n$, there exists a nonnegative integer $q \in R^n$ and a continuous $[A; B]$-valued function h on K such that, for all $\theta \in \mathscr{D}_K$,

$$\langle g, \theta \rangle = \int_K h(t) D^q \theta(t)\, dt. \tag{3}$$

We let $\mathscr{D}_{\mathring{K}}(A)$ be the space of all $\phi \in \mathscr{D}(A)$ whose supports are contained in $\mathring{K}$ and supply $\mathscr{D}_{\mathring{K}}(A)$ with the topology induced by $\mathscr{D}_K(A)$. Also, $\mathscr{D}_{\mathring{K}} \triangleq \mathscr{D}_{\mathring{K}}(C)$. We define a mapping $f_{\mathring{K}}$ on $\mathscr{D}_{\mathring{K}}(A)$ into B by

$$\langle f_{\mathring{K}}, \psi \rangle \triangleq \int_K h(t) D^q \psi(t)\, dt, \qquad \psi \in \mathscr{D}_{\mathring{K}}(A). \tag{4}$$

It follows that $f_{\mathring{K}} \in [\mathscr{D}_{\mathring{K}}(A); B]$. Moreover, if we set $\psi = \theta a$ with $\theta \in \mathscr{D}_{\mathring{K}}$ and $a \in A$, we see from (3) and (4) that $\langle g, \theta \rangle a = \langle f_{\mathring{K}}, \theta a \rangle$, in agreement with (1).

This procedure determines an $f_{\mathring{K}} \in [\mathscr{D}_{\mathring{K}}(A); B]$ for every compact interval K. We now assert that, if the compact interval J contains K, then the restriction of $f_{\mathring{J}}$ to $\mathscr{D}_{\mathring{K}}(A)$ coincides with $f_{\mathring{K}}$. Indeed, for $\theta \in \mathscr{D}_{\mathring{K}}$ and $a \in A$, $\langle f_{\mathring{J}}, \theta a \rangle = \langle g, \theta \rangle a = \langle f_{\mathring{K}}, \theta a \rangle$, and therefore $f_{\mathring{J}}$ coincides with $f_{\mathring{K}}$ on $\mathscr{D}_{\mathring{K}} \odot A$. But, as a consequence of Lemma 3.5-1, $\mathscr{D}_{\mathring{K}} \odot A$ is dense in $\mathscr{D}_{\mathring{K}}(A)$, and this implies our assertion.

It now follows that there exists a unique $f \in [\mathscr{D}(A); B]$ whose restriction to each $\mathscr{D}_{\mathring{K}}(A)$ coincides with $f_{\mathring{K}}$. Thus, every $g \in [\mathscr{D}; [A; B]]$ uniquely determines an $f \in [\mathscr{D}(A); B]$ by means of (1). ◇

3.6. THE ρ-TYPE TESTING FUNCTION SPACES

The rest of this chapter discusses various generalized-function spaces, most (but not all) of which are subspaces of $[\mathscr{D}^m(A); B]$. It is possible to formulate our discussion in such a general fashion that each space of interest to us becomes a special case obtained by making particular choices for certain parameters. The only exceptions to this are the spaces of L_p-type distributions, which are discussed at the end of this chapter. Just as with distributions, generalized functions are continuous linear mappings on certain testing-function spaces. Thus, our first objective is to develop a general formulation for the needed testing-function spaces.

Throughout the rest of this chapter μ, ν, i, j, l, and p denote nonnegative integers in R, whereas $k = \{k_\nu\}_{\nu=1}^n$ is a nonnegative integer in R^n. As before, $m = \{m_\nu\}_{\nu=1}^n$, where each m_ν is either a nonnegative integer or ∞. A sequence $\{K_j\}_{j=1}^\infty$ will be called a *nested closed cover of* R^n if each K_j is a closed subset of R^n, $K_j \subset K_{j+1}$ for every j, $\bigcup_j K_j = R^n$, and each compact subset of R^n is contained in one of the K_j. Let $\{K_j\}_{j=0}^\infty$ and $\{I_p\}_{p=0}^\infty$ be two nested closed covers of R^n. For each j and l, let there be given a continuous function $\xi_{j,l}(t)$ from R^n into R such that $\xi_{j,l}(t) > 0$ for every $t \in R^n$. Also, assume that $\xi_{j,l}(t) \geq \xi_{j+1,l}(t)$ for all j, l, and t. For every j and p, we define the functional $\rho_{j,p}$ on suitably restricted functions $\phi(t)$ from R^n into A by

$$\rho_{j,p}(\phi) \triangleq \max_{\substack{0 \leq l \leq p \\ 0 \leq k_\nu \leq \min(p, m_\nu)}} \sup_{t \in I_p} \|\xi_{j,l}(t)\phi^{(k)}(t)\|_A, \tag{1}$$

where it is understood that the maximum is also taken over $\nu = 1, \ldots, n$. Clearly, $\rho_{j,p}(\phi) \leq \rho_{j,p+1}(\phi)$.

For each j, we define $\mathscr{I}_j^m(A)$ as the linear space of all functions ϕ from R^n into A such that $\phi^{(k)}$ is continuous whenever $0 \leq k \leq m$, $\operatorname{supp} \phi \subset K_j$, and $\rho_{j,p}(\phi) < \infty$. In this case, each $\rho_{j,p}$ is a seminorm on $\mathscr{I}_j^m(A)$ and, if ϕ is not identically zero, then $\rho_{j,p}(\phi) > 0$ for at least one of the $\rho_{j,p}$. We assign to $\mathscr{I}_j^m(A)$ the topology generated by $\{\rho_{j,p}\}_{p=0}^\infty$ and obtain thereby a metrizable locally convex space. That $\mathscr{I}_j^m(A)$ is complete and therefore a Fréchet space can be shown by using the result asserted in Problem 1.6-3. Clearly, the function $\phi \mapsto \phi^{(k)}$, where $k \leq m$, is a continuous linear mapping of $\mathscr{I}_j^m(A)$ into $\mathscr{I}_j^{m-k}(A)$. Both $\mathscr{I}_j^m(A)$ and its topology remain unchanged if the $\rho_{j,p}$ are altered under any one of the following three situations.

(i) If K_j is bounded, all the I_p can be set equal to R^n and $\xi_{j,l}(t)$ can be set equal to 1 for all t and l. In this case, $\mathscr{I}_j^m(A) = \mathscr{D}_{K_j}^m(A)$.

(ii) If any one of the I_p is equal to R^n, all the I_p can be set equal to R^n.

(iii) If every I_p is bounded, $\{I_p\}$ can be replaced by any other nested closed cover of R^n consisting exclusively of bounded sets, and, in addition, $\xi_{j,l}(t)$ can be set equal to 1 for all t and l.

Since $\xi_{j,l}(t) \geq \xi_{j+1,l}(t)$ for all j, l, and t, it follows that $\rho_{j,p}(\phi) \geq \rho_{j+1,p}(\phi)$ for every $\phi \in \mathscr{I}_j^m(A)$. Furthermore, $K_j \subset K_{j+1}$. Consequently, $\mathscr{I}_j^m(A) \subset \mathscr{I}_{j+1}^m(A)$ for all j, and the topology of $\mathscr{I}_j^m(A)$ is stronger than that induced on it by $\mathscr{I}_{j+1}^m(A)$. We define $\mathscr{I}^m(A) \triangleq \bigcup_{j=1}^{\infty} \mathscr{I}_j^m(A)$. This is a linear space, and we equip it with the inductive-limit topology (Appendix E1). We shall call any locally convex space of functions having this form a *ρ-type testing-function space*, and $\mathscr{I}^m(A)$ will always denote such a space.

Clearly, a sufficient condition for $\mathscr{I}^m(A)$ to be strict as an inductive-limit space (Appendix E3) is that $\xi_{j,l} = \xi_{j+1,l}$ for every j and l. If, in addition, $K_j = R^n$ for at least one j, then $\mathscr{I}^m(A) = \mathscr{I}_j^m(A)$, in which case $\mathscr{I}^m(A)$ will be called *degenerate*. As usual, we set $\mathscr{I}^{[\infty]}(A) = \mathscr{I}(A)$, $\mathscr{I}^m(C) = \mathscr{I}^m$, and $\mathscr{I}^{[\infty]}(C) = \mathscr{I}$. This notational convention is followed for every one of the special cases described below.

For subsequent use in defining a topology for $[\mathscr{I}^m(A); B]$, we single out a certain collection $\mathfrak{S}$ of subsets of $\mathscr{I}^m(A)$. The members of $\mathfrak{S}$ will be called *$\mathfrak{S}$-sets*. A subset Ω of $\mathscr{I}^m(A)$ is said to be an *$\mathfrak{S}$-set* if Ω is contained in some $\mathscr{I}_j(A)$ and is a bounded set therein. Thus, every $\mathfrak{S}$-set is a bounded set in $\mathscr{I}^m(A)$, but the converse is not true in general. The converse (namely, the bounded sets in $\mathscr{I}^m(A)$ are $\mathfrak{S}$-sets) is true whenever $\mathscr{I}^m(A)$ is a strict inductive-limit space with the closure property (see Appendix E4).

We shall now list a number of special cases of $\mathscr{I}^m(A)$. In the following, a locally convex space $\mathscr{V}$ of A-valued functions on R^n will be called *normal* if $\mathscr{D}(A)$ is a dense subspace of $\mathscr{V}$ and the canonical injection of $\mathscr{D}(A)$ into $\mathscr{V}$ is continuous.

I. $\mathscr{D}^m(A)$. This occurs when all the K_j are bounded sets. In this case, we can replace $\{K_j\}$ by any other nested closed cover of R^n consisting of bounded sets. By virtue of Lemma 3.2-3, $\mathscr{D}^m(A)$ is normal. For fixed m, $\mathscr{D}^m(A)$ is the smallest ρ-type testing-function space in the following sense. For any other ρ-type testing-function space $\mathscr{I}^m(A)$, we have that $\mathscr{D}^m(A) \subset \mathscr{I}^m(A)$. Moreover, the canonical injection of $\mathscr{D}^m(A)$ into $\mathscr{I}^m(A)$ is continuous.

II. $\mathscr{E}^m(A)$. We obtain this case when every K_j is R^n and every I_p is a bounded set. As was indicated before, we can now set $\xi_{j,l}(t) = 1$ for all j, l, and t. Hence, $\mathscr{E}^m(A)$ is degenerate, and we have

$$\rho_{j,p}(\phi) = \rho_p(\phi) = \max_{0 \leq k_\nu \leq \min(p, m_\nu)} \sup_{t \in I_p} \|\phi^{(k)}(t)\|_A .$$

$\mathscr{E}^m(A)$ is a Fréchet space and is normal. For fixed m, $\mathscr{E}^m(A)$ is the largest of the ρ-type testing-function spaces because $\mathscr{I}^m(A) \subset \mathscr{E}^m(A)$ for any other space $\mathscr{I}^m(A)$. Indeed, every A-valued function ϕ on R^n such that $\phi^{(k)}$ is continuous whenever $0 \leq k \leq m$ is a member of $\mathscr{E}^m(A)$. Thus, $\|\phi^{(k)}(t)\|$ has no restriction on its growth as $|t| \to \infty$. Furthermore, the canonical injection of $\mathscr{I}^m(A)$ into $\mathscr{E}^m(A)$ is continuous.

III. $\mathscr{S}^m(A)$. Now, $K_j = R^n$, $I_p = R^n$, and

$$\xi_{j,l}(t) = (1 + |t|^2)^l$$

for every j, q, l, and t. Thus, $\mathscr{S}^m(A)$ is degenerate, and

$$\rho_{j,p}(\phi) = \rho_p(\phi) = \max_{0 \leq k_\nu \leq \min(p, m_\nu)} \sup_{t \in R^n} \|(1 + |t|^2)^p \phi^{(k)}(t)\|_A .$$

Here, too, $\mathscr{S}^m(A)$ is Fréchet and normal. The members of $\mathscr{S}^{[\infty]}(A) = \mathscr{S}(A)$ are called *A-valued testing functions of rapid descent*. Now, $\|\phi^{(k)}(t)\|$ tends to zero faster than any negative power of $|t|$ as $|t| \to \infty$.

IV. $\mathscr{L}^m_{c,d}(A)$. Let $c = \{c_\nu\}_{\nu=1}^n \in R^n$ and $d = \{d_\nu\}_{\nu=1}^n \in R^n$.

Set

$$\kappa_{c,d}(t) = \prod_{\nu=1}^n \kappa_{c_\nu, d_\nu}(t_\nu),$$

where

$$\kappa_{c_\nu, d_\nu}(t_\nu) = \begin{cases} e^{c_\nu t_\nu}, & 0 \leq t_\nu < \infty, \\ e^{d_\nu t_\nu}, & -\infty < t_\nu < 0. \end{cases}$$

Also, let $K_j = R^n$, $I_p = R^n$, and $\xi_{j,l}(t) = \kappa_{c,d}(t)$ for all j, q, l, and t. So,

$$\rho_{j,p}(\phi) = \rho_{c,d,p}(\phi) = \max_{0 \leq k_\nu \leq \min(p, m_\nu)} \sup_{t \in R^n} \|\kappa_{c,d}(t)\phi^{(k)}(t)\|_A .$$

The resulting ρ-type testing-function space $\mathscr{L}^m_{c,d}(A)$ is degenerate and therefore Fréchet. It is not normal because $\mathscr{D}^m(A)$ is not dense in $\mathscr{L}^m_{c,d}(A)$. This space and the next one arise naturally in the study of the generalized Laplace transformation (Zemanian, 1968a, Chapter 3). The members of these spaces have exponential bounds on their behavior as $|t| \to \infty$.

V. $\mathscr{L}^m(w, z; A)$. Let $w = \{w_\nu\}_{\nu=1}^n$, where each w_ν is either a real number or $-\infty$, and let $z = \{z_\nu\}_{\nu=1}^n$, where each z_ν is either a real number or ∞. Let $\{c_j\}_{j=1}^\infty$ and $\{d_j\}_{j=1}^\infty$ be two sequences in R^n with the following properties: $c_{j+1} < c_j$ and $d_{j+1} > d_j$ for every j. Also, upon denoting the components of c_j by $c_{j,\nu}$ and of d_j by $d_{j,\nu}$, where $\nu = 1, \ldots, n$, assume that, for each ν and as $j \to \infty$, $c_{j,\nu} \to w_\nu$ and $d_{j,\nu} \to z_\nu$. Now, set $K_j = R^n$, $I_p = R^n$, and $\xi_{j,l}(t) = \kappa_{c_j, d_j}(t)$ for every j, p, l, and t. Thus $\xi_{j,l}(t) \geq \xi_{j+1,l}(t)$ as is required. Moreover

$$\rho_{j,p}(\phi) = \rho_{c_j, d_j, p}(\phi) = \max_{0 \leq k_\nu \leq \min(p, m_\nu)} \sup_{t \in R^n} \|\kappa_{c_j, d_j}(t)\phi^{(k)}(t)\|_A .$$

Thus, $\mathscr{L}^m(w, z; A)$ is the inductive-limit space $\bigcup_j \mathscr{L}^m_{c_j, d_j}(A)$. However, it is not strict. On the other hand, it is normal.

VI. $\mathscr{D}_{-}{}^{m}(A)$. In this and the next special case, $n = 1$, so that we will be dealing with functions on the real line R. Also, k is now a nonnegative integer in R, and m is either a nonnegative integer in R or ∞. We set $K_j = (-\infty, j]$, $I_p = [-p, \infty)$, and $\xi_{j,l}(t) = 1$ for all j, p, l, and t. Thus, the functions ϕ in $\mathscr{D}_{-}{}^{m}(A)$ have their supports bounded on the right, and

$$\rho_{j,p}(\phi) = \max_{0 \le k \le p} \sup_{t \in I_p} \|\phi^{(k)}(t)\|_A .$$

However, $\|\phi^{(k)}(t)\|$ has no restriction on its growth as $t \to -\infty$. The space $\mathscr{D}_{-}{}^{m}(A)$ is a normal space as well as a strict inductive-limit space having the closure property.

VII. $\mathscr{D}_{+}{}^{m}(A)$. Here again, $n = 1$. Also k and m are as in the preceding case. We set $K_j = [-j, \infty)$, $I_p = (-\infty, p]$, and $\xi_{j,l}(t) = 1$ for all j, p, l, and t. The functions in $\mathscr{D}_{+}{}^{m}(A)$ have their supports bounded on the left, and

$$\rho_{j,p}(\phi) = \max_{0 \le k \le p} \sup_{t \in I_p} \|\phi^{(k)}(t)\|_A .$$

$D_{+}{}^{m}(A)$ is both normal and strict and has the closure property.

Problem 3.6-1. If $I_q \supset \mathring{K}_j$, then $\rho_{j,p}$ is a norm on $\mathscr{I}_j(A)$ for every $p \ge q$. Conversely, if $\rho_{j,p}$ is a norm on $\mathscr{I}_j(A)$, then $I_p \supset \mathring{K}_j$. Prove these assertions.

Problem 3.6-2. Prove that $I_j{}^m(A)$ is complete.

Problem 3.6-3. Show that the differentiation operator D^k is a continuous linear mapping of $\mathscr{I}^{m+k}(A)$ into $\mathscr{I}^m(A)$.

Problem 3.6-4. Show that $\mathscr{E}^m(A)$, $\mathscr{S}^m(A)$, $\mathscr{L}^m(w, z; A)$, $\mathscr{D}_{-}{}^{m}(A)$, and $\mathscr{D}_{+}{}^{m}(A)$ are all normal. Also, show that the shifting operator σ_τ is an automorphism on each of these spaces.

3.7. GENERALIZED FUNCTIONS

Given any ρ-type testing-function space $\mathscr{I}^m(A)$, a continuous linear mapping of $\mathscr{I}^m(A)$ into B is said to be an *$[A; B]$-valued generalized function on R^n*. As we did with distributions, we will subsequently identify $[\mathscr{I}^m; [A; B]]$ with $[\mathscr{I}^m(A); B]$, and thereby $[\mathscr{I}^m; [C; B]]$ with the space $[\mathscr{I}^m; B]$ of *B-valued generalized functions on R^n*. Since $\mathscr{I}^m(A)$ contains $\mathscr{D}(A)$ and induces a topology on $\mathscr{D}(A)$ weaker than that of $\mathscr{D}(A)$, the restriction of any $f \in [\mathscr{I}^m(A); B]$ to $\mathscr{D}(A)$ is a member of $[\mathscr{D}(A); B]$ (i.e., is an $[A; B]$-valued distribution on R^n). If $\mathscr{I}^m(A)$ happens to be normal, $[\mathscr{I}^m(A); B]$ can be treated as a subspace of $[\mathscr{D}(A); B]$ by identifying each $f \in [\mathscr{I}^m(A); B]$ with its restriction to $\mathscr{D}(A)$.

This is because the said restriction determines f uniquely on all of $\mathscr{I}^m(A)$ by virtue of the density of $\mathscr{D}(A)$ in $\mathscr{I}^m(A)$.

The topology we will usually employ for $[\mathscr{I}^m(A); B]$ is the *topology of uniform convergence on the* $\mathfrak{S}$*-sets in* $\mathscr{I}^m(A)$, which we will simply call the $\mathfrak{S}$-*topology*. This is the topology generated by the collection $\{\gamma_\Phi\}_{\Phi \in \mathfrak{S}}$ of seminorms defined by

$$\gamma_\Phi(f) \triangleq \sup_{\phi \in \Phi} \|\langle f, \phi \rangle\|_B,$$

where $f \in [\mathscr{I}^m(A); B]$ and Φ is an arbitrary $\mathfrak{S}$-set. To indicate that this is' the topology that is being employed, we will use the notation $[\mathscr{I}^m(A); B]^s$. However, when $\mathscr{I}^m(A)$ is either degenerate or more generally a strict inductive limit having the closure property, then the $\mathfrak{S}$-sets are precisely the bounded sets in $\mathscr{I}^m(A)$, so that the $\mathfrak{S}$-topology is the *bounded topology*. In this case, we drop the superscript s on the notation $[\mathscr{I}^m(A); B]$. This is the situation for each of the spaces $[\mathscr{D}^m(A); B]$, $[\mathscr{E}^m(A); B]$, $[\mathscr{S}^m(A); B]$, $[\mathscr{L}^m_{c,d}(A); B]$, $[\mathscr{D}_-{}^m(A); B]$, and $[\mathscr{D}_+{}^m(A); B]$. Still another topology is the *pointwise topology* generated by $\{\gamma_\phi\}$, where ϕ traverses $\mathscr{I}^m(A)$ and

$$\gamma_\phi(f) \triangleq \|\langle f, \phi \rangle\|.$$

When using this topology, we employ the notation $[\mathscr{I}^m(A); B]^\sigma$.

Lemma 3.7-1. *Let F be a bounded set in $[\mathscr{I}^m(A); B]^\sigma$. Then, corresponding to each j, there exists a nonnegative integer $p \in R$ and a constant $M > 0$ such that*

$$\sup_{f \in F} \|\langle f, \phi \rangle\|_B \leq M \rho_{j,p}(\phi) \tag{1}$$

for all $\phi \in \mathscr{I}_j{}^m(A)$. M and p depend in general on F and j.

Proof. Let F_j be the set of restrictions to $\mathscr{I}_j{}^m(A)$ of all $f \in F$. Then, F_j is a bounded set in $[\mathscr{I}_j{}^m(A); B]^\sigma$. Since $\mathscr{I}_j{}^m(A)$ is a Fréchet space, we may invoke Appendix D9 to conclude that F_j is equicontinuous on $\mathscr{I}_j{}^m(A)$, or, in other words, that (1) is satisfied. ◇

Clearly, if F consists of a single element, then F satisfies the hypothesis of the lemma. Another special case arises when F is a bounded set under the $\mathfrak{S}$-topology; for, in this case, F is also bounded under the pointwise topology because the pointwise topology is weaker than the $\mathfrak{S}$-topology.

Generalized differentiation D^k is defined on any generalized function $f \in [\mathscr{I}^m(A); B]$ exactly as it is on any distribution, namely

$$\langle D^k f, \phi \rangle \triangleq (-1)^{|k|} \langle f, D^k \phi \rangle, \qquad \phi \in \mathscr{I}^{m+k}(A).$$

By means of a proof exactly like the one for Theorem 3.3-1, we have the following result.

Theorem 3.7-1. *D^k is a continuous linear mapping of $[\mathscr{I}^m(A); B]^s$ into $[\mathscr{I}^{m+k}(A); B]^s$, as well as of $[\mathscr{I}^m(A); B]^\sigma$ into $[\mathscr{I}^{m+k}(A); B]^\sigma$.*

For future reference, we now list a number of spaces consisting of linear combinations of certain elements.

I. $\mathscr{I}^m \odot A$. This is the span of all elements of the form $\phi a \triangleq a\phi$, where $\phi \in \mathscr{I}$, $a \in A$, and ϕa denotes the mapping $t \mapsto \phi(t)a$. Thus, $\mathscr{I}^m \odot A$ is a subspace of $\mathscr{I}^m(A)$.

II. $[\mathscr{I}^m; C] \odot A$. If $g \in [\mathscr{I}^m; C]$ and $a \in A$, $ga \triangleq ag$ is defined by the equation

$$\langle ga, \phi \rangle \triangleq \langle g, \phi \rangle a, \qquad \phi \in \mathscr{I}^m. \tag{2}$$

It follows that $ga \in [\mathscr{I}^m; A]$. $[\mathscr{I}^m; C] \odot A$ denotes the span of all elements of the form ga and is therefore a subspace of $[\mathscr{I}^m; A]$.

III. $[\mathscr{I}^m; [A; B]] \odot A$. If $g \in [\mathscr{I}^m; [A; B]]$ and $a \in A$, we define ga by Equation (2) again. Now, $ga \in [\mathscr{I}^m; B]$. $[\mathscr{I}^m; [A; B]] \odot A$ denotes the span of all elements of the form ga and is therefore a subspace of $[\mathscr{I}^m; B]$.

Our next objective is to relate $[\mathscr{I}^m(A); B]$ and $[\mathscr{I}^m; [A; B]]$. Through precisely the same proof as that of Lemma 3.3-2, we can establish the following lemma.

Lemma 3.7-2. *Every $f \in [\mathscr{I}^m(A); B]$ uniquely defines a $g \in [\mathscr{I}^m; [A; B]]$ by means of the equation*

$$\langle g, \psi \rangle a \triangleq \langle f, \psi a \rangle, \qquad \psi \in \mathscr{I}^m, \quad a \in A. \tag{3}$$

Theorem 3.7-2. *If $\mathscr{I}^m$ and $\mathscr{I}^m(A)$ are normal spaces, then there exists a bijection from $[\mathscr{I}^m(A); B]$ onto $[\mathscr{I}^m; [A; B]]$ defined by* (3).

PROOF. The steps of this proof are illustrated in Figure 3.7-1. Given $f \in [\mathscr{I}^m(A); B]$, we may replace (3) by

$$\langle f, \phi a \rangle = \langle g, \phi \rangle a, \tag{4}$$

where ϕ is restricted to $\mathscr{D}$, and still obtain a definition for $g \in [\mathscr{I}^m; [A; B]]$. This is because $\mathscr{D}$ is dense in $\mathscr{I}^m$ so that we need merely specify g on $\mathscr{D}$ in order to determine g on all of $\mathscr{I}^m$. Let $F_4: f \mapsto g$ denote this mapping.

Now, let $g \in [\mathscr{I}^m; [A; B]]$ and let $\hat{g}$ be the restriction of g to $\mathscr{D}$. Because of the normality of $\mathscr{I}^m$, the mapping $F_5: g \mapsto \hat{g}$ is a bijection of $[\mathscr{I}^m; [A; B]]$ onto a subspace U of $[\mathscr{D}; [A; B]]$. Moreover, we may write

$$\langle g, \phi \rangle a = \langle \hat{g}, \phi \rangle a, \qquad \phi \in \mathscr{D}, a \in A. \tag{5}$$

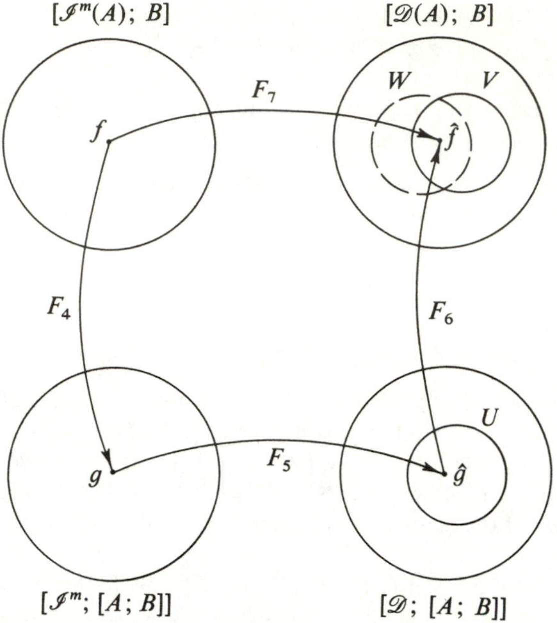

Figure 3.7-1

By Theorem 3.5-1, the equation

$$\langle \hat{g}, \phi \rangle a \triangleq \langle \hat{f}, \phi a \rangle, \tag{6}$$

where $\phi \in \mathscr{D}$, $a \in A$, $\hat{g} \in [\mathscr{D}; [A; B]]$, and $\hat{f} \in [\mathscr{D}(A); B]$, defines a bijection $F_6: \hat{g} \mapsto \hat{f}$ of $[\mathscr{D}; [A; B]]$ onto $[\mathscr{D}(A); B]$. Thus, the subspace U is mapped by F_6 in a one-to-one fashion onto a subspace V of $[\mathscr{D}(A); B]$.

On the other hand, let us denote the restriction of any $f \in [\mathscr{I}^m(A); B]$ to $\mathscr{D}(A)$ by $\hat{f}$. Then, the mapping $F_7: f \mapsto \hat{f}$ is a linear bijection of $[\mathscr{I}^m(A); B]$ onto a subspace W of $[\mathscr{D}(A); B]$ because of the normality of $\mathscr{I}^m(A)$. Moreover, by Lemma 3.5-1, the span of the set of all elements of the form ϕa is dense in $\mathscr{D}(A)$. Therefore, the bijection F_7 is determined by

$$\langle f, \phi a \rangle \triangleq \langle \hat{f}, \phi a \rangle. \tag{7}$$

Upon combining (4)–(7), we see that the composite mapping $F_6 F_5 F_4$ is equal to F_7. Therefore, the subspaces V and W must coincide, and, since F_7 is a bijection of $[\mathscr{I}^m(A); B]$ onto W, so too must be $F_6 F_5 F_4$. But we have already noted that F_5 and F_6 are bijections. Consequently, F_4 must be a bijection. ◇

Because of Theorem 3.7-2, we will customarily denote f and g by the same symbol and will replace (3) by

$$\langle f, \psi\rangle a = \langle f, \psi a\rangle. \tag{8}$$

We end this section by noting certain support conditions possessed by the members of $[\mathscr{E}^m(A); B]$ and $[\mathscr{D}_+{}^m(A); B]$. These members are distributions since both $\mathscr{E}^m(A)$ and $\mathscr{D}_+{}^m(A)$ are normal.

Theorem 3.7-3. *$f \in [\mathscr{E}^m(A); B]$ if and only if $f \in [\mathscr{D}^m(A); B]$ and* supp f *is a compact set.*

PROOF. Given any $f \in [\mathscr{E}^m(A); B]$, there exists a neighborhood Λ of 0 in $\mathscr{E}^m(A)$ such that $\|\langle f, \phi\rangle\|_B \leq 1$ for all $\phi \in \Lambda$. Upon referring to Case II of the preceding section, we see that Λ contains a set Ω consisting of all $\phi \in \mathscr{E}^m(A)$ such that

$$\max_{0 \leq k_\nu \leq \min(p, m_\nu)} \sup_{t \in I_p} \|\phi^{(k)}(t)\|_A \leq \varepsilon,$$

where I_p is a compact set. This implies that supp $f \subset I_p$. Indeed, if f is not equal to the zero distribution on $R^n \backslash I_p$, then we can find a $\theta \in \mathscr{D}(A)$ such that supp $\theta \subset R^n \backslash I_p$ and $\langle f, \theta\rangle \neq 0$. But then, $M\theta \in \Omega$ for all real numbers M, and $\|\langle f, M\theta\rangle\|$ can be made larger than 1 by choosing M appropriately. This is a contradiction.

Conversely, if $f \in [\mathscr{D}^m(A); B]$ and supp f is a compact set, then, by virtue of the paragraph just before Example 3.3-1, f has a unique extension onto any $\phi \in \mathscr{E}^m(A)$ defined by $\langle f, \phi\rangle = \langle f, \lambda\phi\rangle$, where $\lambda \in \mathscr{D}$ is equal to 1 on a neighborhood of supp f. This extension is clearly linear and is continuous because the convergence of $\{\phi_\nu\}$ to 0 in $\mathscr{E}^m(A)$ implies the convergence of $\{\lambda\phi_\nu\}$ to 0 in $\mathscr{D}^m(A)$. ◇

A similar argument establishes the following result.

Theorem 3.7-4. *$f \in [\mathscr{D}_-{}^m(A); B]$ (or $f \in [\mathscr{D}_+{}^m(A); B]$) if and only if $f \in [\mathscr{D}^m(A); B]$ and* supp f *is bounded on the left (or respectively on the right).*

Problem 3.7-1. Show that, when $\mathscr{I}^m(A)$ is normal, the restriction of any $f \in [\mathscr{I}^m(A); B]$ to $\mathscr{D}(A)$ uniquely determines f on all of $\mathscr{I}^m(A)$.

Problem 3.7-2. Prove Theorem 3.7-4.

Problem 3.7-3. Let $\mathscr{I}^m(A)$ be any one of the spaces $\mathscr{E}^m(A)$, $\mathscr{S}^m(A)$, $\mathscr{L}^m(w, z; A)$, $\mathscr{D}_-{}^m(A)$, and $\mathscr{D}_+{}^m(A)$. Show that the shifting operator is an automorphism on $[\mathscr{I}^m(A); B]$.

Problem 3.7-4. Every continuous A-valued function is a regular A-valued distribution (see Example 3.3-2). The canonical injections of $\mathscr{E}^0(A)$ into $[\mathscr{D}^0; A]$ and of $\mathscr{D}^0(A)$ into $[\mathscr{E}^0; A]$ are continuous. Verify these assertions.

Problem 3.7-5. Show that $\mathscr{D}^0(A)$ is a subspace of $[\mathscr{J}^0; A]$ for every ρ-type testing-function space $\mathscr{J}^0$.

Problem 3.7-6. Let $\mathfrak{N}$ be a continuous linear mapping of $\mathscr{D}(A)$ into $[\mathscr{D}; B]$. Define $\mathfrak{M}$ as a mapping on $\mathscr{D}$ by

$$\langle \mathfrak{M}f, \phi \rangle a \triangleq \langle \mathfrak{N}(fa), \phi \rangle, \tag{9}$$

where $f \in \mathscr{D}$, $\phi \in \mathscr{D}$, and $a \in A$. Show that $\mathfrak{M}$ is a continuous linear mapping of $\mathscr{D}$ into $[\mathscr{D}; [A; B]]$.

Problem 3.7-7. Let $\mathscr{J}^m = \bigcup_j \mathscr{J}_j^m$ and $\mathscr{J}^r = \bigcup_j \mathscr{J}_j^r$ be ρ-type testing-function spaces consisting of complex-valued functions. Let $\mathfrak{N}$ be a continuous linear mapping of $[\mathscr{J}^m; A]^s$ into $[\mathscr{J}^r; B]^s$. Define a mapping $\mathfrak{M}$ on $[\mathscr{J}^m; C]$ by (9), where now $f \in [\mathscr{J}^m; C]$, $\phi \in \mathscr{J}^r$, and $a \in A$. Show that $\mathfrak{M}$ is a continuous linear mapping of $[\mathscr{J}^m; C]^s$ into $[\mathscr{J}^r; [A; B]]^s$. Also, show that this result is again valid when the $\mathfrak{S}$-topologies of $[\mathscr{J}^m; A]$, $[\mathscr{J}^r; B]$, $[\mathscr{J}^m; C]$, and $[\mathscr{J}^r; [A; B]]$ are replaced by the pointwise topologies.

3.8. L_p-TYPE TESTING FUNCTIONS AND DISTRIBUTIONS

The L_p-type testing functions do not fit the general formulation for the ρ-type testing functions. They are instead defined as follows. Let $p \in R$ be fixed with $1 \le p < \infty$. The space $\mathscr{D}_{L_p}(A)$ is the linear space of all smooth A-valued functions ϕ on R^n such that, for each integer $k \in R^n$ with $k \ge 0$,

$$\gamma_{p,k}(\phi) \triangleq \left[\int_{R^n} \|\phi^{(k)}(t)\|_A^p \, dt \right]^{1/p} < \infty.$$

We set $\mathscr{D}_{L_p}(C) = \mathscr{D}_{L_p}$. The members of $\mathscr{D}_{L_p}(A)$ are said to be *L_p-type testing functions.* Minkowski's inequality shows that $\gamma_{p,k}$ is a seminorm on $\mathscr{D}_{L_p}(A)$. We assign to $\mathscr{D}_{L_p}(A)$ the topology generated by $\{\gamma_{p,k}\}_{k \ge 0}$. This separates the space because, if $\gamma_{p,0}(\phi) = 0$, it follows from the nonnegativity and continuity of $\|\phi(\cdot)\|$ that $\phi(t) = 0$ for every t. Thus, $\mathscr{D}_{L_p}(A)$ is a metrizable space. The *shifting operator* σ_τ: $\phi(t) \mapsto \phi(t - \tau)$ is an automorphism on $\mathscr{D}_{L_p}(A)$, and differentiation is a continuous linear mapping of $\mathscr{D}_{L_p}(A)$ into $\mathscr{D}_{L_p}(A)$.

The space $\mathscr{D}_{L_p}(A)$ is a subspace of the space $L_p(A)$ of all (equivalence classes of) A-valued Bochner-integrable functions with respect to Lebesgue measure on the Borel subsets of R^n (see Appendix G19). Moreover, the canonical injection of $\mathscr{D}_{L_p}(A)$ into $L_p(A)$ is continuous.

In turn, $L_p(A)$ becomes a subspace of $[\mathscr{D}; A]$ when every $f \in L_p(A)$ is identified with a mapping g of $\mathscr{D}$ into A through the equation

$$\langle g, \phi \rangle \triangleq \int_{R^n} f(t)\phi(t)\, dt.$$

Indeed, g is clearly linear on $\mathscr{D}$. Moreover, for $p > 1$ and $q = p/(p-1)$, we have from Holder's inequality (Appendix G20) that

$$\begin{aligned}\|\langle g, \phi \rangle\| &\leq \left[\int \|f(t)\|^p\, dt\right]^{1/p} \left[\int |\phi(t)|^q\, dt\right]^{1/q} \\ &\leq \left[\int \|f(t)\|^p\, dt\right]^{1/p} [\mathrm{vol}\, K]^{1/q} \sup_{t \in K} |\phi(t)|\end{aligned}$$

for all $\phi \in \mathscr{D}_K$, and hence $g \in [\mathscr{D}; A]$. When $p = 1$, we may write

$$\|\langle g, \phi \rangle\| \leq \int_K \|f(t)\|\, dt \sup_{t \in K} |\phi(t)|$$

for all $\phi \in \mathscr{D}_K$ and thereby conclude once again that $g \in [\mathscr{D}; A]$. These estimates also show that the canonical injection of $L_p(A)$ into $[\mathscr{D}; A]$ is continuous.

We now develop some inequalities that we shall subsequently need. Let 1_+ denote the *n-dimensional unit-step function* defined by

$$1_+(t) = 1_+(t_1) \cdots 1_+(t_n),$$

where in the right-hand side 1_+ is the unit-step function on the real line (see Section 3.4). As usual, $D^{[1]} \triangleq \partial_1 \cdots \partial_n$ denotes the differential operator of order [1]. Also, let $\phi \in \mathscr{E}^{[1]}(A)$ and let $\lambda \in \mathscr{D}$ be such that $\lambda(t) = 1$ for $|t| < 1$ and $\lambda(t) = 0$ for $|t| > 2$. For any fixed $t \in R^n$, we may write

$$\lambda(t - x)\phi(x) = (-1)^n \int_{R^n} 1_+(\tau) D_\tau^{[1]}[\lambda(t - x + \tau)\phi(x - \tau)]\, d\tau.$$

Upon setting $t = x$ and then estimating the result, we get

$$\|\phi(t)\| \leq M \sum_{0 \leq r \leq [1]} \int_E \|\phi^{(r)}(\tau)\|\, d\tau,$$

where M is a constant and $E = \{\tau\colon |t - \tau| < 2\}$. More generally, if k is a nonnegative integer in R^n and if $\phi \in \mathscr{E}^{k+[1]}(A)$, then

$$\|\phi^{(k)}(t)\| \leq M \sum_{0 \leq r \leq [1]} \int_E \|\phi^{(k+r)}(\tau)\|\, d\tau. \tag{1}$$

Also, upon applying Holder's inequality (Appendix G20), we obtain, for $1 < p < \infty$,

$$\|\phi^{(k)}(t)\| \leq N \sum_{0 \leq r \leq [1]} \left[\int_E \|\phi^{(k+r)}(\tau)\|^p\, d\tau\right]^{1/p} \tag{2}$$

where N is still another constant.

Lemma 3.8-1. *$\mathscr{D}_{L_p}(A)$ is a normal space.*

PROOF. Clearly, $\mathscr{D}(A)$ is a subspace of $\mathscr{D}_{L_p}(A)$ and the canonical injection of $\mathscr{D}(A)$ into $\mathscr{D}_{L_p}(A)$ is continuous. To show the density of $\mathscr{D}(A)$ in $\mathscr{D}_{L_p}(A)$, choose arbitrarily a $\phi \in \mathscr{D}_{L_p}(A)$. Also, let $\lambda \in \mathscr{D}$ be as before and set $\lambda_j(t) = \lambda(t/j)$ for $j = 1, 2, \ldots$. Then, $\phi\lambda_j \in \mathscr{D}(A)$. Moreover, for any k, we may write

$$\begin{aligned}
\gamma_{p,k}(\phi - \phi\lambda_j) &= \gamma_{p,0}\{D^k[\phi(1 - \lambda_j)]\} \\
&= \gamma_{p,0}\left[\sum_{0 \le r \le k} \binom{k}{r} (\phi^{(k-r)}(1 - \lambda_j)^{(r)}\right] \\
&\le \sum_{0 \le r \le k} \binom{k}{r} \left[\int_{R^n} \|\phi^{(k-r)}(t)\|^p |D^r[1 - \lambda_j(t)]|^p \, dt\right]^{1/p} \\
&\le M \sum_{0 \le r \le k} \binom{k}{r} \left[\int_{|t| \ge j} \|\phi^{(k-r)}(t)\|^p \, dt\right]^{1/p},
\end{aligned}$$

where M is a positive constant not depending on j. The integral in the last expression tends to zero as $j \to \infty$. Therefore, $\phi\lambda_j \to \phi$ in $\mathscr{D}_{L_p}(A)$, which proves the density of $\mathscr{D}(A)$ in $\mathscr{D}_{L_p}(A)$. ◇

A simpler version of the preceding proof shows that $L_p(A)$ is also a normal space.

Lemma 3.8-2. *Assume that $p, q \in R$ satisfy $1 \le q < p < \infty$. Then, $\mathscr{D}_{L_q}$ is a dense subspace of $\mathscr{D}_{L_p}(A)$, and the canonical injection of $\mathscr{D}_{L_q}(A)$ into $\mathscr{D}_{L_p}(A)$ is continuous.*

PROOF. Let $\phi \in \mathscr{D}_{L_q}(A)$. It follows from (1) that $\|\phi^{(k)}(t)\| \to 0$ as $|t| \to \infty$ for every k. That is, the set $\{t\colon \|\phi^{(k)}(t)\| \ge 1\}$ is bounded. On the complement of that set, $\|\phi^{(k)}(t)\|^q \ge \|\phi^{(k)}(t)\|^p$. Therefore, $\phi \in \mathscr{D}_{L_p}(A)$.

Since $\mathscr{D}(A) \subset \mathscr{D}_{L_q}(A) \subset \mathscr{D}_{L_p}(A)$ and since $\mathscr{D}(A)$ is dense in $\mathscr{D}_{L_p}(A)$, it follows that $\mathscr{D}_{L_q}(A)$ is dense in $\mathscr{D}_{L_p}(A)$.

Finally, let the sequence $\{\phi_j\}$ tend to zero in $\mathscr{D}_{L_q}(A)$. We see from (2) that, for each k, $\{\|\phi_j^{(k)}(t)\|\}_j$ tends to zero uniformly for all $t \in R^n$. Therefore, there exists an integer j_0 such that, for all $j > j_0$,

$$\int_{R^n} \|\phi_j^{(k)}(t)\|^p \, dt \le \int_{R^n} \|\phi_j^{(k)}(t)\|^q \, dt.$$

This implies that $\{\phi_j\}$ tends to zero in $\mathscr{D}_{L_p}(A)$ and completes the proof of the lemma. ◇

As usual, $[\mathscr{D}_{L_p}(A); B]$ is the linear space of all continuous linear mappings of $\mathscr{D}_{L_p}(A)$ into B. Since $\mathscr{D}_{L_p}(A)$ is a normal space, it follows that $[\mathscr{D}_{L_p}(A); B]$ can be identified with a subspace of $[\mathscr{D}(A); B]$ by identifying each $f \in [\mathscr{D}_{L_p}(A);$

$B]$ with its restriction to $\mathscr{D}(A)$. The members of $[\mathscr{D}_{L_p}(A); B]$ are called *$[A; B]$-valued L_p-type distributions.* On the other hand, the members of $[\mathscr{D}_{L_p}; B]$ are said to be *B-valued* since $[C; B]$ can be identified with B. As always, $[\mathscr{D}_{L_p}(A); B]$ is understood to have the bounded topology unless something else is indicated. The shifting operator σ_τ and generalized differentiation D^k are defined on $[\mathscr{D}_{L_p}(A); B]$ in the usual way for distributions (see Section 3.3). σ_τ is an automorphism on $[\mathscr{D}_{L_p}(A); B]$, and D^k is a continuous linear mapping of $[\mathscr{D}_{L_p}(A); B]$ into $[\mathscr{D}_{L_p}(A); B]$. An immediate consequence of Lemma 3.8-2 is the following one.

Lemma 3.8-3. *Assume again that $p, q \in R$ satisfy $1 \leq q < p < \infty$. Then, $[\mathscr{D}_{L_p}(A); B]$ is a subspace of $[\mathscr{D}_{L_q}(A); B]$. Furthermore, the canonical injection of the former space into the latter one is continuous, and the same is true when both spaces possess the pointwise topologies.*

Through the same proof as that of Theorem 3.7-2, we obtain the following result.

Theorem 3.8-1. *There exists a bijection from $[\mathscr{D}_{L_p}(A); B]$ onto $[\mathscr{D}_{L_p}; [A; B]]$ defined by the equation*

$$\langle g, \psi \rangle a = \langle f, \psi a \rangle, \tag{3}$$

where $\psi \in \mathscr{D}_{L_p}$, $a \in A$, $g \in [\mathscr{D}_{L_p}; [A; B]]$, and $f \in [\mathscr{D}_{L_p}(A); B]$.

Because of this bijection, we will usually denote both f and g in (3) by the same symbol.

Chapter 4

Kernel Operators

4.1. INTRODUCTION

This chapter starts our discussion of a realizability theory for continuous linear systems, the next section being devoted to a consideration of various types of systems that generate the operators discussed in this book. The subsequent sections are devoted to our most general class of operators, namely to operators that map $\mathscr{D}_{R^s}(A)$ into $[\mathscr{D}_{R^n}; B]$ linearly and continuously. [$\mathscr{D}_{R^s}(A)$ denotes the space $\mathscr{D}(A)$ with the additional specification that its members are defined on R^s. As usual, $\mathscr{D}_{R^n} \triangleq \mathscr{D}_{R^n}(C)$.] The paramount result in this chapter is the kernel theorem as stated by Theorem 4.4-1. This is an extension to a Banach-space setting of Schwartz's kernel theorem. It provides a kernel representation for the operators at hand, which is the subject of Section 4.5. How causality affects the kernel representation is discussed in Section 4.6, but the implications of other physically motivated assumptions such as time invariance and passivity are reserved for subsequent chapters.

4.2. SYSTEMS AND OPERATORS

Let us start by presenting a physical example of a system whose signals take their instantaneous values in a Hilbert space. This example employs certain results concerning electromagnetic waves and is given only for illustrative purposes. It is not essential to any succeeding discussion.

Example 4.2-1. Consider an electromagnetic cavity resonator that is being excited through a rectangular waveguide as indicated in Figure 4.2-1. The unit vectors for the Cartesian coordinate system $\{x, y, z\}$ are 1_x, 1_y, and 1_z; 1_z is directed along the axis of the waveguide, as shown. $\mathscr{A}$ will denote the cross-sectional surface cut by a fixed transverse plane at some point along the waveguide. Let E be the complex electric field intensity vector and F the complex magnetic field intensity vector at some arbitrary point of $\mathscr{A}$. Thus, E and F are complex vectors depending on $\{x, y, t\}$, where t denotes time.

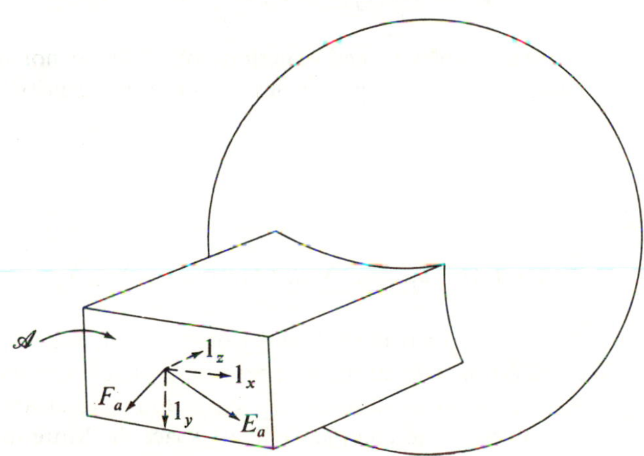

Figure 4.2-1

We let E_a and F_a be the projections of E and F onto $\mathscr{A}$, and therefore E_a and F_a are complex vectors lying in the plane of $\mathscr{A}$. The instantaneous net complex power passing through $\mathscr{A}$ toward the resonator is

$$P(t) = \int_{\mathscr{A}} (\bar{E} \times F) \cdot 1_z \, da = \int_{\mathscr{A}} (\bar{E}_a \times F_a) \cdot 1_z \, da,$$

where the bar denotes the complex conjugate, $\times$ the cross product, $\cdot$ the dot product, and da the incremental area on $\mathscr{A}$. The corresponding real power is $\operatorname{Re} P(t)$.

Now, assume that E_a and F_a are quadratically integrable on $\mathscr{A}$; that is,

$$\int_{\mathscr{A}} \bar{E}_a \cdot E_a \, da < \infty, \qquad \int_{\mathscr{A}} \bar{F}_a \cdot F_a \, da < \infty.$$

According to Jones (1964, p. 246), there exist real basis fields e_j and f_j, where $j = 1, 2, \ldots$, having the following properties. e_j and f_j are real vectors lying in the plane of $\mathscr{A}$ and depending only on $\{x, y\}$; they do not depend on t. Moreover, $f_j = 1_z \times e_j$, $\{e_j\}_{j=1}^{\infty}$ is the orthonormal basis of some separable complex Hilbert space H, and

$$\int_{\mathscr{A}} e_j \cdot e_k \, da = \int_{\mathscr{A}} f_j \cdot f_k \, da = \delta_{j,k},$$

where $\delta_{j,k}$ denotes the Kronecker delta. Finally, E_a and F_a have the unique expansions

$$E_a = \sum_{j=1}^{\infty} v_j e_j, \qquad F_a = \sum_{j=1}^{\infty} u_j f_j,$$

where v_j and u_j are complex-valued functions of t but do not depend on $\{x, y\}$. As a consequence of these properties, we have the identity

$$\int_{\mathscr{A}} (e_j \times f_k) \cdot 1_z \, da = \int_{\mathscr{A}} (f_j \cdot f_k) \, da = \delta_{j,k}$$

and the equation

$$P(t) = \int_{\mathscr{A}} [(\sum \overline{v_j} e_j) \times (\sum u_k f_k)] \cdot 1_z \, da = \sum u_j \bar{v}_j.$$

The point of this example is the following observation. We may view the system of Figure 4.2-1 as generating an operator $\mathfrak{N}$ that maps the quantity $v \triangleq E_a = \sum v_j e_j$ into the the quantity $u \triangleq \sum u_j e_j$. Both u and v are functions of t taking their values in the complex Hilbert space H. Moreover, the instantaneous net complex power $P(t)$ into the system is given by

$$P(t) = \big(u(t), v(t)\big) = \sum u_j(t)\overline{v_j(t)},$$

where $(\cdot, \ \cdot)$ is the inner product for H. Actually, u is a fictitious quantity, which is uniquely related to F_a. Indeed, given any $F_a = \sum u_j f_j$, one need merely replace f_j by e_j to obtain u. Upon borrowing some concepts from electrical network theory, we can consider v to be an H-valued voltage signal and u an H-valued current signal on the system of Figure 4.2-1. Thus, $\mathfrak{N}$ can be taken to be an admittance operator. Were H n-dimensional Euclidean space, our system would be called an n-port by network theorists. In analogy to this, we shall call the present system a *Hilbert port*.

We can view our system in a somewhat different way to obtain a closer analogy to the electrical n-port. We exploit the isomorphism between H and the Hilbert space ℓ_2; that is, we identify v with its sequence $\{v_j\}$ of Fourier coefficients and u with $\{u_j\}$. The coefficient v_j is considered to be the voltage signal at the jth port of an electrical system, and u_j the current at that port, as indicated in Figure 4.2-2. The polarity of v_j and the direction of u_j are so chosen that $u_j(t)\overline{v_j(t)}$ denotes the instantaneous power entering the jth port. Thus, we have arrived at an n-port, where now $n = \infty$. One might call this an *∞-port* (Zemanian, 1972a, Section 8). This ends our discussion of Example 4.2-1. ◇

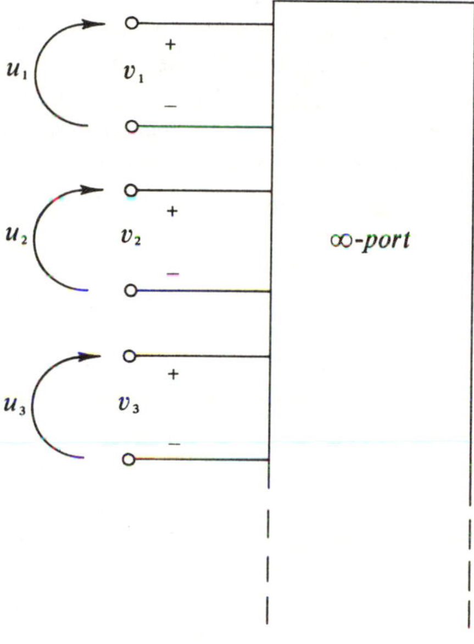

Figure 4.2-2

It is only when considerations of power flow arise that an inner product is needed for the space in which the signals at hand take their values. In this book, this will occur only when we impose the assumption of passivity, which states in effect that the system does not contain energy sources that can transmit energy to the exterior of the system. Thus, in the absence of passivity, we can and shall assume that the signals take their values in Banach spaces. In general, therefore, our analyses will involve signals that are Banach-space-valued functions or distributions. We shall refer to a system having such signals as a *Banach system*. It should also be pointed out that it is usually to transient phenomena that the realizability theories of this book are applied. This means that time is the independent variable for the signals, which are therefore functions or distributions on the real line R. However, there do occur physical phenomena involving signals on R^n that are amenable to some of the subsequent theories. An example of this is the optical system discussed by Meidan (1970). For this reason, we shall allow, at least initially, signals defined on R^n.

A Banach system may have many different Banach spaces associated with it. For example, the signal v representing some physical variable at one location x within the system may be an A-valued distribution, whereas the signal u for another physical variable at a different location y may be a B-valued distribution. Moreover, the system defines the relation $\mathfrak{N}: v \mapsto u$. $\mathfrak{N}$ need not be an operator; that is, more than one u may be assigned to some particular v. (An example of this is the ideal transformer of electrical network theory when u is taken as the current vector and v as the voltage vector.) *However, a basic assumption imposed throughout this book is that every relation with which we shall be concerned is truly an operator.* Furthermore, a Banach system may define many different relations depending on the choices of the locations x and y and the physical variables. The term "Banach system" refers to the entire system and not to any particular relation generated by it. Moreover, it is possible for certain operators generated by a given Banach system to exhibit various properties such as linearity, time invariance, and passivity, while other operators generated by the same system do not. For this reason, the postulates in our subsequent realizability theory will be imposed on particular operators and not on the system as a whole. Indeed, it is such operators and not entire systems that comprise the main concern of this book.

Let us now define the concept of a "Hilbert port." Assume that in a given Banach system we have singled out two physical variables u and v that are complementary in the following sense: When both u and v are ordinary functions taking their values in a (not necessarily separable) complex Hilbert space H, the inner product $(u(t), v(t))$ represents the net complex power entering the Banach system at the instant t. Then, the Banach system with these two variables so singled out is called a *Hilbert port*.

We shall borrow some terminology from electrical network theory by referring to the relation $\mathfrak{N}: v \mapsto u$, when it is truly an operator, as the *admittance operator of the Hilbert port*, and to the relation $\mathfrak{W}: \frac{1}{2}(v + u) \mapsto \frac{1}{2}(v - u)$, when it too is an operator, as the *scattering operator of the Hilbert port*. This agrees with the usual terminology for electrical n-ports when v is identified with the voltage vector and u with the current vector. For the scattering operator, certain normalizations of v and u are also implied by this (Carlin and Giordano, 1964, p. 225). It is the admittance and scattering operators with which we shall be concerned when the passivity hypothesis is imposed.

Still another operator we could consider is the impedance operator $\mathfrak{Z}: u \mapsto v$, where again u is current and v is voltage. However, everything we shall say about $\mathfrak{N}$ can be applied equally well to $\mathfrak{Z}$ by interchanging the roles of u and v.

4.3. THE SPACE $\mathscr{H} = \mathscr{D}(\mathscr{V})$

Schwartz's kernel theorem (Schwartz, 1957, p. 93) characterizes separately continuous bilinear mappings of $\mathscr{D}_{R^n} \times \mathscr{D}_{R^s}$ into C in terms of complex-valued distributions on $R^n \times R^s$. There now exist a number of alternative proofs for it (Bogdanowicz, 1961; Ehrenpreis, 1956; Gask, 1960; Gelfand and Vilenkin, 1964, Section 1.3). For our purposes, we need an extension of this theorem to separately continuous bilinear mappings of $\mathscr{D}_{R^n} \times \mathscr{D}_{R^s}(A)$ into B. Actually, Bogdanowicz's proof establishes the kernel theorem for mappings of $\mathscr{D}_{R^n} \times \mathscr{D}_{R^s}(A)$ into C, and with some obvious modifications, we can replace C by B. His proof is the subject of this and the next section. The present section is devoted to a generalization of the space $\mathscr{D}_{R^s}(A)$ resulting from the replacement of A by a more general type of space $\mathscr{V}$.

In the following, we let $\mathscr{V}$ be the strict inductive limit of a sequence $\{\mathscr{V}_j\}_{j=1}^{\infty}$ of Fréchet spaces. Since every Fréchet space is separated and complete, we can conclude that $\mathscr{V}$ possesses the closure property (Appendix E4). For each j, we let Z_j denote a sequence $\{\zeta_{j,q}\}_{q=0}^{\infty}$ of seminorms that generates the topology of $\mathscr{V}_j$. We can always choose the multinorm Z_j such that $\zeta_{j,0} \leq \zeta_{j,1} \leq \zeta_{j,2} \leq \cdots$; this we do. As a consequence, a base of neighborhoods of 0 in $\mathscr{V}_j$ consists of all sets of the form

$$\{v \in \mathscr{V}_j : \zeta_{j,q}(v) < \varepsilon\},$$

where $\varepsilon \in R_+$ and q are arbitrary.

Now, let $\{K_j\}_{j=1}^{\infty}$ be a sequence of compact intervals in R^n such that $K_j \subset \mathring{K}_{j+1}$ for every j and $\bigcup K_j = R^n$. We let $\mathscr{H} \triangleq \mathscr{D}_{R^n}(\mathscr{V})$ denote the linear space of all smooth $\mathscr{V}$-valued functions on R^n having compact supports. It follows from Theorem 1.3-1 that, for any $h \in \mathscr{H}$, $h(R^n)$ is a bounded subset of $\mathscr{V}$. Consequently, according to Appendix E4, $h(R^n) \subset \mathscr{V}_j$ for some j depending on h. We now let $\mathscr{H}_j \triangleq \mathscr{D}_{K_j}(\mathscr{V}_j)$ be the linear space of all $h \in \mathscr{H}$ such that $h(R^n) \subset \mathscr{V}_j$ and $\operatorname{supp} h \subset K_j$. Thus, $\mathscr{H}_j \subset \mathscr{H}_{j+1}$ for every j, and $\mathscr{H} = \bigcup \mathscr{H}_j$.

Fix j and consider $\mathscr{H}_j$. For any $h \in \mathscr{H}_j$ and any nonnegative integer $p \in R^n$, $h^{(p)}$ is a continuous function from R^n into $\mathscr{V}$ and its range is contained in $\mathscr{V}_j$. Since the topology of $\mathscr{V}_j$ is identical to the topology induced on $\mathscr{V}_j$ by $\mathscr{V}$, $h^{(p)}$ is also continuous from R^n into $\mathscr{V}_j$. This means that, for any given $\zeta \in Z_j$ and p, we can define a finite-valued functional $\varkappa_{p,\zeta}$ on $\mathscr{H}_j$ by means of

$$\varkappa_{p,\zeta}(h) \triangleq \sup_{t \in R^n} \zeta[h^{(p)}(t)], \qquad h \in \mathscr{H}_j.$$

Each $\varkappa_{p,\zeta}$ is a seminorm on $\mathscr{H}_j$. We equip $\mathscr{H}_j$ with the topology generated by the collection $\Gamma_j \triangleq \{\varkappa_{p,\zeta}\}$ of all such seminorms. This makes makes $\mathscr{H}_j$ a metrizable locally convex space.

A useful and easily shown fact is the following. Let k and p be integers in R^n with $0 \leq k \leq p$, and let ω and ζ be seminorms in Z_j with $\omega \leq \zeta$. Finally, let $T = \operatorname{diam} K_j$. Then, for all $h \in \mathscr{H}_j$,

$$\varkappa_{k,\omega}(h) \leq T^{|p-k|} \varkappa_{p,\zeta}(h). \tag{1}$$

Lemma 4.3-1. *The topology of $\mathscr{H}_j$ is identical to the topology induced on $\mathscr{H}_j$ by $\mathscr{H}_{j+1}$.*

PROOF. Since the topology of $\mathscr{V}_j$ is the same as the topology induced on $\mathscr{V}_j$ by $\mathscr{V}_{j+1}$, every set of the form $\{v \in \mathscr{V}_j : \zeta(v) < \varepsilon, \zeta \in Z_j\}$ contains a set of the form $\{v \in \mathscr{V}_j : \hat{\zeta}(v) < \hat{\varepsilon}, \hat{\zeta} \in Z_{j+1}\}$, and conversely. Now, consider the following neighborhood of zero in $\mathscr{H}_j$:

$$\left\{h \in \mathscr{H}_j : \sup_t \zeta[h^{(p)}(t)] < \varepsilon, \quad \zeta \in Z_j\right\}. \tag{2}$$

By our first observation, the set (2) contains a neighborhood of zero in the induced topology of $\mathscr{H}_j$ of the form

$$\left\{h \in \mathscr{H}_j : \sup_t \hat{\zeta}[h^{(p)}(t)] < \hat{\varepsilon}, \quad \hat{\zeta} \in Z_{j+1}\right\}. \tag{3}$$

But a base of neighborhoods of zero in $\mathscr{H}_j$ consists of all intersections of finite collections of sets of the form (2), and similarly for the induced topology with respect to (3). Therefore, we can conclude that the topology of $\mathscr{H}_j$ is weaker than the topology induced on it by $\mathscr{H}_{j+1}$. The same kind of argument proves that the topology of $\mathscr{H}_j$ is stronger than the induced topology. ◇

Henceforth, we assign to $\mathscr{H} = \bigcup \mathscr{H}_j$ the inductive-limit topology. In view of Lemma 4.3-1, this makes $\mathscr{H}$ the strict inductive limit of $\{\mathscr{H}_j\}$.

Lemma 4.3-2. *Given any $h \in \mathscr{H}_{j-1}$ with $j > 1$, any seminorm $\varkappa_{p,\zeta} \in \Gamma_j$, and any $\varepsilon \in R_+$, there exists an $h_0 \in \mathscr{H}_j$ such that $\varkappa_{p,\zeta}(h - h_0) < \varepsilon$ and h_0 is the following sum of a finite number of terms:*

$$h_0 = \phi_1 v_1 + \cdots + \phi_k v_k. \tag{4}$$

Here, $\phi_\nu \in \mathscr{D}_{K_j}$ and $v_\nu \in \mathscr{V}_j$, where $\nu = 1, \ldots, k$.

PROOF. We can choose a function $\psi \in \mathscr{D}_{K_j}$ such that $\psi(t) = 1$ for all $t \in K_{j-1}$. For $t, \tau \in K_j$, set $g(t) = D^{p+[1]}h(t)$,

$$Q(t, \tau) = \sum_{0 \leq \mu \leq p} \binom{p}{\mu} \frac{(t-\tau)^{\mu}}{\mu!} D^{\mu}\psi(t),$$

and

$$L = \sup_{t, \tau \in K_j} |Q(t, \tau)|.$$

Here, μ and p are nonnegative integers in R^n, $\binom{p}{\mu}$ is the n-dimensional binomial coefficient, and

$$\mu! \triangleq (\mu_1!) \cdots (\mu_n!),$$
$$(t-\tau)^{\mu} \triangleq (t_1 - \tau_1)^{\mu_1} \cdots (t_n - \tau_n)^{\mu_n}.$$

Now, let $\varepsilon \in R_+$ be given. Since g has a compact support, g is uniformly continuous on R^n, and therefore there exists an $\varepsilon_1 \in R_+$ such that

$$\zeta[g(t) - g(\tau)] < \frac{\varepsilon}{L \operatorname{vol} K_j} \tag{5}$$

whenever $|t - \tau| < \varepsilon_1$.

Next, we can find a finite collection $\{\Omega_i\}_{i=1}^m$ of open sets such that $\bigcup \Omega_i \supset K_{j-1}$, $\Omega_i \subset K_j$, and diam $\Omega_i < \varepsilon_1$ for every i. Moreover, we can find a finite collection $\{\lambda_i\}_{i=1}^m$ of smooth nonnegative functions on R^n such that supp $\lambda_i \subset \Omega_i$ for every i, $\sum \lambda_i(t) = 1$ for all $t \in K_{j-1}$, and $\sum \lambda_i(t) \leq 1$ for all $t \in R^n$. [That all this can be done is shown, for example, in the volume by Zemanian (1965, Section 1.8).] For each i, choose a $t_i \in \Omega_i$ and set

$$g_0(t) \triangleq \sum_{i=1}^{m} \lambda_i(t) g(t_i).$$

Then, for every $t \in R^n$, we may write

$$\zeta[g(t) - g_0(t)] = \zeta\{\sum \lambda_i(t)[g(t) - g(t_i)]\}$$
$$\leq \sum \lambda_i(t)\zeta[g(t) - g(t_i)] < \frac{\varepsilon}{L \operatorname{vol} K_j} \tag{6}$$

because of (5).

Let us define the integration operator $I^{p+[1]}$ on any $\chi \in \mathscr{H}$ by

$$(I^{p+[1]}\chi)(t) \triangleq \int_{-\infty}^{t} \frac{(t-\tau)^p}{p!} \chi(\tau)\, d\tau.$$

Also, set

$$h_0(t) \triangleq \psi(t)(I^{p+[1]}g_0)(t).$$

h_0 is of the form asserted in the lemma. Since $h = I^{p+[1]}g \in \mathscr{D}_{K_{j-1}}(\mathscr{V})$ and $\psi(t) = 1$ for $t \in K_{j-1}$, we also have that

$$D^p h(t) = D^p[\psi(t) I^{p+[1]} g].$$

Hence, by differentiating under the integral sign, we get

$$D^p[h(t) - h_0(t)] = D^p\{\psi(t)[I^{p+[1]}(g - g_0)](t)\}$$
$$= \int_{-\infty}^{t} Q(t, \tau)[g(\tau) - g_0(\tau)]\, d\tau.$$

Thus,

$$\varkappa_{p,\zeta}(h - h_0) = \sup_{t \in K_j} \zeta\{D^p[h(t) - h_0(t)]\}$$
$$\leq L \operatorname{vol} K_j \sup_{\tau \in K_j} \zeta[g(\tau) - g_0(\tau)] < \varepsilon,$$

where the last inequality follows from (6). ◇

Example 4.3-1. We end this section by describing a special case of $\mathscr{H}$ obtained by setting $\mathscr{V} = \mathscr{D}_{R^s}(A)$ and $\mathscr{V}_j = \mathscr{D}_{L_j}(A)$, where $j = 1, 2, \ldots$ and the L_j are compact intervals in R^s such that $L_j \subset \mathring{L}_{j+1}$ for every j and $\bigcup L_j = R^s$. In this case, one multinorm $Z = \{\zeta_q\}_{q=0}^{\infty}$ will serve for every $\mathscr{D}_{L_j}(A)$ when we set

$$\zeta_q(v) \triangleq \max_{0 \leq k \leq [q]} \sup_{x \in R^s} \|v^{(k)}(x)\|_A, \qquad v \in \mathscr{D}_{R^s}(A). \tag{7}$$

We can identify $\mathscr{H}$ with the strict inductive-limit space $\mathscr{D}_{R^n \times R^s}(A) = \mathscr{D}_{R^{n+s}}(A)$ as follows.

Let $t \in R^n$ and $x \in R^s$. Furthermore, let $\phi \in \mathscr{D}_{R^{n+s}}(A)$ be given. Then, the mapping $h\colon t \mapsto \phi(t, \cdot)$ is a continuous function from R^n into $\mathscr{D}_{R^s}(A)$ because, for each nonnegative integer $p \in R^s$, $\{t, x\} \mapsto D_x{}^p\phi(t, x)$ is a uniformly continuous function on $R^n \times R^s$ and has a compact support therein. Moreover, $(D^k h)(t) = D_t{}^k\phi(t, \cdot)$. So, the same considerations show that h is a smooth function from R^n into $\mathscr{D}_{R^s}(A)$. Clearly, $\operatorname{supp} h$ is bounded. Thus, to each $\phi \in \mathscr{D}_{R^{n+s}}(A)$ there corresponds a unique $h \in \mathscr{H}$.

Conversely, let $h \in \mathscr{H}$ be given. We define an A-valued function ϕ on R^{n+s} by setting $\phi(t, x) = [h(t)](x)$. The function ϕ has a compact support in R^{n+s} because h has a compact support in R^n and the range of h is contained in some $D_{L_j}(A)$. That ϕ is smooth can be shown in the following way.

The fact that h is continuous from R^n into $\mathscr{D}_{R^s}(A)$ implies that, for any nonnegative integer $k \in R^s$, for a fixed $t \in R^n$, and as $\Delta t \to 0$ in R^n, $D_x{}^k\phi(t + \Delta t, x)$ tends to $D_x{}^k\phi(t, x)$ uniformly for all $x \in R^s$. But $D_x{}^k\phi(t, x)$ is a continuous function of x for each fixed t. We can conclude therefore that $D_x{}^k\phi(t, x)$ is a continuous function of $\{t, x\}$ whatever be k.

Next, let $\partial_i \triangleq \partial_{t_i}$ denote differentiation with respect to the ith component t_i of t. By setting up the incremental definition of $\partial_i h$ and taking the limit, we see that $\partial_i \phi(t, x) = [\partial_i h(t)](x)$. After proceeding exactly as in the preceding paragraph, we can conclude that $D_x{}^k \partial_{t_i} \phi$ is a continuous function on R^{n+s}, whatever be k. Moreover, Theorem 1.6-2 implies that the order of differentiation in $D_x{}^k \partial_{t_i} \phi$ can be changed in any fashion without altering the result.

These arguments can be applied to all the derivatives of ϕ, which leads to the conclusion that ϕ is a smooth A-valued function on R^{n+s}. Since supp ϕ is bounded, $\phi \in \mathscr{D}_{R^{n+s}}(A)$.

We have so far shown that the equation

$$h(t) = \phi(t, \cdot) \tag{8}$$

sets up a bijection from $\mathscr{H}$ onto $\mathscr{D}_{R^{n+s}}(A)$. Clearly, this bijection and its inverse are linear. On the other hand, we have from (7) that

$$\varkappa_{p, \zeta_q}(h) = \max_{0 \le k \le [q]} \sup_{t \in R^n;\, x \in R^s} \|D_x{}^k D_t{}^p \phi(t, x)\|_A .$$

This shows that, given the compact intervals $K_j \subset R^n$ and $L_j \subset R^s$, the bijection defined by (8) is an isomorphism from $\mathscr{H}_j$ onto $\mathscr{D}_{K_j \times L_j}$. Consequently, the bijection is also an isomorphism from $\mathscr{H}$ onto $\mathscr{D}_{R^{n+s}}(A)$.

Problem 4.3-1. Prove (1).

Problem 4.3-2. Show that $\mathscr{H}$ possesses the closure property; i.e., for each j, $\mathscr{H}_j$ is a closed subspace of $\mathscr{H}_{j+1}$.

4.4. THE KERNEL THEOREM

We continue to use the notation defined in the last section. Moreover, we can let $\{\rho_l\}_{l=0}^{\infty}$, where

$$\rho_l(\phi) \triangleq \max_{0 \le k \le [l]} \sup_{t \in R^n} |\phi^{(k)}(t)|,$$

be the multinorm for $\mathscr{D}_{K_j}$ whatever be j. Clearly, $\rho_0 \le \rho_1 \le \rho_2 \le \cdots$.

The primary result of this chapter is the following vector version of the kernel theorem.

Theorem 4.4-1. *Corresponding to every separately continuous bilinear mapping $\mathfrak{M}$ of $\mathscr{D}_{R^n} \times \mathscr{V}$ into B there exists one and only one $\mathfrak{T} \in [\mathscr{H}; B]$ such that*

$$\mathfrak{M}(\phi, v) = \mathfrak{T}(\phi v) \tag{1}$$

for all $\phi \in \mathscr{D}_{R^n}$ and $v \in \mathscr{V}$.

Note. A much easier fact to establish is the converse: Any given $\mathfrak{T}$ determines a unique $\mathfrak{M}$ by means of (1).

PROOF. For every pair of positive integers j and i, $\mathfrak{M}$ is a separately continuous bilinear mapping of $\mathscr{D}_{K_{j+1}} \times \mathscr{V}_i$ into B because $\mathscr{D}_{R^n}$ and $\mathscr{V}$ are inductive-limit spaces (Appendix E2). Moreover, both $\mathscr{D}_{K_{j+1}}$ and $\mathscr{V}_i$ are Fréchet spaces, and their respective multinorms $\{\rho_l\}$ and $\{\zeta_{i,q}\}$ are monotonic increasing. Therefore, we can invoke Appendix F2 to conclude the following. With $\mathscr{D}_{K_{j+1}}$ and $\mathscr{V}_i$ fixed, there exists a ρ_l and a $\zeta_{i,q}$ such that

$$\|\mathfrak{M}(\phi, v)\|_B \le M\rho_l(\phi)\zeta_{i,q}(v) \tag{2}$$

for all $\phi \in \mathscr{D}_{K_{j+1}}$ and all $v \in \mathscr{V}_i$.

Now, let θ be any nonnegative member of $\mathscr{D}_{R^n}$ such that $\theta(t) = 0$ when $|t| \ge 1$ and

$$\int_{R^n} \theta(t)\, dt = 1.$$

Also, for each j, let $\varepsilon_j \in R$ be such that

$$0 < \varepsilon_j < \inf\{d \colon d = |\xi - \chi|, \quad \xi \in K_j, \quad \chi \in R^n \backslash K_{j+1}\}$$

and, as $j \to \infty$, ε_j tends monotonically to zero. Set $\eta_j(t) = \varepsilon_j^{-n}\theta(t/\varepsilon_j)$. Then, $\{\eta_j\}_{j=1}^{\infty}$ tends to the delta functional, and supp η_j is contained in the n-dimensional sphere $\{t \colon |t| < \varepsilon_j\}$. As a result, for any $\phi \in \mathscr{D}_{K_j}$, the convolution product $\eta_i * \phi$ is a member of $\mathscr{D}_{K_{j+1}}$ for every $i \ge j$. Also, as $i \to \infty$, $\eta_i * \phi$ tends to ϕ in $\mathscr{D}_{K_{j+1}}$.

Observe that $\tau \mapsto \eta_j(\cdot - \tau)$ is a continuous mapping of K_j into $\mathscr{D}_{K_{j+1}}$. Also, any given $h \in \mathscr{H}$ belongs to some $\mathscr{H}_i$, and therefore $\tau \mapsto h(\tau)$ is a continuous mapping of K_j into $\mathscr{V}_i$. By virtue of (2), $\tau \mapsto \mathfrak{M}(\eta_j(\cdot - \tau), h(\tau))$ is a continuous mapping of K_j into B. Hence,

$$\mathfrak{T}_j(h) \triangleq \int_{K_j} \mathfrak{M}\big(\eta_j(\cdot - \tau), h(\tau)\big)\, d\tau \tag{3}$$

exists as a Riemann integral, and, in view of (2), satisfies the following estimate:

$$\|T_j(h)\|_B \le M\varkappa_{0,\zeta_{i,q}}(h) \sup_{\tau \in K_j} \rho_l[\eta(\cdot - \tau)] \operatorname{vol} K_j. \tag{4}$$

This shows that the linear mapping $\mathfrak{T}_j \colon \mathscr{H}_i \to B$ is continuous. Since this is true for every i, $\mathfrak{T}_j$ is continuous and linear from $\mathscr{H}$ into B as well.

We set $h(\tau) = \phi(\tau)v$, where $\phi \in \mathscr{D}_{K_j}$, $v \in \mathscr{V}_j$, and j is arbitrary. Also, let $i \ge j$. Because of the bilinearity of $\mathfrak{M}$, we may write

$$\mathfrak{T}_i(\phi v) = \int_{K_j} \mathfrak{M}\big(\eta_i(\cdot - \tau)\phi(\tau), v\big)\, d\tau.$$

But $\tau \mapsto \eta_i(\cdot - \tau)\phi(\tau)$ is a continuous mapping of K_i into $\mathscr{D}_{K_{i+1}}$ and $\psi \mapsto \mathfrak{M}(\psi, v)$ is a continuous linear mapping of $\mathscr{D}_{K_{i+1}}$ into B. So, by note II of Section 1.4,

$$\mathfrak{T}_i(\phi v) = \mathfrak{M}\left(\int_{K_j} \eta_i(\cdot - \tau)\phi(\tau)\, d\tau, v\right)$$
$$= \mathfrak{M}(\eta_i * \phi, v). \tag{5}$$

We have noted that $\eta_i * \phi \to \phi$ in $\mathscr{D}_{K_{j+1}}$ as $i \to \infty$. Consequently, (5) tends to $\mathfrak{M}(\phi, v)$ in B whatever be $\phi \in \mathscr{D}_{K_j}$ and $v \in \mathscr{V}_j$. Since j can be any positive integer here,

$$\lim_{i\to\infty} \mathfrak{T}_i(\phi v) = \mathfrak{M}(\phi, v), \qquad \phi \in \mathscr{D}, \quad v \in \mathscr{V}. \tag{6}$$

We now state a lemma but postpone its proof.

Lemma 4.4-1. *Given any j, there exists a constant $M_1 > 0$ and a seminorm $\varkappa_{p,\zeta}$ in the multinorm Γ_j for $\mathscr{H}_j$ such that*

$$\|\mathfrak{T}_i(h)\|_B \le M_1 \varkappa_{p,\zeta}(h) \tag{7}$$

for all $h \in \mathscr{H}_j$ and all i. That is, $\{\mathfrak{T}_i\}$ is an equicontinuous set of mappings on $\mathscr{H}_j$.

We shall now show that $\{\mathfrak{T}_i\}$ converges in $[\mathscr{H}; B]^\sigma$ to a limit $\mathfrak{T} \in [\mathscr{H}; B]$. Consider any $h \in \mathscr{H}$; h will be in $\mathscr{H}_{j-1}$ for some j. Choose any $\varepsilon_1 > 0$ and let M_1 and $\varkappa_{p,\zeta}$ be as in Lemma 4.4-1. By virtue of Lemma 4.3-2, we can select an h_0 of the form (4) in Section 4.3 such that

$$\varkappa_{p,\zeta}(h - h_0) < \varepsilon_1/3M_1. \tag{8}$$

Moreover,

$$\|\mathfrak{T}_i(h) - \mathfrak{T}_l(h)\| \le \|\mathfrak{T}_i(h - h_0)\| + \|\mathfrak{T}_i(h_0) - \mathfrak{T}_l(h_0)\| + \|\mathfrak{T}_l(h_0 - h)\|.$$

For all i and l, the first and third terms on the right-hand side are both bounded by $\varepsilon_1/3$ by virtue of Lemma 4.4-1 and (8). Moreover, there exists a constant $N > 0$ such that, for all $i, l > N$, the second term is bounded by $\varepsilon_1/3$ because of (6). Hence, by the completeness of B, $\mathfrak{T}_i(h)$ converges to a limit, say $\mathfrak{T}(h)$. This defines a mapping $\mathfrak{T}$ from $\mathscr{H}_{j-1}$ into B, which is linear since each $\mathfrak{T}_i$ is linear. Moreover,

$$\|\mathfrak{T}(h)\|_B \le M_1 \varkappa_{p,\zeta}(h)$$

for all $h \in \mathscr{H}_{j-1}$. But, Lemma 4.3-1 implies that $\varkappa_{p,\zeta}$ is a continuous seminorm on $\mathscr{H}_{j-1}$. Thus, $\mathfrak{T}$ is continuous on every $\mathscr{H}_{j-1}$. This being true for every j, we have that $\mathfrak{T} \in [\mathscr{H}; B]$.

Equation (6) now shows that (1) holds true. Clearly, given $\mathfrak{T}$, $\mathfrak{M}$ is uniquely determined by (1). On the other hand, Lemma 4.3-2 coupled with (1) of Section 4.3 states in effect that the set Θ of all elements of the form ϕv, where $\phi \in \mathscr{D}$ and $v \in \mathscr{V}$, is total in $\mathscr{H}$. Hence, any member of $[\mathscr{H}; B]$ that coincides with $\mathfrak{T}$ on Θ must be identical to $\mathfrak{T}$ on $\mathscr{H}$. This completes the proof of Theorem 4.4-1 except for the proof of Lemma 4.4-1. In order to establish that, we shall need still another lemma.

Lemma 4.4-2. *Given any nonnegative integer $p \in R^n$, define the complex-valued function r_i on R^n by*

$$r_i(t) = \int_{-\infty}^{t} \frac{(t-\tau)^p}{p!} \eta_i(\tau)\, d\tau,$$

where η_i is the function defined in the preceding proof. Let $j > 1$ and let $\psi \in \mathscr{D}_{K_{j+2}}$ be such that $\psi(t) = 1$ on K_{j+1}. Then, for all $i > j$ and all $h \in \mathscr{H}_{j-1}$,

$$\mathfrak{T}_i(h) = \int_{K_{j-1}} \mathfrak{M}(\psi(\cdot)r_i(\cdot - \tau), D^{p+[1]}h(\tau))\, d\tau. \tag{9}$$

PROOF. For any $\phi \in \mathscr{D}_{K_j}$ and $i > j$, we have that $\operatorname{supp} \eta_i * \phi \subset K_{j+1}$ and therefore

$$\begin{aligned}(\eta_i * \phi)(t) &= \int_{R^n} \eta_i(t-\tau)\phi(\tau)\, d\tau \\ &= \int_{R^n} \psi(t) r_i(t-\tau) D^{p+[1]}\phi(\tau)\, d\tau.\end{aligned}$$

Here, we have used repeated integration by parts. Consider the function

$$\tau \mapsto \psi(\cdot) r_i(\cdot - \tau) D^{p+[1]}\phi(\tau).$$

It is smooth on R^n with values in $\mathscr{D}_{K_{j+2}}$, and its support is contained in K_j. By appealing to (5), we may write, for any $\phi \in \mathscr{D}_{K_j}$, any $v \in \mathscr{V}_j$, and all $i > j$,

$$\begin{aligned}\mathfrak{T}_i(\phi v) &= \mathfrak{M}(\eta_i * \phi, v) \\ &= \mathfrak{M}\left(\int_{K_j} \psi(\cdot) r_i(\cdot - \tau) D^{p+[1]}\phi(\tau)\, d\tau, v\right) \\ &= \int_{K_j} \mathfrak{M}(\psi(\cdot) r_i(\cdot - \tau), D^{p+[1]}\phi(\tau) v)\, d\tau.\end{aligned}$$

In the last step, we have used Note II of Section 1.4 and the separate continuity and bilinearity of $\mathfrak{M}$.

With $i > j$ still, define the linear operator $\hat{\mathfrak{T}}_i$ on $\mathscr{H}_j$ by

$$\hat{\mathfrak{T}}_i(h) \triangleq \int_{K_j} \mathfrak{M}(\psi(\cdot)r_i(\cdot - \tau), D^{p+[1]}h(\tau))\, d\tau, \qquad h \in \mathscr{H}_j. \tag{10}$$

Thus, from (2), we have

$$\|\hat{\mathfrak{T}}_i(h)\|_B \leq N\varkappa_{s,\omega}(h) \sup_{\tau \in K_j} \rho_l(\psi(\cdot)r_i(\cdot - \tau)) \operatorname{vol} K_j, \tag{11}$$

where $s = p + [1]$, $\omega = \zeta_{j,v}$ for some integer v, and N is a constant. This shows that $\hat{\mathfrak{T}}_i$ is continuous. Obviously, $\mathfrak{T}_i$ and $\hat{\mathfrak{T}}_i$ coincide on the set $\{\phi v: \phi \in \mathscr{D}_{K_j}, v \in \mathscr{V}_j\}$ and therefore on its span S. Now, let $h \in \mathscr{H}_{j-1}$ and $h_0 \in S$. We may write

$$\|\mathfrak{T}_i(h) - \hat{\mathfrak{T}}_i(h)\|_B \leq \|\mathfrak{T}_i(h - h_0)\| + \|\hat{\mathfrak{T}}_i(h_0 - h\|.$$

Upon referring to (1) of Section 4.3, to (4) and (11), and to Lemma 4.3-2, we see that $\mathfrak{T}_i$ and $\hat{\mathfrak{T}}_i$ coincide on $\mathscr{H}_{j-1}$. Finally, note that the interval K_j of integration in (10) may be replaced by K_{j-1} when $h \in \mathscr{H}_{j-1}$. This proves (9). ◇

PROOF OF LEMMA 4.4-1. We may apply the estimate (2) (with $\mathscr{D}_{K_{j+1}}$ replaced by $\mathscr{D}_{K_{j+2}}$ and $\mathscr{V}_i$ by $\mathscr{V}_{j-1}$) to (9), where p is chosen equal to $[l]$. This yields, for all $i > j$,

$$\|\mathfrak{T}_i(h)\|_B \leq M\varkappa_{s,\beta}(h)\left\{\sup_{\tau \in K_{j-1}} \rho_l[\psi(\cdot)r_i(\cdot - \tau)]\right\} \operatorname{vol} K_{j-1},$$

where $s = p + [1]$ and $\beta = \zeta_{j-1,q}$. It is not difficult to show that the quantity within the braces is bounded by a constant not depending on i. (This is a result of the fact that diam supp η_i is bounded for all i.) Thus, $\{\mathfrak{T}_i\}_{i>j}$ is an equicontinuous set of mappings on $\mathscr{H}_{j-1}$. Moreover, for $0 \leq i \leq j$, each $\mathfrak{T}_i$ is continuous on $\mathscr{H}_{j-1}$. Consequently, $\{\mathfrak{T}_i\}_{i=1}^{\infty}$ is also equicontinuous on $\mathscr{H}_{j-1}$. ◇

4.5. KERNEL OPERATORS

Theorem 4.4-1 provides a characterization of the continuous linear operators from $\mathscr{D}_{R^s}(A)$ into $[\mathscr{D}_{R^n}; B]^\sigma$; this is the content of the present section. We start with a special case of Theorem 4.4-1.

Theorem 4.5-1. *Corresponding to every separately continuous bilinear mapping $\mathfrak{M}$ of $\mathscr{D}_{R^n} \times \mathscr{D}_{R^s}(A)$ into B there exists one and only one distribution $f \in [\mathscr{D}_{R^{n+s}}(A); B]$ such that*

$$\mathfrak{M}(\phi, v) = \langle f(t, x), \phi(t)v(x)\rangle, \qquad \phi \in \mathscr{D}_{R^n}, \quad v \in \mathscr{D}_{R^s}(A). \tag{1}$$

PROOF. We choose $\mathscr{V} = \mathscr{D}_{R^s}(A)$, as was done in Example 4.3-1, and then employ (8) of Section 4.3 to set up an isomorphism $\mathfrak{J}$ from $\mathscr{H} = \mathscr{D}_{R^n}(\mathscr{D}_{R^s}(A))$ onto $\mathscr{D}_{R^{n+s}}(A)$. This induces a bijection from $[\mathscr{H}; B]$ onto $[\mathscr{D}_{R^{n+s}}(A); B]$ by means of the equation

$$\mathfrak{T}(\phi v) = \langle f(t, x), \phi(t)v(x)\rangle,$$

where $\mathfrak{T} \in [\mathscr{H}; B]$. (See Figure 4.5-1). But, according to Theorem 4.4-1, every $\mathfrak{M}$ can be identified with one and only one such $\mathfrak{T}$ by setting $\mathfrak{M}(\phi, v) = \mathfrak{T}(\phi v)$. Thus, this theorem is established. ◇

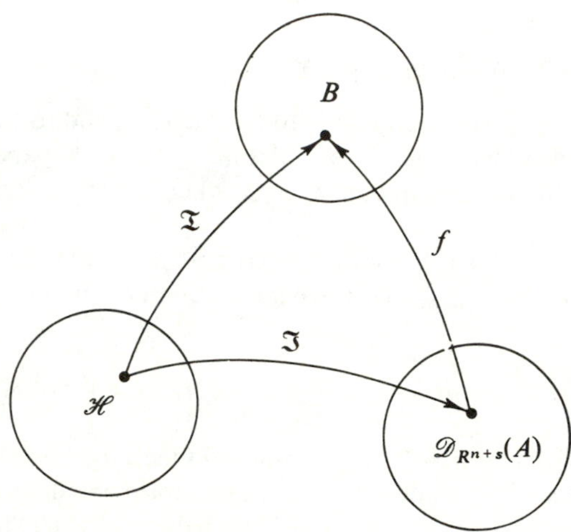

Figure 4.5-1

We shall use the right-hand side of (1) to define an operator $f\cdot$, which we call a *kernel operator* or alternatively a *composition operator*. Given $f \in [\mathscr{D}_{R^{n+s}}(A); B]$ and any $v \in \mathscr{D}_{R^s}(A)$, we define the *composition product* $f\cdot v$ as a mapping on all $\phi \in \mathscr{D}_{R^n}$ by

$$\langle f\cdot v, \phi\rangle \triangleq \langle f(t, x), \phi(t)v(x)\rangle, \qquad t \in R^n, \quad x \in R^s. \tag{2}$$

Thus, $f\cdot v$ maps $\mathscr{D}_{R^n}$ into B linearly and continuously. Hence, the kernel operator $f\cdot: v \mapsto f\cdot v$ maps $\mathscr{D}_{R^s}(A)$ into $[\mathscr{D}_{R^n}; B]$.

Clearly, $f\cdot$ is linear. To show its continuity, let Φ be a bounded set in $\mathscr{D}_{R^n}$. There exists a compact interval K in R^n such that $\Phi \subset \mathscr{D}_K$. Also, let J be any compact interval in R^s. Then, for all $v \in \mathscr{D}_J(A)$, the mapping $\{t, x\} \mapsto \phi(t)v(x)$ is a member of $\mathscr{D}_{K\times J}(A)$. Therefore, there exists a constant Q and a nonnegative integer $r = \{r_1, r_2\} \in R^{n+s}$ such that

$$\sup_{\phi\in\Phi}\|\langle f\cdot v, \phi\rangle\| \le \sup_{\phi\in\Phi} Q \max_{0\le k\le r} \sup_{t,x}\|D^k[\phi(t)v(x)]\|.$$

But the right-hand side is bounded by

$$P \max_{0 \le k_2 \le r_2} \sup_x \|D^{k_2} v(x)\| = P\rho_{r_2}(v),$$

where P does not depend on v. Thus, $f\cdot$ is continuous on $\mathscr{D}_J(A)$ for every J. This implies that $f\cdot$ is continuous on $\mathscr{D}_{R^s}(A)$.

We summarize these results as follows.

Theorem 4.5-2. *For any given $f \in [\mathscr{D}_{R^{n+s}}(A); B]$, the kernel operator $f\cdot$ is a continuous linear mapping of $\mathscr{D}_{R^s}(A)$ into $[\mathscr{D}_{R^n}; B]$.*

Our next objective is to develop a converse to Theorem 4.5-2. We first state a lemma whose proof is quite straightforward.

Lemma 4.5-1. *Let $\mathfrak{N}$ be a continuous linear mapping of $\mathscr{D}_{R^s}(A)$ into $[\mathscr{D}_{R^n}; B]^\sigma$. Define $\mathfrak{M}$ from $\mathfrak{N}$ by*

$$\mathfrak{M}(\phi, v) \triangleq \langle \mathfrak{N}v, \phi \rangle, \qquad v \in \mathscr{D}_{R^s}(A), \quad \phi \in \mathscr{D}_{R^n}. \tag{3}$$

Then, $\mathfrak{M}$ is a uniquely defined, separately continuous bilinear mapping of $\mathscr{D}_{R^n} \times \mathscr{D}_{R^s}(A)$ into B.

Theorem 4.5-3. *If $\mathfrak{N}$ is a sequentially continuous linear mapping of $D_{R^s}(A)$ into $[\mathscr{D}_{R^n}; B]^\sigma$, then there exists a unique $f \in [\mathscr{D}_{R^{n+s}}(A); B]$ such that $\mathfrak{N} = f\cdot$ on $\mathscr{D}_{R^s}(A)$.*

Note. $f\cdot$ is called the *kernel representation for* $\mathfrak{N}$ and f is called the *kernel of* $\mathfrak{N}$. This theorem implies the precise converse of Theorem 4.5-2 since continuity implies sequential continuity and the pointwise topology is weaker than the bounded topology.

PROOF. As was noted in Section 3.2, the sequential continuity of $\mathfrak{N}$ on $\mathscr{D}_{R^s}(A)$ implies its continuity on $\mathscr{D}_{R^s}(A)$. Therefore, $\mathfrak{N}$ satisfies the hypothesis of Lemma 4.5-1. By virtue of Theorem 4.5-1, the $\mathfrak{M}$ that is uniquely defined by (3) determines a unique $f \in [\mathscr{D}_{R^{n+s}}(A); B]$ such that (1) is satisfied. Upon combining (1)–(3), we see that $\mathfrak{N}v = f\cdot v$ for all $v \in \mathscr{D}_{R^s}(A)$. ◇

Example 4.5-1. We determine the kernel representation $f\cdot$ for the operator $c\sigma_\tau D^p$, where c is an $[A; B]$-valued continuous function on R^n, $\tau \in R^n$, σ_τ is the shifting operator defined in Section 3.3, and p is a nonnegative integer in R^n. Note that $c\sigma_\tau D^p$ is a sequentially continuous linear mapping of $\mathscr{D}_{R^n}(A)$ into $[\mathscr{D}_{R^n}; B]^\sigma$, so that it must have a kernel representation.

Throughout the following, both t and x are variables in R^n. Define the distribution $g \in [\mathscr{D}^0_{R^{2n}}(B); B]$ by

$$\langle g, \theta \rangle \triangleq \int_{R^n} \theta(t, t)\, dt \in B, \qquad \theta \in \mathscr{D}^0_{R^{2n}}(B).$$

Then, define the distribution f by

$$f(t, x) \triangleq c(t)\sigma_{-\tau}(x)(-1)^{|p|} D_x{}^p g(t, x), \tag{4}$$

where $\sigma_{-\tau}(x)$ denotes a shift of $-\tau$ in the x direction. Actually, $f \in [\mathscr{D}^q_{R^{2n}}(A); B]$, where q is the $(2n)$-tuple whose first n components are zero and whose last n components are the components of p. This can be seen from the following equation, where $\psi \in \mathscr{D}^q_{R^{2n}}(A)$:

$$\begin{aligned} \langle f, \psi \rangle &= \langle g(t, x), D_x{}^p \sigma_\tau(x) c(t) \psi(t, x) \rangle \\ &= \int_{R^n} c(t)[D_x{}^p \psi(t, x - \tau)]_{x=t}\, dt. \end{aligned}$$

Moreover, for any $v \in \mathscr{D}_{R^n}(A)$ and $\phi \in \mathscr{D}_{R^n}$,

$$\begin{aligned} \langle f \cdot v, \phi \rangle &= \langle f(t, x), \phi(t) v(x) \rangle \\ &= \int_{R^n} c(t)\phi(t) D_t{}^p v(t - \tau)\, dt \\ &= \langle c\sigma_\tau D^p v, \phi \rangle. \end{aligned}$$

Thus, we have shown that $c\sigma_\tau D^p = f\cdot$ on $\mathscr{D}_{R^n}(A)$, where f is defined by (4). ◇

There are a number of extensions of the kernel operators discussed heretofore. Schwartz (1957, pp. 124–126) discusses kernel operators that map $\mathscr{D}_{R^s}$ into $\mathscr{D}_{R^n}(\mathscr{V})$, where $\mathscr{V}$ is a separated locally convex space. Meidan (1972) considers kernel operators that map $\mathscr{D}_{R^s}$ into $\mathscr{E}^m_{R^n}$, and, by going to the adjoint operators, he obtains mappings of $[\mathscr{E}^m_{R^n}; C]$ into $[\mathscr{D}_{R^s}; C]$. An extension onto various spaces of Banach-space-valued distributions is given by Zemanian (1970c); for example, representations for mappings of $[\mathscr{E}^m_{R^s}; A]$ into $[\mathscr{D}^j_{R^n}; B]$ are presented there. The latter two works are related to still another method of representing certain continuous linear mappings of one space of distributions into another; it was first introduced by Cristescu (1964) and subsequently developed by Cristescu and Marinescu (1966), Sabac (1965), Wexler (1966), Cioranescu (1967), Pondelicek (1969), Dolezal (1970), and Zemanian (1972a). The basic idea is to assume that we are given a family $\{y_x\}$ of distributions in $[\mathscr{D}_{R^n}; C]$ depending on the parameter $x \in R^s$ such that the following condition is satisfied. For $\psi_\phi(x) \triangleq \langle y_x, \phi \rangle$, $\phi \mapsto \psi_\phi$ is a mapping of $\mathscr{D}_{R^n}$ into $\mathscr{E}_{R^s}$. Then, define the product $v \circ y_x$, where $v \in [\mathscr{E}_{R^s}; C]$ by

$\langle v \circ y_x, \phi \rangle \triangleq \langle v, \psi_\phi \rangle$. It can be shown that $v \mapsto v \circ y_x$ is a continuous linear mapping of $[\mathscr{E}_{R^s}; C]$ into $[\mathscr{D}_{R^n}; C]$. This implies that it must also be a kernel operator on $\mathscr{D}_{R^s}$.

Problem 4.5-1. Prove Lemma 4.5-1.

4.6. CAUSALITY AND KERNEL OPERATORS

Theorems 4.5-2 and 4.5-3 characterize any continuous linear operator $\mathfrak{N}$ that maps $\mathscr{D}_{R^s}(A)$ into $[\mathscr{D}_{R^n}; B]$. The facts that the domain $\mathscr{D}_{R^s}(A)$ of $\mathfrak{N}$ is a small space with a strong topology [as compared to other spaces, say $L_2(A)$ or $\mathscr{E}^0(A)$ that one might choose for a realizability theory] and that the range of $\mathfrak{N}$ is contained in a large space $[\mathscr{D}_{R^n}; B]$ with a weak topology implies that a wide class of operators is encompassed by Theorems 4.5-2 and 4.5-3. Any expansion of the domain space or weakening of its topology and similarly any diminution of the range space or strengthening of its topology will in general decrease the class of continuous linear operators under consideration. This is one reason why a distributional approach using testing-function spaces as domains and distribution spaces as range spaces is such a powerful tool in realizability theory.

There are many physical phenomena that can be modeled by continuous linear operators on $\mathscr{D}_{R^s}(A)$. However, many of those phenomena possess still another property, namely, causality. It can be stated loosely by saying that a physical system cannot respond to an excitation until that excitation has been imposed. Or, alternatively, systems cannot predict the future behavior of their excitations. In this section, we define causality and show how it can be characterized in terms of a condition on the support of the kernel f of any kernel operator $\mathfrak{N} = f\cdot$. Although our results can be formulated for signals on R^n (see Problem 4.6-1), the physically significant situation arises when $n = 1$. In the latter case, the independent variable $t \in R = R^1$ is taken to be time and the signals at hand are distributions on the t-axis. We therefore restrict ourselves to this situation and set $\mathscr{D} = \mathscr{D}_{R^1}$. Also, $t, x \in R$ throughout.

Definition 4.6-1. Let $\mathfrak{N}$ be an operator mapping a set $\mathscr{X} \subset [\mathscr{D}; A]$ into $[\mathscr{D}; B]$. $\mathfrak{N}$ is said to be *causal on* $\mathscr{X}$ if, for every $T \in R$, we have that $\mathfrak{N}v_1 = \mathfrak{N}v_2$ on the open interval $(-\infty, T)$ (in the sense of equality in $[\mathscr{D}; B]$) whenever $v_1, v_2 \in \mathscr{X}$ and $v_1 = v_2$ on $(-\infty, T)$.

It follows that a linear operator $\mathfrak{N}$ on a linear space $\mathscr{X}$ is causal if and only if $\mathfrak{N}v = 0$ on $(-\infty, T)$ whenever $v \in \mathscr{X}$ and $v = 0$ on $(-\infty, T)$.

Theorem 4.6-1. *Let* $f \in [\mathscr{D}_{R^2}(A); B]$ *and let* $\Omega \triangleq \{\{t, x\}: t \geq x\}$. *The kernel operator* $f\cdot$ *is causal on* $\mathscr{D}(A)$ *if and only if* $\operatorname{supp} f \subset \Omega$.

PROOF. Assume $\operatorname{supp} f \subset \Omega$. Let $v \in \mathscr{D}(A)$ be such that $v = 0$ on $(-\infty, T)$. The causality of $f\cdot$ will follow once we show that $f \cdot v = 0$ on $(-\infty, T)$ in the sense of equality in $[\mathscr{D}; B]$. Choose an arbitrary $\phi \in \mathscr{D}$ with supp ϕ contained in $(-\infty, T)$. Then,

$$\langle f \cdot v, \phi \rangle = \langle f(t, x), \phi(t)v(x) \rangle = 0$$

because the support of the function $\{t, x\} \mapsto \phi(t)v(x)$ does not intersect Ω and is therefore contained in the null set of f. So, truly, $f \cdot v = 0$ on $(-\infty, T)$.

Conversely, assume that $f\cdot$ is causal on $\mathscr{D}(A)$. Consequently, for any $v \in \mathscr{D}(A)$ with $\operatorname{supp} v \subset [T, \infty)$ and any $\phi \in \mathscr{D}$ with $\operatorname{supp} \phi \subset (-\infty, T)$, we have

$$\langle f(t, x), \phi(t)v(x) \rangle = \langle f \cdot v, \phi \rangle = 0. \tag{1}$$

Choose any $\psi \in \mathscr{D}_{R^{2n}}(A)$ such that supp ψ does not intersect Ω. We shall prove that $\langle f, \psi \rangle = 0$ and conclude thereby that $\operatorname{supp} f \subset \Omega$.

Since supp ψ is a compact set and Ω is a closed set, the distance between supp ψ and Ω is a positive quantity; that is,

$$\inf\{|w - z| : w \in \operatorname{supp} \psi, \quad z \in \Omega\} > 0.$$

Now, we can choose two finite collections $\{K_\mu\}$ and $\{J_\mu\}$ of closed intervals in R^{2n} and a finite collection $\{\psi_\mu\}$ of members of $\mathscr{D}_{R^{2n}}(A)$ such that the following five conditions hold: $K_\mu \subset \mathring{J}_\mu$ for each μ; $\bigcup K_\mu \supset \operatorname{supp} \psi$; $\bigcup J_\mu$ does not meet Ω; $\operatorname{supp} \psi_\mu \subset K_\mu$; and finally, $\psi = \sum \psi_\mu$. (See Zemanian, 1965, Section 1.8.) For any fixed μ and all $\theta \in \mathscr{D}_{J_\mu}(A)$, we may write

$$\|\langle f, \theta \rangle\|_B \leq M \sup_{t, x \in R^n} \|D_t{}^q D_x{}^r \theta(t, x)\|_A, \tag{2}$$

where the constant M and the integers q and r do not depend on θ. By Lemma 4.3-2 under the special case of Example 4.3-1, given any $\varepsilon \in R_+$, we can choose

$$h_\mu(t, x) = \sum_\nu \phi_{\mu,\nu}(t) v_{\mu,\nu}(x)$$

such that the summation is over a finite number of terms, $\phi_{\mu,\nu} \in \mathscr{D}$, $v_{\mu,\nu} \in \mathscr{D}(A)$, the support of the function $\{t, x\} \mapsto \phi_{\mu,\nu}(t)v_{\mu,\nu}(x)$ is contained in J_μ, and

$$\sup_{t, x \in R^n} \|D_t{}^q D_x{}^r [\psi_\mu(t, x) - h_\mu(t, x)]\| < \varepsilon/M.$$

Hence, by (1) and (2),

$$\|\langle f, \psi_\mu \rangle\| = \|\langle f, \psi_\mu - h_\mu \rangle\| < \varepsilon$$

so that $\langle f, \psi_\mu \rangle = 0$. In view of the fact that the collection $\{\psi_\mu\}$ is finite, we can conclude that

$$\langle f, \psi \rangle = \sum_\mu \langle f, \psi_\mu \rangle = 0.$$

Thus, $\operatorname{supp} f \subset \Omega$. ◇

Problem 4.6-1. Definition 4.6-1 and Theorem 4.6-1 continue to hold when we let $\mathscr{D} = \mathscr{D}_{R^n}$, $T \in R^n$, and $f \in [\mathscr{D}_{R^{2n}}(A); B]$. Show this.

Chapter 5

Convolution Operators

5.1. INTRODUCTION

Time invariance is a property possessed by many physical systems. It arises when the structure of the system and the values of its parameters remain fixed with time. As a result, if $u(t)$ is the responding signal to some driving signal $v(t)$, then $u(t - \tau)$ will be the response to $v(t - \tau)$, whatever be the real number τ. In fact, for any operator $\mathfrak{N}$ of a given Banach system, time invariance is characterized by saying that $\mathfrak{N}$ commutes with the shifting operator σ_τ. Thus, this property is also called translation invariance.

The objectives of the present chapter are to discuss the convolution process in the context of Banach systems and to establish the following basic result: Translation-invariant kernel operators are convolution operators, and conversely.

5.2. CONVOLUTION

Here is a theory for the convolution of an $[A; B]$-valued generalized function y on R^n with an A-valued generalized function v on R^n. The resulting convolution product $y * v$ will be a B-valued generalized function on R^n.

Let there be given three ρ-type testing-function spaces: $\mathscr{I}(A)$, $\mathscr{I} = \mathscr{I}(C)$, and $\mathscr{K} = \mathscr{K}(C)$. Also, assume that the following three conditions are satisfied.

Conditions E

E1. *If $\phi \in \mathscr{K}$, then, for each fixed $t \in R^n$, $\phi(t + \cdot)$ is a member of $\mathscr{I}$.*

E2. *With $v \in [\mathscr{I}; A]$ and $\phi \in \mathscr{K}$, define*

$$\psi(t) \triangleq \langle v(x), \phi(t + x)\rangle. \tag{1}$$

Then, with v fixed $\phi \mapsto \psi$ is a continuous linear mapping of $\mathscr{K}$ into $\mathscr{I}(A)$.

E3. *Under the preceding notation and with $\phi \in \mathscr{K}$ fixed, $v \mapsto \psi$ is a continuous linear mapping of $[\mathscr{I}; A]^s$ into $\mathscr{I}(A)$; moreover, it is uniformly continuous with respect to the $\mathfrak{S}$-sets in $\mathscr{K}$.* (By this uniform continuity, we mean the following. Given any $\mathfrak{S}$-set Φ in $\mathscr{K}$ and any neighborhood Λ of zero in $\mathscr{I}(A)$, there exists a neighborhood Ξ of zero in $[\mathscr{I}; A]^s$ such that $\psi \in \Lambda$ for all $\phi \in \Phi$ and all $v \in \Xi$. See Figure 5.2-1.)

Definition 5.2-1. Under Conditions E, the *convolution product $y * v$ of any $y \in [\mathscr{I}(A); B]$ and any $v \in [\mathscr{I}; A]$* is defined as a mapping on $\mathscr{K}$ by

$$\langle y * v, \phi\rangle \triangleq \langle y(t), \langle v(x), \phi(t + x)\rangle\rangle, \qquad \phi \in \mathscr{K}. \tag{2}$$

Note that Conditions E1 and E2 ensure that the right-hand side of (2) has a sense and is a member of B.

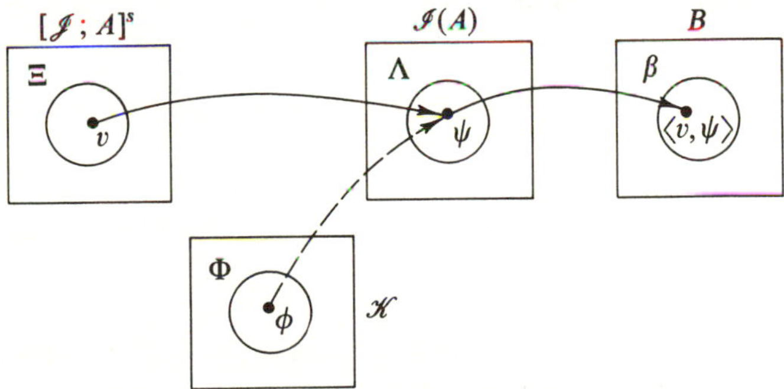

Figure 5.2-1

Theorem 5.2-1. *Given the complex Banach spaces A and B and the ρ-type testing-function spaces $\mathscr{I}(A)$, $\mathscr{J}$, and $\mathscr{K}$, assume that Conditions E are satisfied. If $y \in [\mathscr{I}(A); B]$, then the operator $y *: v \mapsto y * v$ is a continuous linear mapping of $[\mathscr{J}; A]^s$ into $[\mathscr{K}; B]^s$.*

PROOF. We have already noted that $y * v$ maps $\mathscr{K}$ into B. Now, observe that $y * v$ is the composite mapping $\phi \mapsto \psi \mapsto \langle y, \psi \rangle$. By virtue of Condition E2, we conclude that $y * v \in [\mathscr{K}; B]$. As is evident from (2), the mapping $v \mapsto y * v$ is linear. To show that it is continuous, let Φ be an arbitrary $\mathfrak{S}$-set in $\mathscr{K}$ and consider

$$\gamma_\Phi(y * v) \triangleq \sup_{\phi \in \Phi} \|\langle y * v, \phi \rangle\|_B = \sup_{\phi \in \Phi} \|\langle y, \psi \rangle\|_B .$$

Given any $\varepsilon \in R_+$, set $\beta \triangleq \{b \in B: \|b\| < \varepsilon\}$. Since y is continuous on $\mathscr{I}(A)$, there exists a neighborhood Λ of zero in $\mathscr{I}(A)$ such that y maps Λ into β. (See Figure 5.2-1.) But, by Condition E3, there exists a neighborhood Ξ of zero in $[\mathscr{J}; A]^s$ such that the composite mapping $v \mapsto \psi \mapsto \langle v, \psi \rangle$ carries Ξ into β, whatever be the choice of $\phi \in \Phi$. Therefore, $\gamma_\Phi(y * v) < \varepsilon$ for all $v \in \Xi$. This proves the asserted continuity because any neighborhood Υ of zero in $[\mathscr{K}; B]^s$ contains the intersection of a finite collection of sets of the form

$$\{u \in [\mathscr{K}; B]: \gamma_\Phi(u) < \varepsilon\}$$

and we need merely take the intersection of the corresponding Ξ's to get a neighborhood of zero in $[\mathscr{J}; A]^s$ that is mapped by $y *$ into Υ. ◇

The mapping $y *: v \mapsto y * v$ will be called a *convolution operator*. Any given $y \in [\mathscr{I}(A); B]$ generates such an operator so long as Conditions E are satisfied.

Problem 5.2-1. Show that Theorem 5.2-1 remains true when the $\mathfrak{S}$-topologies of $[\mathscr{J}; A]$ and $[\mathscr{K}; B]$ are replaced by their pointwise topologies and Condition E3 is replaced by the following statement: $v \mapsto \psi$ is a continuous linear mapping of $[\mathscr{J}; A]^\sigma$ into $\mathscr{I}(A)$.

5.3. SPECIAL CASES

We now list several specific choices of the triplet $\{\mathscr{I}(A), \mathscr{J}, \mathscr{K}\}$ for which Conditions E and therefore the hypothesis of Theorem 5.2-1 are satisfied.

I. $\mathscr{I}(A) = \mathscr{D}(A)$, $\mathscr{J} = \mathscr{E}$, $\mathscr{K} = \mathscr{D}$.

As in the preceding section, all testing functions are defined on R^n, and so it is understood that $t \in R^n$, $\mathscr{D} = \mathscr{D}_{R^n}$, and $\mathscr{E} = \mathscr{E}_{R^n}$. Up to now, we have viewed the topology of $\mathscr{D}(A)$ as an inductive-limit topology. But, since this topology is locally convex, it must be obtainable from a generating family of seminorms. We shall now determine such a family and use it subsequently.

Let $\varepsilon \triangleq \{\varepsilon_\nu\}_{\nu=0}^{\infty}$ be a sequence of positive numbers tending monotonically to zero and let $l \triangleq \{l_\nu\}_{\nu=0}^{\infty}$ be a sequence of nonnegative integers tending monotonically to ∞. $\Lambda(l, \varepsilon)$ is defined as the set of all $\phi \in \mathscr{D}(A)$ such that, for each $\nu = 0, 1, 2, \ldots$, we have

$$\|\phi^{(k)}(t)\|_A < \varepsilon_\nu$$

whenever $|k| \leq l_\nu$ and $|t| \geq \nu$. By taking the collection of all such $\Lambda(l, \varepsilon)$ as a basis of neighborhoods for $\mathscr{D}(A)$, we obtain a locally convex topology $\mathscr{O}$ for $\mathscr{D}(A)$. (See Appendix C4.) Note that $\mathscr{O}$ is generated by the collection $\{\gamma_{l,\varepsilon}\}$ of all seminorms on $\mathscr{D}(A)$ defined by

$$\gamma_{l,\varepsilon}(\phi) \triangleq \sup_{\nu} \sup_{\substack{|t| \geq \nu \\ |k| \leq l_\nu}} \|\varepsilon_\nu^{-1} \phi^{(k)}(t)\|_A. \tag{1}$$

We shall show that $\mathscr{O}$ is precisely the inductive-limit topology that was previously assigned to $\mathscr{D}(A)$. Indeed, this is an immediate consequence of the following lemma. We let K be an arbitrary compact subset of R^n.

Lemma 5.3-1. *A convex set $\Omega \subset \mathscr{D}(A)$ is a neighborhood of zero with respect to $\mathscr{O}$ if and only if it intersects each $\mathscr{D}_K$ in a neighborhood of zero in $\mathscr{D}_K$.*

PROOF. Any neighborhood of zero with respect to $\mathscr{O}$ contains a set of the form $\Lambda(l, \varepsilon)$, and the latter in turn clearly intersects every $\mathscr{D}_K$ in a neighborhood of zero in $\mathscr{D}_K$.

Conversely, assume that the convex set Ω intersects each $\mathscr{D}_K$ in a neighborhood of zero in $\mathscr{D}_K$. Let $K_\nu \triangleq \{t \in R^n: |t| \leq \nu + 2\}$. For each ν, there exists an integer $l_\nu \geq 0$ and a positive number η_ν such that every $\phi \in \mathscr{D}(A)$ satisfying supp $\phi \subset K_\nu$ and $\|\phi^{(k)}(t)\| \leq \eta_\nu$ for $|k| \leq l_\nu$ is a member of Ω. We can choose the sequence $l = \{l_\nu\}$ to be monotonically increasing to ∞. Furthermore, we can choose a sequence $\{\lambda_\nu\}_{\nu=0}^{\infty} \subset \mathscr{D}$ such that $\lambda_\nu(t) \geq 0$ and $\sum \lambda_\nu(t) = 1$ for all t and supp $\lambda_\nu \subset \{t: \nu \leq |t| \leq \nu + 2\}$. (See Zemanian 1965, Section 1.8.) Then, for every $\phi \in \mathscr{D}(A)$,

$$\phi = \sum_{\nu=0}^{\infty} 2^{-\nu-1} 2^{\nu+1} \lambda_\nu \phi,$$

where the summation contains only a finite number of nonzero terms. Because of the convexity of Ω, $\phi \in \Omega$ whenever $2^{\nu+1}\lambda_\nu \phi \in \Omega$ for every ν.

Now, if $\|\phi^{(k)}(t)\| \leq \varepsilon_\nu$ for $|k| \leq l_\nu$ and $|t| \geq \nu$, then, by virtue of Leibniz's rule for the differentiation of a product,

$$\|2^{\nu+1} D^k[\lambda_\nu(t)\phi(t)]\| \leq c_\nu \varepsilon_\nu$$

for all t and $|k| \leq l_\nu$, where the c_ν are constants not depending on the choice of $\phi \in \mathscr{D}(A)$. We can choose $\varepsilon \triangleq \{\varepsilon_\nu\}$ to be monotonically decreasing to zero and such that $c_\nu \varepsilon_\nu < \eta_\nu$ for every ν. Thus, the fact that $\phi \in \Lambda(l, \varepsilon)$ implies that $2^{\nu+1}\lambda_\nu \phi \in \Omega$, which, as was noted above, implies that $\phi \in \Omega$. So, truly, Ω is a neighborhood of zero under the topology $\mathcal{O}$. ◇

We now show that Conditions E are satisfied when $\mathscr{I}(A) = \mathscr{D}(A)$, $\mathscr{J} = \mathscr{E}$, and $\mathscr{K} = \mathscr{D}$. Condition E1 is obviously fulfilled. To verify Condition E2, we first note that supp ψ is a bounded set since supp v and supp ϕ are. It is also true that, for every nonnegative integer $k \in R^n$,

$$\psi^{(k)}(t) = \langle v(x), \phi^{(k)}(t + x)\rangle. \tag{2}$$

Indeed, this is true for $k = 0$ by definition of ψ. So, assume it is true for any other k, fix $t \in R^n$, and let $\Delta t_\nu \in R$ with $\Delta t_\nu \neq 0$. Then,

$$\frac{\psi^{(k)}(t + \Delta t\,|_\nu) - \psi^{(k)}(t)}{\Delta t_\nu} - \langle v(x), \partial_\nu \phi^{(k)}(t + x)\rangle = \langle v(x), \theta_{\Delta t_\nu}(x)\rangle, \tag{3}$$

where

$$\theta_{\Delta t_\nu}(x) = \frac{1}{\Delta t_\nu} \int_0^{\Delta t_\nu} d\tau_\nu \int_0^{\tau_\nu} \partial_\nu{}^2 \phi^{(k)}(t + x + z\,|_\nu)\, dz_\nu. \tag{4}$$

As usual, $z|_\nu$ denotes an n-tuple all of whose components are zero except possibly for the νth component z_ν. A straightforward manipulation shows that $\theta_{\Delta t_\nu} \to 0$ in $\mathscr{E}$ as $\Delta t_\nu \to 0$. Since $v \in [\mathscr{E}; A]$, (3) tends to zero. We can conclude by induction that (2) holds true for all k. This implies that ψ is a smooth function and is therefore a member of $\mathscr{D}(A)$.

Thus, $\phi \mapsto \psi$ is a mapping of $\mathscr{D}$ into $\mathscr{D}(A)$, which is clearly linear. Furthermore, for any compact set K, supp ψ is contained in another fixed compact set J for all $\phi \in \mathscr{D}_K$. Also, since $v \in [\mathscr{E}; A]$,

$$\begin{aligned} \|\psi^{(k)}(t)\| &\leq M \max_{0 \leq k_\nu \leq p} \sup_{x \in I_p} |\phi^{(k)}(t + x)| \\ &\leq M \max_{0 \leq k_\nu \leq p} \sup_{x \in R^n} |\phi^{(k)}(x)|. \end{aligned}$$

(See Case II of Section 3.6.) This implies that $\phi \mapsto \psi$ is continuous from $\mathscr{D}_K$ into $\mathscr{D}_J(A)$ for every K and corresponding J and therefore from $\mathscr{D}$ into $\mathscr{D}(A)$. So, truly, Condition E2 is fulfilled.

Finally, Condition E3 asserts that $v \mapsto \psi$ is linear from $[\mathscr{E}; A]$ into $\mathscr{D}(A)$ and uniformly continuous with respect to the bounded sets in $\mathscr{D}$. The linearity is clear. To show the uniform continuity, choose an arbitrary seminorm $\gamma_{l,\varepsilon}$ for the topology of $\mathscr{D}(A)$ as defined by (1) and let Φ be a bounded set in $\mathscr{D}$. Thus, Φ is contained in some $\mathscr{D}_K$ and is bounded therein. We have that

$$\sup_{\phi \in \Phi} \gamma_{l,\varepsilon}(\psi) = \sup_{\phi \in \Phi} \sup_{\nu} \sup_{\substack{|t| \geq \nu \\ |k| \leq l_\nu}} \|\langle v(x), \varepsilon_\nu^{-1} \phi^{(k)}(t + x)\rangle\|. \tag{5}$$

Now, observe that

$$\{\varepsilon_\nu^{-1}\phi^{(k)}(t+\cdot):\ \phi\,\varepsilon\,\Phi,\quad |k|\le l_\nu,\quad |t|\ge\nu,\quad \nu=0,1,2,\ldots\}$$

is a bounded set in $\mathscr{E}$. Indeed, let I be any compact set in R^n. Then, there exists an integer ν_0 such that, for all $|t|\ge\nu_0$ and all $\phi\in\Phi$,

$$I\cap\operatorname{supp}\phi^{(k)}(t+\cdot)$$

is a void set. Thus, for only a finite number of ν's are there functions $\phi^{(k)}(t+\cdot)$, where $|t|\ge\nu$, that are not identically equal to zero on I. Our assertion follows from this fact. Thus, (5) is the same as

$$\sup_{\phi\in\Phi}\gamma_{l,\varepsilon}(\psi)=\sup_{\theta\in\Theta}\|\langle v,\theta\rangle\|,$$

where Θ is a bounded set in $\mathscr{E}$. This establishes the uniform continuity of $\phi\to\psi$ and completes the proof of Condition E3.

By virtue of Theorem 5.2-1, we have established the following.

Theorem 5.3-1. *If $y\in[\mathscr{D}(A);B]$, then $v\mapsto y*v$ is a continuous linear mapping of $[\mathscr{E};A]$ into $[\mathscr{D};B]$.*

II. $\mathscr{I}(A)=\mathscr{D}_-(A)$, $\mathscr{J}=\mathscr{D}_-$, $\mathscr{K}=\mathscr{D}_-$.

In this part, we take $R^n=R$ so that $t\in R$. We first present a direct characterization of the topology of $\mathscr{D}_-(A)$. Once again, let $\varepsilon\triangleq\{\varepsilon_\nu\}_{\nu=0}^\infty$ be a sequence of positive numbers tending monotonically to zero and let $l=\{l_\nu\}_{\nu=0}^\infty$ be a sequence of nonnegative integers tending monotonically to infinity. Also, let p be an arbitrary positive integer. $\Lambda(l,\varepsilon,p)$ denotes the set of all $\phi\in\mathscr{D}_-(A)$ such that

$$\|\phi^{(k)}(t)\|_A\le\varepsilon_0,\qquad -p\le t,\quad 0\le k\le l_0$$

and

$$\|\phi^{(k)}(t)\|_A\le\varepsilon_\nu,\qquad \nu\le t,\quad 0\le k\le l_\nu,\quad \nu=1,2,\ldots.$$

The collection of all such $\Lambda(l,\varepsilon,p)$ is a basis of neighborhoods of zero for a topology $\mathscr{Q}$ of $\mathscr{D}_-(A)$. A family $\{\gamma_{l,\varepsilon,p}\}$ of seminorms generating $\mathscr{Q}$ is defined by

$$\gamma_{l,\varepsilon,p}(\phi)\triangleq\max\{\alpha(\phi),\beta(\phi)\},\tag{6}$$

where

$$\alpha(\phi)=\sup_{\substack{-p\le t\\0\le k\le l_0}}\|\varepsilon_0^{-1}\phi^{(k)}(t)\|_A$$

and

$$\beta(\phi)=\sup_{1\le\nu}\ \sup_{\substack{\nu\le t\\0\le k\le l_\nu}}\|\varepsilon_\nu^{-1}\phi^{(k)}(t)\|_A.$$

The next lemma shows that $\mathscr{Q}$ is identical to the inductive-limit topology previously assigned to $\mathscr{D}_-(A)$. We let K be an interval of the form $(-\infty, T]$, where $T \in R$, and let $\mathscr{D}_K$ be the space of all $\phi \in \mathscr{D}_-(A)$ with supp $\phi \subset K$ and supplied with the topology generated by $\{\rho_{j,p}\}_{p=0}^{\infty}$, where $\rho_{j,p}$ is defined in Case VI of Section 3.6.

Lemma 5.3-2. *A convex set $\Omega \subset \mathscr{D}_-(A)$ is a neighborhood of zero with respect to $\mathscr{Q}$ if and only if it intersects each $\mathscr{D}_K$ in a neighborhood of zero in $\mathscr{D}_K$.*

The proof is just a modification of the proof of Lemma 5.3-1.

We turn to a verification of Conditions E. It is obvious that Condition E1 is fulfilled. For Condition E2, we first note that supp v is bounded on the left (Theorem 3.7-4) and that supp ϕ is bounded on the right. It follows that ψ is an A-valued function whose support is bounded on the right. In fact, for all $\phi \in \mathscr{D}_K$, where $K = (-\infty, T]$, we have that supp $\psi \subset N \triangleq (-\infty, T - \tau]$, where $\tau = \inf \operatorname{supp} v$. That (2) holds and that ψ is smooth follows as in Case I. (Now, however, we have to show that $\theta_{\Delta t}$ tends to zero in $\mathscr{D}_-$ as $\Delta t \to 0$.) Thus, $\phi \mapsto \psi$ is a linear mapping of $\mathscr{D}_K$ into $\mathscr{D}_N(A)$. It is also continuous. Indeed, for any $\phi \in \mathscr{D}_K$, for any nonnegative integers k and q, and for each $t \in I_q$, where $I_q \triangleq [-q, \infty)$, we have that $\phi(t + \cdot) \in \mathscr{D}_J$, where $J = (-\infty, T + q]$. Since $v \in [\mathscr{D}_J; A]$, there exists a constant M and a nonnegative integer p such that

$$\sup_{t \in I_q} \|\psi^{(k)}(t)\| \le M \max_{0 \le r \le p} \sup_{\substack{t \in I_q \\ x \in I_p}} |\phi^{(k+r)}(t + x)|.$$

Hence, $\phi \mapsto \psi$ is a continuous linear mapping of $\mathscr{D}_K$ into $\mathscr{D}_N(A)$ and therefore of $\mathscr{D}_-$ into $\mathscr{D}_-(A)$.

To verify Condition E3, we first note that $v \mapsto \psi$ is a linear mapping of $[\mathscr{D}_-; A]$ into $\mathscr{D}_-(A)$. That this mapping is uniformly continuous with respect to all ϕ in any bounded set Φ in $\mathscr{D}_-$ is shown as follows. We know that Φ is contained in some space $\mathscr{D}_K$, where $K = (-\infty, T]$ as above, and is bounded therein. It is also true that

$$\begin{aligned} \sup_{\phi \in \Phi} \gamma_{l,\varepsilon,p}(\psi) &= \sup_{\phi \in \Phi} \gamma_{l,\varepsilon,p}(\langle v(x), \phi(t + x)\rangle) \\ &= \sup_{\theta \in \Theta} \|\langle v, \theta\rangle\|, \end{aligned} \tag{7}$$

where Θ is a bounded set in $\mathscr{D}_-$. This is a consequence of (2) and the fact that the collections

$$\{\varepsilon_0^{-1}\phi^{(k)}(t + x) \colon \phi \in \Phi, \quad -p \le t, \quad 0 \le k \le l_0\}$$

and

$$\{\varepsilon_\nu^{-1}\phi^{(k)}(t+x): \phi \in \Phi, \quad \nu \le t, \quad 0 \le k \le l_\nu, \quad \nu = 1, 2, \ldots\}$$

are both bounded sets in $\mathscr{D}_-$. (Note that, given any interval of the form $I = [q, \infty)$, there is a ν_0 such that $I \cap \operatorname{supp} \phi(t + \cdot)$ is void for every $t \ge \nu_0$ and all $\phi \in \Phi$.) Equation (7) proves the uniform continuity of $v \mapsto \psi$ with respect to all $\phi \in \Phi$. Thus, we have established the following theorem.

Theorem 5.3-2. *If $y \in [\mathscr{D}_-(A); B]$ then $v \mapsto y * v$ is a continuous linear mapping of $[\mathscr{D}_-; A]$ into $[\mathscr{D}_-; B]$.*

III. $\mathscr{I}(A) = \mathscr{S}(A)$, $\mathscr{J} = \mathscr{L}(w, z)$, where $w < 0 < z$, $\mathscr{K} = \mathscr{S}$.

We shall examine this case under the restriction that our testing functions are defined on R rather than on R^n. This, in fact, is all we shall need subsequently. Essentially the same arguments can be used for R^n, but the notation becomes considerably more complicated.

Every $\phi \in \mathscr{S}$ is a member of $\mathscr{L}(w, z)$ because $\phi(t)$ tends to zero faster than any negative power of $|t|$ as $|t| \to \infty$, whereas the members of $\mathscr{L}(w, z)$ are allowed to grow exponentially as $|t| \to \infty$ by virtue of the condition $w < 0 < z$. Moreover, it is easy to see that $\mathscr{L}(w, z)$ is closed under the shifting operator. This implies that Condition E1 is satisfied.

Turning to Condition E2, we set up Equation (3) for the A-valued function ψ, where now Δt_ν is replaced by Δt and ∂_ν by D. From (4) and the fact that $\phi \in \mathscr{S}$, it follows that, for each nonnegative integer p, $\theta_{\Delta t}^{(p)}$ converges uniformly to zero on R as $\Delta t \to 0$. In other words, $\theta_{\Delta t} \to 0$ in $\mathscr{L}_{0,0}$ and therefore in $\mathscr{L}(w, z)$ as well. But $v \in [\mathscr{L}(w, z); A]$. Hence, (3) tends to zero, which by induction establishes (2) and the smoothness of ψ.

To show that ψ is of rapid descent, we first observe that $\phi \in \mathscr{S}$ implies the existence of positive constants $C_{q,p}$ such that

$$|\phi^{(p)}(t)| \le C_{q,p}/(1 + t^2)^q, \qquad q, p = 0, 1, \ldots. \tag{8}$$

Now, let $w < c < 0 < d < z$, so that $\mathscr{S} \subset \mathscr{L}_{c,d}$. Since the restriction of v to $\mathscr{L}_{c,d}$ is continuous and linear,

$$\|(1 + t^2)^l \psi^{(j)}(t)\|_A \le M \max_{0 \le k \le r} \sup_{x \in R} |(1 + t^2)^l \kappa_{c,d}(x) \phi^{(k+j)}(t + x)|. \tag{9}$$

The quantity within the magnitude signs in the right-hand side is bounded on the $\{t, x\}$ plane. Indeed, for $t \ge 0$ and $x \le -t/2$,

$$|(1 + t^2)^l \kappa_{c,d}(x) \phi^{(p)}(t + x)| \le (1 + t^2)^l e^{-dt/2} C_{0,p}$$

and the right-hand side is bounded for all $t \ge 0$. On the other hand, for $t \ge 0$ and $x > -t/2$,

$$|(1 + t^2)^l \kappa_{c,d}(x) \phi^{(p)}(t + x)| \le (1 + t^2)^l \frac{C_{q,p}}{(1 + \frac{1}{4}t^2)^q}$$

and the right-hand side is bounded for $t \geq 0$ when $q \geq l$. A similar argument for $t \leq 0$ establishes our assertion concerning the right-hand side of (9). This proves that $\psi \in \mathscr{S}(A)$.

To prove that the linear mapping $\phi \mapsto \psi$ of $\mathscr{S}$ into $\mathscr{S}(A)$ is continuous, assume that $\{\phi_\nu\}_{\nu=1}^\infty$ converges in $\mathscr{S}$ to zero. This means that there exist positive constants $C_{q,p,\nu}$ which replace $C_{q,p}$ in (8) when ϕ is replaced by ϕ_ν and which tend to zero as $\nu \to \infty$, q and p being fixed. Then, (9) and the argument following it show that $\{\psi_\nu\}_{\nu=1}^\infty$ converges in $\mathscr{S}(A)$ to zero.

Finally, consider Condition E3. Let Φ be a bounded set in $\mathscr{S}$ and write

$$\sup_{\phi \in \Phi} \sup_{t \in R} \|(1 + t^2)^l \psi^{(j)}(t)\|_A = \sup_{\phi \in \Phi} \sup_{t \in R} \|\langle v(x), (1 + t^2)^l \phi^{(j)}(t + x)\rangle\|_A.$$

Condition E3 will be verified once we show that, as ϕ traverses Φ and t traverses R, $(1 + t^2)^l \phi^{(j)}(t + \cdot)$ traverses an $\mathfrak{S}$-set in $\mathscr{L}(w, z)$. To prove the latter, let $w < c < 0 < d < z$, so that $\mathscr{S} \subset \mathscr{L}_{c,d}$. By Liebniz's rule for the differentiation of a product and by the argument following (9), we have that, for every k,

$$\sup_{x \in \mathrm{R}} |\kappa_{c,d}(x) D_x^k[(1 + t^2)^l \phi^{(j)}(t + x)]|$$

is uniformly bounded for all $t \in R$ and all $\phi \in \Phi$, which is what we want.

The next theorem has hereby been established.

Theorem 5.3-3. *If* $y \in [\mathscr{S}(A); B]$, *then* $v \mapsto y * v$ *is a continuous linear mapping of* $[\mathscr{L}(w, z); A]^s$ *into* $[\mathscr{S}; B]$ *when* $w < 0 < z$.

Problem 5.3-1. Prove Lemma 5.3-2.

Problem 5.3-2. Show that, if $y \in [\mathscr{E}(A); B]$, then $v \mapsto y * v$ is a continuous linear mapping of $[\mathscr{D}; A]$ into $[\mathscr{D}; B]$.

Problem 5.3-3. Show that, if $y \in [\mathscr{L}(w, z; A); B]$, where $w < z$, then $v \mapsto y * v$ is a continuous linear mapping of $[\mathscr{L}(w, z); A]^s$ into $[\mathscr{L}(w, z); B]^s$.

5.4. THE COMMUTATIVITY OF CONVOLUTION OPERATORS WITH SHIFTING AND DIFFERENTIATION

In each of the three cases considered in the preceding section, the convolution operator $y *$ commutes with the shifting operator and differentiation. For instance, in case I, we have the following result.

Theorem 5.4-1. *If* $y \in [\mathscr{D}(A); B]$, $v \in [\mathscr{E}; A]$, $\tau \in R^n$, *and* k *is a nonnegative integer in* R^n, *then*

$$\sigma_\tau(y * v) = y * (\sigma_\tau v) \tag{1}$$

and

$$D^k(y * v) = y * (D^k v) \tag{2}$$

in the sense of equality in $[\mathscr{D}; B]$.

PROOF. Since $\phi \mapsto \sigma_\tau \phi$ is an isomorphism on $\mathscr{D}$, we may write

$$\begin{aligned} \langle \sigma_\tau(y * v), \phi \rangle &= \langle y * v, \sigma_{-\tau}\phi \rangle = \langle y(t), \langle v(x), \phi(t + x + \tau)\rangle\rangle \\ &= \langle y(t), \langle v(x - \tau), \phi(t + x)\rangle\rangle = \langle y * (\sigma_\tau v), \phi \rangle. \end{aligned}$$

This establishes (1). A similar argument with σ_τ replaced by D^k establishes (2). ◇

That $y *$ commutes with σ_τ is an important fact. As was indicated in the introduction, it means that $y *$ is a translation-invariant operator. Moreover, $\mathscr{D}(A)$ can be identified as a subspace of $[\mathscr{E}; A]$ in accordance with Example 3.3-2, and the canonical injection of $\mathscr{D}(A)$ into $[\mathscr{E}; A]$ is continuous. Consequently, the restriction of $y*$ to $\mathscr{D}(A)$ is a translation-invariant kernel operator. A major objective of this chapter is to demonstrate the converse; namely, every translation-invariant kernel operator is a convolution operator.

For Cases II and III of the preceding section, we have the next theorem. Its proof is quite similar to that of Theorem 5.4-1.

Theorem 5.4-2. *Let* $\tau \in R$ *and let* k *be a nonnegative integer in* R. *Then,* (1) *and* (2) *are equalities in* $[\mathscr{K}; B]$ *under either one of the following conditions*:

(i) $y \in [\mathscr{D}_-(A); B]$, $v \in [\mathscr{D}_-; A]$, $\mathscr{K} = \mathscr{D}_-$.
(ii) $y \in [\mathscr{S}(A); B]$, $v \in [\mathscr{L}(w, z); A]$, *where* $w < 0 < z$, $\mathscr{K} = \mathscr{S}$.

Problem 5.4-1. Establish Theorem 5.4-2. Do the same for the convolutions of Problems 5.3-2 and 5.3-3.

5.5. REGULARIZATION

When the convolution operator $y *$, where $y \in [\mathscr{D}(A); B]$, is applied to any $v \in \mathscr{D}(A)$, the result $y * v$ is a smooth B-valued function. This is called the *regularization of* y *by* v, a process we shall now investigate.

Lemma 5.5-1. *Let* $y \in [\mathscr{D}(A); B]$ *and* $v \in \mathscr{D}(A)$. *Set*

$$u(x) \triangleq \langle y(t), v(x-t)\rangle. \tag{1}$$

Then, $v \mapsto u$ *is a continuous linear mapping of* $\mathscr{D}(A)$ *into* $\mathscr{E}(B)$.

PROOF. Clearly, u maps R^n into B. The argument that was applied to (2) of Section 5.3 can again be used to show that u is smooth [i.e., $u \in \mathscr{E}(B)$] and that

$$u^{(k)}(x) = \langle y(t), v^{(k)}(x-t)\rangle \tag{2}$$

for every nonnegative integer $k \in R^n$. To show the continuity of the linear mapping $v \mapsto u$, let K and N be arbitrary compact sets in R^n. Then,

$$\{v(x-\cdot): x \in K, \quad v \in \mathscr{D}_N(A)\} \subset \mathscr{D}_J(A),$$

where J is some other compact set in R^n. Since the restriction of y to $\mathscr{D}_J(A)$ is continuous and linear,

$$\begin{aligned}\sup_{x \in K} \|u^{(k)}(x)\|_B &= \sup_{x \in K} \|\langle y(t), v^{(k)}(x-t)\rangle\| \\ &\leq \sup_{x \in K} M \max_{0 \leq l \leq r} \sup_{t \in J} \|v^{(k+l)}(x-t)\| \\ &= M \max_{0 \leq l \leq r} \sup_{t \in R^n} \|v^{(k+l)}(t)\|.\end{aligned}$$

This shows that $v \mapsto u$ is continuous from $\mathscr{D}_N(A)$ into $\mathscr{E}(B)$ for every N and therefore is continuous on $\mathscr{D}(A)$. ◇

Theorem 5.5-1. *If* $y \in [\mathscr{D}(A); B]$ *and* $v \in \mathscr{D}(A)$, *then, in the sense of equality in* $[\mathscr{D}; B]$,

$$y * v = \langle y(t), v(\cdot - t)\rangle. \tag{3}$$

Moreover, $v \mapsto y * v$ *is a continuous linear mapping of* $\mathscr{D}(A)$ *into* $\mathscr{E}(B)$.

PROOF. Let $\phi \in \mathscr{D}$ and consider the manipulations

$$\begin{aligned}\langle y * v, \phi\rangle &= \left\langle y(t), \int_{R^n} v(x)\phi(t+x)\,dx\right\rangle \\ &= \left\langle y(t), \int_{R^n} v(x-t)\phi(x)\,dx\right\rangle \qquad (4)\\ &= \int_{R^n} \langle y(t), v(x-t)\rangle\phi(x)\,dx \qquad (5)\\ &= \langle\langle y(t), v(x-t)\rangle, \phi(x)\rangle.\end{aligned}$$

All these equalities are obvious except for the one between (4) and (5). To establish that one, let K be a compact interval containing supp ϕ. Approximate

$$\int_K v(x-t)\phi(x)\,dx \tag{6}$$

by the Riemann sum

$$\sum_{[1]\le\mu\le l} v(x_\mu - t)\phi(x_\mu)\operatorname{vol} I_\mu, \tag{7}$$

where

$$\pi \triangleq \{I_\mu\}_{[1]\le\mu\le l}$$

is a rectangular partition of K and $x_\mu \in I_\mu$. (We are using the notation of Section 1.4.) Upon applying y to (7), we get

$$\langle y(t), \textstyle\sum v(x_\mu - t)\phi(x_\mu)\operatorname{vol} I_\mu\rangle = \textstyle\sum \langle y(t), v(x_\mu - t)\rangle\phi(x_\mu)\operatorname{vol} I_\mu. \tag{8}$$

Because of the continuity of $\langle y(t), v(\cdot - t)\rangle$ and ϕ, the right-hand side of (8) tends to (5) as $|\pi| \to 0$. To show that the left-hand side of (8) tends to (4), we need merely show that the function of t defined by (7) tends in $\mathscr{D}(A)$ to the function of t defined by (6). This can be done in a straightforward manner using the facts that the function

$$\{t, x\} \mapsto v(x-t)\phi(x)$$

has a compact support and that, for every nonnegative integer $k \in R^n$,

$$\{t, x\} \mapsto D^k v(x-t)\phi(x)$$

is a uniformly continuous function on R^n. Thus, (3) holds in the sense of equality in $[\mathscr{D}; B]$.

The second sentence of our theorem has already been established by Lemma 5.5-1. ◇

There is another form of regularization that we shall need at one point in Chapter 7. It states in effect that, if y is an $[A; B]$-valued L_1-type distribution and v is a smooth A-valued function such that it and all its derivatives decrease exponentially as the magnitude of its independent variable increases indefinitely, then $y * v$ is a member of $\mathscr{B}(B)$. Here, $\mathscr{B}(B)$ is the space of B-valued smooth functions ϕ on R^n such that, for each nonnegative integer $k \in R^n$,

$$\gamma_k(\phi) \triangleq \sup_{t\in R^n} \|\phi^{(k)}(t)\|_B < \infty.$$

We assign to $\mathscr{B}(B)$ the topology generated by $\{\gamma_k\}$, and this makes $\mathscr{B}(B)$ a Fréchet space.

Theorem 5.5-2. *If* $y \in [\mathscr{D}_{L_1}(A); B]$ *and* $v \in \mathscr{L}_{a,b}(A)$, *where* $b < 0 < a$, *then, in the sense of equality in* $[\mathscr{S}; B]$,

$$y * v = \langle y(t), v(\cdot - t)\rangle,$$

and $v \mapsto y * v$ *is a continuous linear mapping of* $\mathscr{L}_{a,b}(B)$ *into* $\mathscr{B}(B)$.

The proof follows the same scheme as that of Theorem 5.5-1, but the details are rather more complicated. They may be found in the work by Zemanian (1970a, pp. 112–114).

5.6. PRIMITIVES

In this and the next three sections, we discuss four concepts concerning A-valued distributions which we shall use in proving that translation-invariant kernel operators are convolution operators. They are the primitives of a distribution, the direct product of an A-valued distribution with a complex-valued distribution, distributions that are independent of certain coordinates, and a change-of-variable formula. These discussions are much the same as those for complex-valued distributions.

In our discussion of primitives, we assume once again that $t \in R \triangleq R^1$ and that $\mathscr{D} = \mathscr{D}_{R^1}$. Let H denote that subspace of $\mathscr{D}$ whose elements χ have the form $\chi = \psi^{(1)}$, where $\psi \in \mathscr{D}$. Given any $\phi_0 \in \mathscr{D}$ such that $\int \phi_0(t)\, dt = 1$, every $\phi \in \mathscr{D}$ has the unique decomposition

$$\phi = c\phi_0 + \chi, \tag{1}$$

where $\chi \in H$ and $c = \int \phi(t)\, dt$. Indeed, χ is clearly a uniquely determined member of $\mathscr{D}$, and so too is $\psi(t) \triangleq \int_{-\infty}^{t} \chi(x)\, dx$ because $\psi(t) = 0$ for all sufficiently large t.

A *primitive of* $f \in [\mathscr{D}; A]$ is any $g \in [\mathscr{D}; A]$ such that $g^{(1)} = f$. As we shall see, any $f \in [\mathscr{D}; A]$ has an infinity of primitives, any two of which differ by a constant member of $[\mathscr{D}; A]$. (A *constant member of* $[\mathscr{D}; A]$ is a regular member of $[\mathscr{D}; A]$ generated by a function of the form $t \mapsto a$, where a is a fixed member of A. We let a also denote the corresponding constant member of $[\mathscr{D}; A]$.)

To determine a particular primitive $f^{(-1)}$ of a given $f \in [\mathscr{D}; A]$, choose ϕ_0 as above and assign some value in A to $\langle f^{(-1)}, \phi_0\rangle$. We then define $f^{(-1)}$ on $\phi = c\phi_0 + \chi$ by

$$\langle f^{(-1)}, \phi\rangle \triangleq c\langle f^{(-1)}, \phi_0\rangle - \langle f, \psi\rangle. \tag{2}$$

It is not difficult to show that, as $\{\phi_j\} \to 0$ in $\mathscr{D}$, the corresponding numerical sequence $\{c_j\}$ tends to zero, whereas $\{\psi_j\}$ tends to zero in $\mathscr{D}$. This implies that $f^{(-1)}$ is truly a member of $[\mathscr{D}; A]$.

Next, we verify that $f^{(-1)}$ is a primitive of f; we do this by showing that $(f^{(-1)})^{(1)} = f$. For any $\phi \in \mathscr{D}$, the decomposition (1) of $\phi^{(1)}$ yields $c = 0$ and $\chi = \phi^{(1)}$. Therefore,

$$\langle (f^{(-1)})^{(1)}, \phi \rangle = \langle f^{(-1)}, -\phi^{(1)} \rangle = \left\langle f(t), \int_{-\infty}^{t} \phi^{(1)}(x)\,dx \right\rangle$$
$$= \langle f, \phi \rangle,$$

which is what we wished to establish.

Finally, we demonstrate that any primitive g of f differs from the primitive $f^{(-1)}$ defined by (2) by a constant member of $[\mathscr{D}; A]$. For any $\psi \in \mathscr{D}$,

$$\langle g, \psi^{(1)} \rangle = \langle g^{(1)}, -\psi \rangle = -\langle f, \psi \rangle.$$

So, upon applying g to the decomposition (1) of any $\phi \in \mathscr{D}$, we may write

$$\langle g, \phi \rangle = c\langle g, \phi_0 \rangle + \langle g, \psi^{(1)} \rangle = c\langle g, \phi_0 \rangle - \langle f, \psi \rangle.$$

Subtracting (2) from this, we get

$$\langle g - f^{(1)}, \phi \rangle = a \int \phi(t)\,dt = \langle a, \phi \rangle,$$

where $a = \langle g, \phi_0 \rangle - \langle f^{(-1)}, \phi_0 \rangle$ and is therefore a constant member of $[\mathscr{D}; A]$.

Here is a fact we shall need later on. As usual, σ_τ represents the shifting operator.

Lemma 5.6-1. *If $g \in [\mathscr{D}; A]$ and if $\sigma_\tau g = g$ for all $\tau \in R$, then*

$$\langle g, \phi \rangle = a \int \phi(t)\,dt, \tag{3}$$

where $a \in A$ is uniquely determined by g.

PROOF. We first note that $g^{(1)} = 0$ in $[\mathscr{D}; A]$. Indeed, for every $\phi \in \mathscr{D}$,

$$\langle g^{(1)}, \phi \rangle = \langle g, -\phi^{(1)} \rangle = \lim_{\tau \to 0} \langle g, (\sigma_\tau \phi - \phi)/\tau \rangle$$
$$= \lim 0 = 0.$$

Now, one primitive of zero is zero. Since g is also a primitive of zero, it must differ from zero by a constant member of $[\mathscr{D}; A]$. In other words, (3) holds. Obviously, there cannot be two different values for a satisfying (3) for all ϕ. ◇

Problem 5.6-1. Prove that $f^{(-1)}$ as defined by (2) is a member of $[\mathscr{D}; A]$.

5.7. DIRECT PRODUCTS

In this section, $t \in R^n$ and $x \in R^s$. Moreover, $\phi(t, x)$ will now denote a function on R^{n+s} (and not the value of ϕ at $\{t, x\}$, which is our usual interpretation). Let $f \in [\mathscr{D}_{R^n}; A]$ and $g \in [\mathscr{D}_{R^s}; C]$. We define the *direct product* $f(t) \times g(x)$ as a mapping on any $\phi(t, x) \in \mathscr{D}_{R^{n+s}}$ by

$$\langle f(t) \times g(x), \phi(t, x)\rangle \triangleq \langle f(t), \langle g(x), \phi(t, x)\rangle\rangle. \tag{1}$$

The right-hand side has a meaning since, as a function of t, $\langle g(x), \phi(t, x\rangle$ is a member of $\mathscr{D}_{R^n}$. This can be shown through the same argument as that used in regard to (2)–(4) of Section 5.3. That argument also establishes the following equation:

$$D_t^k\langle g(x), \phi(t, x)\rangle = \langle g(x), D_t^k\phi(t, x)\rangle. \tag{2}$$

It follows from (1) that $f(t) \times g(x)$ is a linear mapping of $\mathscr{D}_{R^{n+s}}$ into A. Moreover, a direct estimate of the right-hand side of (1) with the use of (2) shows that $f(t) \times g(x)$ is continuous on $\mathscr{D}_J$ for every compact set $J \subset R^{n+s}$. Thus,

$$f(t) \times g(x) \in [\mathscr{D}_{R^{n+s}}; A].$$

A useful fact is the following. The restriction of $f(t) \times g(x)$ to the set Ω of all testing functions of the form $\theta(t)\psi(x)$, where $\theta \in \mathscr{D}_{R^n}$ and $\psi \in \mathscr{D}_{R^s}$, uniquely determines $f(t) \times g(x)$ on all of $\mathscr{D}_{R^{n+s}}$. This is because Ω is total in $\mathscr{D}_{R^{n+s}}$. (See Schwartz, 1966, pp. 108–109.) For this restriction, we have

$$\langle f(t) \times g(x), \theta(t)\psi(x)\rangle = \langle f, \theta\rangle\langle g, \psi\rangle.$$

5.8. DISTRIBUTIONS THAT ARE INDEPENDENT OF CERTAIN COORDINATES

Let

$$j \in [\mathscr{D}_{R^{n+s}}; A],$$

$$\xi = \{\xi_1, \ldots, \xi_n\} \in R^n,$$

$$\eta = \{\eta_1, \ldots, \eta_s\} \in R^s,$$

$$\{\xi, \eta\} = \{\xi_1, \ldots, \xi_n, \eta_1, \ldots, \eta_s\} \in R^{n+s}.$$

We say that *j is independent of* η if, for every $\tau \in R^{n+s}$ with

$$\tau = \{0, \ldots, 0, \tau_1, \ldots, \tau_s\},$$

we have that $\sigma_\tau j = j$ [that is, if $j(\xi, \eta)$ is independent of shifts through the η coordinates]. It will now be shown that such a distribution can be written as the direct product $y(\xi) \times 1(\eta)$, where $y(\xi)$ denotes a member of $[\mathscr{D}_{R^n}; A]$ and $1(\eta)$ denotes the regular distribution corresponding to the function that equals 1 everywhere on R^s.

Let

$$\phi(\xi, \eta) = \theta(\xi_1, \ldots, \xi_n, \eta_1, \ldots, \eta_{s-1})\psi(\eta_s),$$

where $\theta \in \mathscr{D}_{R^{n+s-1}}$ and $\psi \in \mathscr{D}_{R^1}$. Now, j is independent of η_s. Therefore, $\psi \mapsto \langle j, \theta\psi \rangle$ is a member of $[\mathscr{D}_{R^1}; A]$, which is also independent of η_s. We may therefore invoke Lemma 5.6-1 to write

$$\langle j, \theta\psi \rangle = a(\theta) \int \psi(\eta_s)\, d\eta_s = a(\theta)\langle 1(\eta_s), \psi(\eta_s)\rangle,$$

where $a(\theta)$ is a member of A depending on θ but not on ψ. Upon fixing ψ such that $\int \psi(\eta_s)\, d\eta_s = 1$, we see that $a(\theta) = \langle q, \theta \rangle$, where $q \in [\mathscr{D}_{R^{n+s-1}}; A]$ is uniquely determined by j. In view of the last paragraph of the preceding section, we can conclude that j is equal to the following direct product:

$$j(\xi, \eta) = q(\xi_1, \ldots, \xi_n, \eta_1, \ldots, \eta_{s-1}) \times 1(\eta_s).$$

Next, we observe that, since j is independent of shifts through η_{s-1}, so, too, is q. Indeed, let σ_τ be a shift through the coordinate η_{s-1} only and let θ and ψ be as above. Then,

$$\langle q, \theta \rangle \langle 1, \psi \rangle = \langle j, \theta\psi \rangle = \langle \sigma_\tau j, \theta\psi \rangle = \langle q, \sigma_{-\tau}\theta \rangle \langle 1, \psi \rangle.$$

Since this holds for all such θ and ψ, $q = \sigma_\tau q$. Therefore, by applying the argument of the preceding paragraph to q, we see that there exists a unique $p \in [\mathscr{D}_{R^{n+s-2}}; A]$ such that

$$j(\xi, \eta) = p(\xi_1, \ldots, \xi_n, \eta_1, \ldots, \eta_{s-2}) \times 1(\eta_{s-1}) \times 1(\eta_s).$$

We may also write $1(\eta_{s-1}) \times 1(\eta_s) = 1(\eta_{s-1}, \eta_s)$.

Continuing in this way, we arrive at the following result.

Theorem 5.8-1. *Let $j \in [\mathscr{D}_{R^{n+s}}; A]$ be independent of η. Then, there exists a unique $y \in [\mathscr{D}_{R^n}; A]$ such that*

$$j(\xi, \eta) = y(\xi) \times 1(\eta).$$

5.9. A CHANGE-OF-VARIABLE FORMULA

In this section, $\mathscr{D} = \mathscr{D}_{R^n}$. Let $z, \zeta \in R^n$ and set $z = U\zeta$, where U is a nonsingular linear transformation on R^n. Thus, U can be represented by a nonsingular $n \times n$ matrix of real numbers and has an inverse U^{-1}. Moreover, $|U|^{-1} = |U^{-1}|$, where $|U|$ denotes the magnitude of the determinant of U.

Given any $f \in [\mathscr{D}; A]$, we let $f(U\zeta)$ denote the linear mapping from $\mathscr{D}$ into A defined by

$$\langle f(U\zeta), \theta(\zeta)\rangle \triangleq \langle f(z), |U|^{-1}\theta(U^{-1}z)\rangle, \qquad \theta \in \mathscr{D}. \tag{1}$$

$f(U\zeta)$ is continuous. Indeed, if the sequence $\{\theta_\nu\}$ tends to zero in D, then, clearly, $\{|U|^{-1}\theta_\nu(U^{-1}z)\}$ tends to zero uniformly for all $z \in R^n$ and the supports of these functions remain contained within a fixed compact set. Moreover, the chain rule for differentiation (Kaplan, 1952, p. 86) shows that the same is true for each fixed-order differentiation of these functions. Thus, $|U|^{-1}\theta_\nu(U^{-1}z) \to 0$ in $\mathscr{D}$ as $\nu \to \infty$. The continuity of $f(U\zeta)$ now follows from (1). We conclude that $f(U\zeta) \in [\mathscr{D}; A]$.

An alternative form for (1) can be obtained by setting $\psi(z) = \theta(U^{-1}z)$:

$$\langle f(z), \psi(z)\rangle = \langle f(U\zeta), |U|\psi(U\zeta)\rangle. \tag{2}$$

5.10. CONVOLUTION OPERATORS

We are at last ready to establish the condition (namely, translation invariance) under which a kernel operator from $\mathscr{D}(A)$ into $[\mathscr{D}; B]$ becomes a convolution operator. Once again, it is understood that $\mathscr{D} = \mathscr{D}_{R^n}$ and $\mathscr{D}(A) = \mathscr{D}_{R^n}(A)$.

Definition 5.10-1. Let $\mathscr{X}$ and $\mathscr{Y}$ be spaces of functions or distributions on R^n and assume that $\mathscr{X}$ and $\mathscr{Y}$ are closed under the shifting operator σ_τ. A mapping $\mathfrak{N}$ from $\mathscr{X}$ into $\mathscr{Y}$ is called *translation invariant* if $\sigma_\tau \mathfrak{N} f = \mathfrak{N}\sigma_\tau f$ for every $f \in \mathscr{X}$ and every $\tau \in R^n$.

In certain subsequent discussions, we shall call the members of a given family of operators *translation varying* to indicate that the condition of translation invariance is not imposed even though certain members of that family may be translation invariant. Thus, translation-invariant operators are considered to be a special case of translation-varying operators.

When $n = 1$ and the space R^1 on which the members of $\mathscr{X}$ and $\mathscr{Y}$ are given is interpreted as the time axis, the adjective translation-invariant is commonly replaced by *time-invariant* and translation-varying by *time-varying*.

Assume now that $\mathfrak{N}$ is a continuous linear mapping of $\mathscr{D}(A)$ into $[\mathscr{D}; B]^\sigma$. By Theorem 4.5-3, $\mathfrak{N}$ is a kernel operator; that is, $\mathfrak{N} = f\,\cdot$ on $\mathscr{D}(A)$, where $f \in [\mathscr{D}_{R^{2n}}(A); B]$. If, in addition, $\mathfrak{N}$ is translation invariant, we may write for every $\phi \in \mathscr{D}$, $v \in \mathscr{D}(A)$, and $\tau \in R^n$,

$$\begin{aligned}\langle f(t, x), \phi(t + \tau)v(x)\rangle &= \langle \mathfrak{N}v, \sigma_{-\tau}\phi\rangle = \langle \sigma_\tau \mathfrak{N}v, \phi\rangle \\ &= \langle \mathfrak{N}\sigma_\tau v, \phi\rangle = \langle f(t, x), \phi(t)v(x - \tau)\rangle.\end{aligned}$$

By Theorem 3.5-1, f is also a member of $[\mathscr{D}_{R^{2n}}; [A; B]]$. So, we may replace v by any $w \in \mathscr{D}$ to obtain

$$\langle f(t, x), \phi(t + \tau)w(x)\rangle = \langle f(t, x), \phi(t)w(x - \tau)\rangle.$$

Since the set of functions of the form $\phi(t)w(x)$ is total in $\mathscr{D}_{R^{2n}}$, we can conclude that f is independent of translations along the subspace of R^{2n} defined by $t = x$. In symbols,

$$f(t, x) = f(t - \tau, x - \tau) \tag{1}$$

for every $\tau \in R^n$.

We now make the change of variables $t = \xi + \eta$, $x = \eta$. In matrix notation, this can be written as $z = U\zeta$, where

$$z = \begin{bmatrix} t \\ x \end{bmatrix}, \qquad \zeta = \begin{bmatrix} \xi \\ \eta \end{bmatrix}, \qquad U = \begin{bmatrix} 1_n & 1_n \\ 0 & 1_n \end{bmatrix}.$$

Here, 1_n denotes the $n \times n$ matrix, U is nonsingular, and $|U| = 1$. Thus, by Equation (2) of Section 5.9,

$$\langle f(t, x), \psi(t, x)\rangle = \langle j(\xi, \eta), \psi(\xi + \eta, \eta)\rangle, \qquad \psi \in \mathscr{D}_{R^{2n}} \tag{2}$$

where $j(\xi, \eta) \triangleq f(\xi + \eta, \eta)$ represents a member of $[\mathscr{D}_{R^{2n}}; [A; B]]$. We now shift j in the η direction through the increment $\tau \in R^n$ and appeal to (1) and (2):

$$\begin{aligned} \langle j(\xi, \eta - \tau), \psi(\xi + \eta, \eta)\rangle &= \langle f(t - \tau, x - \tau), \psi(t, x)\rangle \\ &= \langle f(t, x), \psi(t, x)\rangle = \langle j(\xi, \eta), \psi(\xi + \eta, \eta)\rangle. \end{aligned}$$

This shows that $j(\xi, \eta)$ is independent of η because, for $\theta(\xi, \eta) \triangleq \psi(\xi + \eta, \eta)$, θ traverses $\mathscr{D}_{R^{2n}}$ as ψ traverses $\mathscr{D}_{R^{2n}}$.

According to Theorem 5.8-1, $j(\xi, \eta) = y(\xi) \times 1(\eta)$ for some unique $y \in [\mathscr{D}; [A; B]]$ or, equivalently, $y \in [\mathscr{D}(A); B]$. Thus, by (2) again and the definition of the direct product,

$$\begin{aligned} \langle f(t, x), \psi(t, x)\rangle &= \langle y(\xi) \times 1(\eta), \psi(\xi + \eta, \eta)\rangle \\ &= \left\langle y(\xi), \int_{R^n} \psi(\xi + \eta, \eta)\, d\eta \right\rangle. \end{aligned} \tag{3}$$

Set $\psi(t, x) = \phi(t)w(x)$, where $\phi, w \in \mathscr{D}$, and let $a \in A$. Then, (3) can be re-written as

$$\begin{aligned} \langle \mathfrak{N}(wa), \phi\rangle &= \langle f \cdot (wa), \phi\rangle = \langle f(t, x), \phi(t)w(x)a\rangle \\ &= \left\langle y(\xi), \int_{R^n} w(\eta)a\phi(\xi + \eta)\, d\eta \right\rangle = \langle y * (wa), \phi\rangle. \end{aligned}$$

But Lemma 3.5-1 asserts that the set of elements of the form wa is total in $\mathscr{D}(A)$. Moreover, both $\mathfrak{N}$ and $y\,*$ are continuous and linear on $\mathscr{D}(A)$. Thus, under our assumptions on $\mathfrak{N}$, we can conclude that $\mathfrak{N} = y\,*$ on $\mathscr{D}(A)$, where y is unique.

Conversely, Theorems 5.4-1 and 5.5-1 show that every convolution operator $y\,*$ possesses the properties assigned to $\mathfrak{N}$. We summarize all of this as follows.

Theorem 5.10-1. *$\mathfrak{N}$ is a continuous, linear translation-invariant mapping of $\mathscr{D}(A)$ into $[\mathscr{D}; B]^\sigma$ if and only if there exists a $y \in [\mathscr{D}(A); B]$ such that $\mathfrak{N} = y\,*$ on $\mathscr{D}(A)$. y is uniquely determined by $\mathfrak{N}$, and conversely.*

An alternative proof of this that does not make use of the kernel theorem is given by Zemanian (1970a, pp. 118–119). Also, we again point out that a linear mapping $\mathfrak{N}$ on $\mathscr{D}(A)$ is continuous if and only if it is sequentially continuous.

Corollary 5.10-1a. *A continuous linear translation-invariant mapping $\mathfrak{N}$ from $\mathscr{D}(A)$ into $[\mathscr{D}; B]^\sigma$ can be extended by means of its convolution representation $\mathfrak{N} = y\,*$ onto the space $[\mathscr{J}; A]$ and the range of the extended mapping will be contained in $[\mathscr{K}; B]$ so long as $y \in [\mathscr{J}(A); B]$ and the spaces $\mathscr{J}(A)$, $\mathscr{J}$, and $\mathscr{K}$ are ρ-type testing-function spaces satisfying the Conditions E stated in Section 5.2. Furthermore, if $\mathscr{D}(A)$ is dense in $[\mathscr{J}; A]$, this extension is unique in the sense that no other continuous linear mapping of $[\mathscr{J}; A]$ into $[\mathscr{K}; B]$ can coincide with $\mathfrak{N} = y\,*$ on $\mathscr{D}(A)$.*

It is a fact that $\mathscr{D}(A)$ is dense in $[\mathscr{E}; A]$ (see Problem 5.10-1). Consequently, for every $y \in [\mathscr{D}(A); B]$, $y\,*$ has a unique extension onto $[\mathscr{E}; A]$. Moreover, with $\delta \in [\mathscr{E}; C]$ denoting the delta functional (δ is also called the *unit impulse*), we can write $y * \delta = y$ in the sense of equality in $[\mathscr{D}; [A; B]]$. Thus, we may interpret y as the response of $y\,*$ to the input δ. Because of this, we shall refer to y as the *unit-impulse response of* $\mathfrak{N} = y\,*$. The Laplace transform Y of y (which is defined in the next chapter) is called the *system function for* $\mathfrak{N}$.

If it happens that supp y is bounded on the left or, equivalently, if $y \in [\mathscr{D}_-(A); B]$ (see Theorem 3.7-4), $y\,*$ can be extended onto $[\mathscr{D}_-; A]$. Moreover, $\mathscr{D}(A)$ is dense in $[\mathscr{D}_-; A]$, and therefore this extension of $y\,*$ is unique. Similarly, if supp y is bounded or, equivalently, if $y \in [\mathscr{E}(A); B]$ (see Theorem 3.7-3), $y\,*$ has a unique extension onto $[\mathscr{D}; A]$ according to Problem 5.3-2 and the density of $\mathscr{D}(A)$ in $[\mathscr{D}; A]$.

Problem 5.10-1. Prove that $\mathscr{D}(A)$ is dense in $[\mathscr{E}; A]$. *Hint*: Choose a sequence $\{\theta_j\} \subset \mathscr{D}$ such that $\int \theta_j\, dt = 1$, $\theta_j \geq 0$, and $\operatorname{supp} \theta_j \subset [-j^{-1}, j^{-1}]$.

Show that $\theta_j \to \delta$ in $[\mathscr{E}; C]$, $f * \theta_j \in \mathscr{D}(A)$ for $f \in [\mathscr{E}; A]$, and there exists a compact set K such that $\operatorname{supp} f * \theta_j \subset K$ for every j. Conclude from this that $f * \theta_j \to f$ in $[\mathscr{E}; A]$.

Problem 5.10-2. Prove that $\mathscr{D}(A)$ is dense in both $[\mathscr{D}_-; A]$ and $[\mathscr{D}; A]$.

5.11. CAUSALITY AND CONVOLUTION OPERATORS

As was mentioned in Section 4.6, the concept of causality (see Definition 4.6-1) has physical significance when the signals at hand are defined on the real line. For this reason, we again restrict t and x to $R \triangleq R^1$ and set $\mathscr{D} = \mathscr{D}_{R^1}$. Our objective now is to establish a condition on supp y that characterizes the causality of the convolution operator $y *$.

Theorem 5.11-1. *Let $y \in [\mathscr{D}(A); B]$. The convolution operator $y *$ is causal on $\mathscr{D}(A)$ if and only if* $\operatorname{supp} y \subset [0, \infty)$.

PROOF. Assume $y *$ is causal on $\mathscr{D}(A)$. The regularization formula (Theorem 5.5-1) is

$$(y * v)(t) = \langle y(x), v(t - x)\rangle, \qquad v \in \mathscr{D}(A). \tag{1}$$

If $\operatorname{supp} v \subset (0, \infty)$, then, by causality, this regularization is equal to zero for $t < 0$. But, given any $\phi \in \mathscr{D}(A)$ with $\operatorname{supp} \phi \subset (-\infty, 0)$, we can choose v and t such that $\phi(x) = v(t - x)$. Hence, $\langle y, \phi\rangle = 0$ for every such ϕ, which means that $\operatorname{supp} y \subset [0, \infty)$.

Conversely, assume that $\operatorname{supp} y \subset [0, \infty)$. If $\operatorname{supp} v \subset [T, \infty)$, then $\operatorname{supp} v(t - \cdot) \subset (-\infty, t - T]$. So, $\operatorname{supp} v(t - \cdot)$ does not meet supp y when $t < T$. Hence, $y * v = 0$ on $(-\infty, T)$, and by the linearity of $y *$, this implies the causality of $y *$ on $\mathscr{D}(A)$. ◇

Theorem 5.11-2. *Assume that $\mathscr{I}(A)$, $\mathscr{J}$, and $\mathscr{K}$ are normal ρ-type testing-function spaces satisfying the Conditions E given in Section 5.2. Let $y \in [\mathscr{I}(A); B]$. If $y *$ is causal on $\mathscr{D}(A)$, then $y *$ is causal on $[\mathscr{J}; A]$.*

PROOF. Since $\mathscr{I}(A)$, $\mathscr{J}$, and $\mathscr{K}$ are normal, it follows that $y \in [\mathscr{D}(A); B]$, $v \in [\mathscr{J}; A]$, and $y * v \in [\mathscr{K}; B]$ are distributions. According to the preceding theorem, $\operatorname{supp} y \subset [0, \infty)$. Assume that $\operatorname{supp} v \subset [T, \infty)$. We wish to show that $\operatorname{supp} y * v \subset [T, \infty)$ also. Let $\phi \in \mathscr{K}$ with $\operatorname{supp} \phi \subset (-\infty, T)$. Then,

$$\langle v(x), \phi(\cdot + x)\rangle \in \mathscr{I}(A).$$

Moreover,

$$\operatorname{supp}\langle v(x), \phi(\cdot + x)\rangle \subset (-\infty, 0)$$

because, for any fixed $t > 0$, $\phi(x) \mapsto \phi(t + x)$ is a shift to the left, so that the supports of v and $\phi(t + \cdot)$ do not meet. Consequently,

$$\langle y * v, \phi\rangle = \langle y(t), \langle v(x), \phi(t + x)\rangle\rangle = 0.$$

Since this is true for all ϕ as chosen above, $\operatorname{supp} y * v \subset [T, \infty)$. ◇

Problem 5.11-1. Do the results of this section extend to convolution operators whose signals are given on R^n?

Problem 5.11-2. Construct another proof of Theorem 5.11-1 by deriving it as a special case of Theorem 4.6-1.

Chapter 6

The Laplace Transformation

6.1. INTRODUCTION

Much of realizability theory for a translation-invariant operator consists of criteria imposed upon its system function, which by definition is the Laplace transform of the unit-impulse response of the operator. Consequently, the Laplace transformation on operator-valued distributions is essential to our purposes and is our next topic. The scalar version of the results of this chapter is given by Zemanian (1968a, Chapter 3).

Throughout this chapter, we restrict our attention to testing functions and distributions on the real line R. Thus, $t \in R$ and $\mathscr{D} = \mathscr{D}_{R^1}$.

6.2. THE DEFINITION OF THE LAPLACE TRANSFORMATION

A natural testing-function space to use in discussing the Laplace transformation of $[A; B]$-valued distributions is $\mathscr{L}(w, z; A)$ (see parts IV and V of

Section 3.6 with m set equal to $[\infty]$). We repeat its definition here for the case where its members are defined on R rather than on R^n.

Given any $c, d \in R$, set

$$\kappa_{c,d}(t) = \begin{cases} e^{ct}, & 0 \le t < \infty \\ e^{dt}, & -\infty < t < 0. \end{cases}$$

$\mathscr{L}_{c,d}(A)$ is the linear space of all A-valued smooth functions on R such that

$$\gamma_{c,d,k}(\phi) \triangleq \sup_{t \in R} \|\kappa_{c,d}(t)\phi^{(k)}(t)\| < \infty, \qquad k = 0, 1, \ldots .$$

$\mathscr{L}_{c,d}(A)$ is a Fréchet space under the topology generated by the multinorm $\{\gamma_{c,d,k}\}_{k=0}^{\infty}$.

Now, let $\{c_j\}_{j=1}^{\infty}$ be a strictly decreasing sequence in R tending to w, where either $w \in R$ or $w = -\infty$. Similarly, let $\{d_j\}_{j=1}^{\infty}$ be a strictly increasing sequence in R tending to z, where either $z \in R$ or $z = \infty$. By definition, $\mathscr{L}(w, z; A)$ is the inductive limit of the $\mathscr{L}_{c_j, d_j}(A)$. It is not a strict inductive limit but is, on the other hand, a normal ρ-type testing-function space. As usual, we set $\mathscr{L}(w, z; C) \triangleq \mathscr{L}(w, z)$. For each fixed $\zeta \in C$ and nonnegative integer p, $t \mapsto t^p e^{-\zeta t}$ is a member of $\mathscr{L}(w, z)$ if and only if $w < \operatorname{Re} \zeta < z$.

As was indicated in Section 3.7, $[\mathscr{L}(w, z; A); B]$ is a subspace of $[\mathscr{D}(A); B]$ and is therefore a space of $[A; B]$-valued distributions. According to Theorem 3.7-2, $[\mathscr{L}(w, z; A); B]$ can be identified with $[\mathscr{L}(w, z); [A; B]]$ by means of the bijection $f \mapsto g$ defined by

$$\langle g, \psi \rangle a = \langle f, \psi a \rangle, \tag{1}$$

where $f \in [\mathscr{L}(w, z; A); B]$, $g \in [\mathscr{L}(w, z);\ [A; B]]$, $\psi \in \mathscr{L}(w, z)$, and $a \in A$. This identification is henceforth understood, and we will use the same symbol to denote both f and g.

A $y \in [\mathscr{D}(A); B]$ is said to be *Laplace-transformable* if there exist two elements η_1 and η_2 in the extended real line $[-\infty, \infty]$ such that $\eta_1 < \eta_2$, $y \in [\mathscr{L}(\eta_1, \eta_2; A); B]$, and $y \notin [\mathscr{L}(w, z; A); B]$ if either $w < \eta_1$ or $z > \eta_2$. The open strip

$$\Omega_y \triangleq \{\zeta \in C; \quad \eta_1 < \operatorname{Re} \zeta < \eta_2\}$$

will be called the *strip of definition for the Laplace transform of* y. By the aforementioned identification, $y \in [\mathscr{L}(\eta_1, \eta_2); [A; B]]$ also. Thus, we may define the *Laplace transform* Y *of* y as a mapping of Ω_y into $[A; B]$ by

$$Y(\zeta) \triangleq \langle y(t), e^{-\zeta t} \rangle, \qquad \zeta \in \Omega_y. \tag{2}$$

The *Laplace transformation* $\mathfrak{L}$ is by definition the mapping $y \mapsto Y$. Whenever, we write "$y \in [\mathscr{D}(A); B]$ and $(\mathfrak{L}y)(\zeta) = Y(\zeta)$ for $\zeta \in \Omega_y$," it is understood that y is a Laplace-transformable member of $[\mathscr{D}(A); B]$ and Ω_y is the strip of definition for $\mathfrak{L}y \triangleq Y$.

We may replace A by C and identify $[C; B]$ with B, in which case (2) becomes the definition of the Laplace transform Y of a B-valued distribution and Y becomes a B-valued function on Ω_y.

One further comment should be made here. Our definition of a Laplace-transformable distribution can be simplified somewhat. We could simply say that $g \in [\mathscr{D}(A); B]$ is Laplace-transformable if $f \in [\mathscr{L}(\sigma_1, \sigma_2; A); B]$ for some $\sigma_1, \sigma_2 \in [-\infty, \infty]$ such that $\sigma_1 < \sigma_2$. This is because there will then exist a unique pair $\eta_1, \eta_2 \in [-\infty, \infty]$ with $\eta_1 \leq \sigma_1$ and $\sigma_2 \leq \eta_2$ and a unique extension f of g that satisfies our previous definition for a Laplace-transformable distribution (Zemanian, 1968a, pp. 55–56). It will tacitly be assumed throughout this book that every such g has been replaced by its extension f, and we will on occasion use the simpler definition of a Laplace-transformable distribution.

Problem 6.2-1. Show that, if $u \leq w$ and $z \leq v$, then $\mathscr{L}(w, z)$ is a dense subspace of $\mathscr{L}(u, v)$ and the canonical injection of $\mathscr{L}(w, z)$ into $\mathscr{L}(u, v)$ is continuous. As a result, $[\mathscr{L}(u, v; A); B]$ is a subspace of $[\mathscr{L}(w, z; A); B]$.

Problem 6.2-2. Show that the shifting operator is an automorphism on $[\mathscr{L}(w, z; A); B]$ under either the pointwise topology or the $\mathfrak{S}$-topology.

6.3. ANALYTICITY AND THE EXCHANGE FORMULA

Theorem 6.3-1. *If* $y \in [\mathscr{D}(A); B]$ *and* $(\mathfrak{L}y)(\zeta) = Y(\zeta)$ *for* $\zeta \in \Omega_y$, *then* $Y \triangleq \mathfrak{L}y$ *is an* $[A; B]$*-valued analytic function on* Ω_y, *and, for each nonnegative integer* k,

$$Y^{(k)}(\zeta) = \langle y(t), (-t)^k e^{-\zeta t}\rangle, \qquad \zeta \in \Omega_y. \tag{1}$$

This is proven by induction on k in the usual way. Upon fixing $\zeta \in \Omega_y$, we write

$$\frac{Y^{(k)}(\zeta + \Delta\zeta) - Y^{(k)}(\zeta)}{\Delta\zeta} - \langle y(t), (-t)^{k+1} e^{-\zeta t}\rangle = \langle y(t), \psi_{\Delta\zeta}(t)\rangle$$

and then show that $\psi_{\Delta\zeta}$ tends to zero in $\mathscr{L}(\eta_1, \eta_2)$ as $\Delta\zeta \to 0$.

The Laplace transformation converts convolution into multiplication. This is indicated by Equation (2) in the next theorem, which is called the *exchange formula.*

Theorem 6.3-2. *If* $y \in [\mathscr{D}(A); B]$ *and* $(\mathfrak{L}y)(\zeta) = Y(\zeta)$ *for* $\zeta \in \Omega_y$, *if* $v \in [\mathscr{D}; A]$ *and* $(\mathfrak{L}v)(\zeta) = V(\zeta)$ *for* $\zeta \in \Omega_v$, *and if* $\Omega_y \cap \Omega_v$ *is not empty, then* $y * v$ *exists in* *accordance with Problem* 5.3-3, *where now the open interval* (w, z) *is the intersection of* $\Omega_y \cap \Omega_v$ *with the real axis. Moreover, for each* $\zeta \in \Omega_y \cap \Omega_v$,

$$(\mathfrak{L}y * v)(\zeta) = Y(\zeta)(V(\zeta)). \tag{2}$$

Note. The right-hand side is a B-valued analytic function of $\zeta \in \Omega_y \cap \Omega_v$ because Y is $[A; B]$-valued and V is A-valued.

PROOF. According to Problem 6.2-1, $y \in [\mathscr{L}(w, z; A); B]$ and $v \in [\mathscr{L}(w, z); A]$. Hence, $y * v$ exists in the sense of Problem 5.3-3. Moreover, for each $\zeta \in \Omega_y \cap \Omega_v$, we may write

$$\begin{aligned}(\mathfrak{L}y * v)(\zeta) &= \langle y(t), \langle v(x), e^{-\zeta(t+x)}\rangle\rangle \\ &= \langle y(t), e^{-\zeta t}V(\zeta)\rangle.\end{aligned}$$

Since $V(\zeta) \in A$, we may invoke (1) of Section 6.2 to equate the last expression to

$$\langle y(t), e^{-\zeta t}\rangle V(\zeta) = Y(\zeta)V(\zeta). \quad \diamond$$

Problem 6.3-1. Prove Theorem 6.3-1.

Problem 6.3-2. Again let $y \in [\mathscr{D}(A); B]$ and $(\mathfrak{L}y)(\zeta) = Y(\zeta)$ for $\zeta \in \Omega_y$. Establish the following two operation-transform formulas, where k is a nonnegative integer and σ_τ is the shifting operator:

$$(\mathfrak{L}y^{(k)})(\zeta) = \zeta^k Y(\zeta), \qquad \zeta \in \Omega_y, \tag{3}$$

$$(\mathfrak{L}\sigma_\tau y)(\zeta) = e^{-\zeta\tau} Y(\zeta), \qquad \zeta \in \Omega_y. \tag{4}$$

6.4. INVERSION AND UNIQUENESS

Two $[A; B]$-valued distributions having the same Laplace transforms with identical strips of definition, say $\{\zeta \in C: \eta_1 < \operatorname{Re} \zeta < \eta_2\}$, must be equal as members of $[\mathscr{L}(\eta_1, \eta_2; A); B]$. This is the uniqueness property of the Laplace transformation. It is an almost immediate consequence of the following inversion formula.

Theorem 6.4-1 (Inversion). *If* $y \in [\mathscr{D}(A); B]$ *and* $(\mathfrak{L}y)(\zeta) = Y(\zeta)$ *for* $\zeta \in \Omega_y = \{\zeta \in C: \eta_1 < \operatorname{Re} \zeta < \eta_2\}$, *then, in the sense of convergence in* $[\mathscr{D}(A); B]^\sigma$,

$$y(t) = \lim_{r\to\infty} (1/2\pi) \int_{-r}^{r} Y(\zeta)e^{\zeta t}\, d\omega, \tag{1}$$

where $\omega = \operatorname{Im} \zeta$ *and* $\operatorname{Re} \zeta$ *is fixed such that* $\eta_1 < \operatorname{Re} \zeta < \eta_2$.

The proof of this theorem is the same as that of Theorem 3.5-1 given by Zemanian (1968a). Even though we are now dealing with A-valued testing functions and $[A; B]$-valued distributions, no alteration in the proof is needed except for the replacement of certain magnitude signs by norm symbols.

Theorem 6.4-2 (Uniqueness). *Let y_1 and y_2 be members of $[\mathscr{D}(A); B]$ and let $(\mathfrak{L}y_j)(\zeta) = Y_j(\zeta)$ for $s \in \Omega_{y_j}$ and $j = 1, 2$. Also, assume that $\Omega_{y_1} \cap \Omega_{y_2}$ is not void and that $Y_1 = Y_2$ on $\Omega_{y_1} \cap \Omega_{y_2}$. Then, $y_1 = y_2$ in the sense of equality in $[\mathscr{L}(w, z; A); B]$, where $(w, z) = \Omega_{y_1} \cap \Omega_{y_2} \cap R$.*

PROOF. By Theorem 6.4-1, y_1 coincides with y_2 on $\mathscr{D}(A)$. But $\mathscr{D}(A)$ is dense in $\mathscr{L}(w, z; A)$, and the restrictions of y_1 and y_2 to $\mathscr{L}(w, z; A)$ are continuous. Hence, y_1 and y_2 coincide on $\mathscr{L}(w, z; A)$ as well. ◇

6.5. A CAUSALITY CRITERION

If the unit-impulse response y of a convolution operator happens to be Laplace-transformable, then the causality of $y\, *$ can be characterized by a growth condition on the Laplace transform Y of y.

Theorem 6.5-1. *Necessary and sufficient conditions for Y to be the Laplace transform of a Laplace-transformable distribution $y \in [\mathscr{D}(A); B]$ with* supp *$y \subset [0, \infty)$ are that there be a half-plane $C_\sigma \triangleq \{\zeta \in C\colon \operatorname{Re}\zeta > \sigma\}$ on which Y is an $[A; B]$-valued analytic function and there be a polynomial P for which*

$$\| Y(\zeta)\|_{[A;\, B]} \le P(|\zeta|), \qquad \operatorname{Re}\zeta > \sigma. \tag{1}$$

When this is the case, $y \in [\mathscr{L}(\sigma, \infty; A); B]$ and the strip of definition of $\mathfrak{L}y$ contains the half-plane C_σ.

PROOF. *Necessity*. Since supp $y \subset [0, \infty)$ and y is Laplace-transformable, it follows that $y \in [\mathscr{L}_{\sigma, d}(A); B]$ for some σ and every d. We are free to choose $\sigma > 0$ because, for $\sigma < \eta$, $\mathscr{L}_{\sigma, d}(A) \supset \mathscr{L}_{\eta, d}(A)$. Thus, $Y \triangleq \mathfrak{L}y$ exists on some half-plane bounded on the left and containing C_σ.

Now, let λ be a smooth, real-valued function such that $\lambda(t) = 0$ for $t < -1$ and $\lambda(t) = 1$ for $t > -\frac{1}{2}$. Arbitrarily choose a $\zeta \in C$ such that $\operatorname{Re}\zeta > \sigma > 0$. Then, for every $d > \operatorname{Re}\zeta$, we have that $t \mapsto e^{-\zeta t}$ is a member of $\mathscr{L}_{\sigma, d}(A)$ and $\kappa_{\sigma, d}(t) \le e^{\sigma t}$ for all t. Therefore, there exists a constant $M > 0$ and a non-negative integer r such that

$$\begin{aligned} \| Y(\zeta)\| &= \|\langle y(t), e^{-\zeta t}\rangle\| \\ &\le M \sup_{0 \le k \le r} \sup_{t \ge -1/|\zeta|} |e^{-\sigma t} D^k[\lambda(|\zeta| t) e^{-\zeta t}]|. \end{aligned} \tag{2}$$

Note that the right-hand side is independent of d and that (2) holds whenever $\operatorname{Re} \zeta > \sigma$. Moreover, $e^{(\sigma-\zeta)t}$ remains bounded for all t and ζ such that $\operatorname{Re} \zeta > \sigma > 0$ and $t \geq -1/|\zeta|$. As a result, the right-hand side of (2) is bounded by a polynomial in $|\zeta|$.

Sufficiency. We first establish a lemma concerning classical Laplace transforms.

Lemma 6.5-1. *Assume that, on the half-plane* $C_\sigma \triangleq \{\zeta \in C: \operatorname{Re} \zeta > \sigma\}$, *$G$ is an $[A; B]$-valued analytic function and*

$$\|G(\zeta)\| \leq M|\zeta|^{-2}, \tag{3}$$

where M is a constant. Set

$$g(t) \triangleq (1/2\pi) \int_{-\infty}^{\infty} G(c + i\omega)e^{(c+i\omega)t}\, d\omega, \qquad c > \sigma. \tag{4}$$

Then, g is a continuous $[A; B]$-valued function for all t and is independent of the choice of $c > \sigma$. Moreover, $g(t) = 0$ for $t < 0$, and

$$G(\zeta) = \int_0^{\infty} g(t)e^{-\zeta t}\, dt, \qquad \zeta \in C_\sigma. \tag{5}$$

In fact, G is the Laplace transform of the regular distribution $f \in [\mathscr{L}(\sigma, \infty; A); B]$ generated by g.

PROOF. That g does not depend on the choice of $c > \sigma$ follows from (3) and Cauchy's theorem (Theorem 1.8-1). Its continuity follows from the continuity of $t \mapsto e^{-ct}g(t)$, which in turn can be shown by writing

$$\begin{aligned}\|e^{-c(t+x)}g(t + x) - e^{-ct}g(t)\| &\leq \frac{1}{2\pi} \int_{-\infty}^{\infty} \|G(c + i\omega)\|\, |e^{i\omega t}(e^{i\omega x} - 1)|\, d\omega \\ &\leq \frac{1}{2\pi} \int_{-\infty}^{\infty} \|G(c + i\omega)\| \left| 2 \sin \frac{\omega x}{2} \right| d\omega\end{aligned}$$

and noting that the right-hand side tends to zero as $x \to 0$.

Similarly, it can be seen that $e^{-ct}g(t)$ is bounded on the domain

$$\{\{t, c\}: t \in R, c > \sigma\}.$$

This implies that $g(t) = 0$ for $t < 0$. Indeed, if $g(t) \neq 0$ at some $t < 0$, then $e^{-ct}\|g(t)\|$ can be made arbitrarily large by choosing c large enough.

We now observe that $\omega \mapsto G(c + i\omega)$ is a smooth $[A; B]$-valued function. Hence, by virtue of (3), the standard proof for the Fourier-inversion formula (see, for example, Zemanian, 1965, Section 7.2) can be applied to obtain

(5) from (4). Finally, g generates a regular distribution $f \in [\mathscr{L}(\sigma, \infty; A); B]$ by means of the definition

$$\langle f, \phi \rangle \triangleq \int_{-\infty}^{\infty} g(t)\phi(t)\, dt, \qquad \phi \in \mathscr{L}(\sigma, \infty; A)$$

because, for any $a, b \in R$ with $\sigma < a < b < \infty$, we can choose c with $\sigma < c < a$ and then write, for all $\phi \in \mathscr{L}_{a,b}(A)$,

$$\|\langle f, \phi \rangle\|_B = \left\| \int_0^{\infty} e^{-ct} g(t) e^{(c-a)t} e^{at} \phi(t)\, dt \right\|$$

$$\leq \sup_t \|e^{-ct} g(t)\| \sup_t \|\kappa_{a,b}(t)\phi(t)\| \int_0^{\infty} e^{(c-a)t}\, dt.$$

Thus, $f \in [\mathscr{L}_{a,b}(A); B]$, and therefore $f \in [\mathscr{L}(\sigma, \infty; A); B]$. Hence, (5) is equivalent to $G(\zeta) = (\mathfrak{L}f)(\zeta)$ for $\zeta \in C_\sigma$. Our lemma is established.

The sufficiency part of Theorem 6.5-1 now follows readily. By the condition (1), there exists a positive integer m such that $G(\zeta) \triangleq Y(\zeta)/\zeta^m$ satisfies the hypothesis of the lemma. Therefore, $G = \mathfrak{L}f$, where $f \in [\mathscr{L}(\sigma, \infty; A); B]$. According to Problem 6.3-2, $Y(\zeta) = \zeta^m G(\zeta) = (\mathfrak{L}f^{(m)})(\zeta)$ for $\zeta \in C_\sigma$. We conclude by noting that

$$\operatorname{supp} f^{(m)} \subset \operatorname{supp} f = \operatorname{supp} g \subset [0, \infty). \quad \diamond$$

Since the condition $\operatorname{supp} y \subset [0, \infty)$ is equivalent to the causality of $y\, *$ (Theorem 5.11-1), the last theorem provides the following causality criterion promised in the title of this section.

Corollary 6.5-1a. *Let $y \in [\mathscr{D}(A); B]$ and $Y(\zeta) = (\mathfrak{L}y)(\zeta)$ for $\zeta \in \Omega_y$. The convolution operator $y\, *$ is causal on $\mathscr{D}(A)$ if and only if there exists a half-plane $\{\zeta \in C: \operatorname{Re} \zeta > \sigma\}$ on which Y is analytic and $\|Y(\zeta)\|$ is bounded by a polynomial in $|\zeta|$. In this case, y is causal on $[\mathscr{D}_-; A]$ as well.*

For the last sentence, see Theorem 5.11-2.

Chapter 7

The Scattering Formulism

7.1. INTRODUCTION

So far, we have seen that the linearity and continuity of an operator $\mathfrak{N}$ from $\mathscr{D}(H)$ into $[\mathscr{D}; H]$ is equivalent to a kernel representation for $\mathfrak{N}$ (Theorems 4.5-2 and 4.5-3), that $\mathfrak{N}$ is in addition translation invariant if and only if it has a convolution representation (Theorem 5.10-1), and that the causality of $\mathfrak{N}$ is characterized by a support condition on the kernel or unit-impulse response of these representations (Theorems 4.6-1 and 5.11-1). However, we have not as yet developed any results arising from energy considerations. This is our last objective.

A Hilbert space is the natural framework in which to examine questions concerning energy and power flow, and consequently we will be concerned henceforth with Hilbert ports. The net energy $e(I)$ absorbed by a Hilbert port over some time interval $I \subset R$ is described in two distinct ways, depending on whether the scattering formulism or the admittance formulism is chosen. In the former case,

$$e(I) = \int_I [\|q(t)\|^2 - \|\mathfrak{W}q(t)\|^2]\, dt, \tag{1}$$

where $\mathfrak{W}$ is the scattering operator for the Hilbert port. In the latter case,

$$e(I) = \operatorname{Re} \int_I (\mathfrak{N}v(t), v(t))\, dt, \tag{2}$$

where $\mathfrak{N}$ is the admittance operator. If the Hilbert port has no energy sources within it, it cannot impart more energy to its surroundings during the time interval $I_T \triangleq \{t\colon -\infty < t < T\}$, where $T \leq \infty$, than it has received, and thus, $e(I_T) \geq 0$. This property is called passivity. However, one should (and we will) distinguish between the passivity conditions arising from (1) and (2) because they affect the representations for $\mathfrak{W}$ and $\mathfrak{N}$ in very different ways. Also, the cases where $T = \infty$ and T is finite but arbitrary will also be treated separately, the former one being a weaker assumption than the latter. We will take up the scattering formulism in this chapter and the admittance formulism in the next.

Passivity is a strong assumption. For example, linearity and passivity imply causality and continuity. This is fairly obvious in the scattering formulism (see Section 7.3) but is by no means obvious in the admittance formulism (see Sections 8.2 and 8.3). Similarly, linearity, continuity, time invariance, and causality do not ensure that the unit-impulse response g of the operator at hand is Laplace-transformable. However, g is indeed Laplace-transformable when the assumption of passivity is added. Much of the subsequent discussion will be directed toward the Laplace transform of g and will result in the so-called frequency-domain formulation for our realizability theory.

Throughout our discussion of Hilbert ports, we adopt the natural assumption that the signals at hand are functions or distributions on the real time axis. Thus, in this and the next chapter, $t \in R$ and $\mathscr{D} = \mathscr{D}_{R^1}$.

7.2. PRELIMINARY CONSIDERATIONS CONCERNING L_p-TYPE DISTRIBUTIONS

We start by establishing certain properties of the distributions in $[\mathscr{D}_{L_1}(A); B]$ and $[\mathscr{D}_{L_2}(A); B]$. These two spaces were discussed in Section 3.8.

Lemma 7.2-1. *If* $f \in [\mathscr{D}_{L_1}(A); B]$ *and if* $\operatorname{supp} f \subset [0, \infty)$, *then* $f \in [\mathscr{L}(0, \infty; A); B]$.

Note. According to Lemma 3.8-3 $[\mathscr{D}_{L_2}(A); B] \subset [\mathscr{D}_{L_1}(A); B]$, and hence this lemma also holds for all $f \in [\mathscr{D}_{L_2}(A); B]$.

PROOF. Let $c, d \in R$ with $0 < c < d < \infty$. Let $\phi \in \mathscr{L}_{c,d}(A)$. Finally, let λ be a smooth, real-valued function on R such that $\lambda(t) = 0$ for $-\infty < t < -1$ and $\lambda(t) = 1$ for $-\frac{1}{2} < t < \infty$. Then, $\lambda\phi \in \mathscr{D}_{L_1}(A)$. Indeed

$$\int_{-\infty}^{\infty} \|D^k[\lambda(t)\phi(t)]\|_A \, dt \le \sum_{l=0}^{k} \binom{k}{l} \int_{-1}^{\infty} |\lambda^{(k-l)}(t)| \; \|\phi^{(l)}(t)\|_A \, dt, \tag{1}$$

and the right-hand side is finite because $\|\phi^{(l)}(t)\| \le Ne^{-ct}$ on $-1 < t < \infty$ for some constant N.

As was indicated in Section 3.3, $\langle f, \phi \rangle$ depends only on the values that ϕ assumes on some arbitrarily small neighborhood of supp f. Therefore, we can extend the definition of f onto $\mathscr{L}_{c,d}(A)$ by means of the equation

$$\langle f, \phi \rangle = \langle f, \lambda\phi \rangle, \qquad \phi \in \mathscr{L}_{c,d}(A).$$

Since $f \in [\mathscr{D}_{L_1}(A); B]$, there exists a constant $M > 0$ and a nonnegative integer r such that

$$\|\langle f, \phi \rangle\|_B = \|\langle f, \lambda\phi \rangle\| \le M \max_{0 \le k \le r} \int_{-\infty}^{\infty} \|D^k[\lambda(t)\phi(t)]\|_A \, dt.$$

In view of (1), the right-hand side is bounded by

$$M \max_{0 \le k \le r} \sum_{l=0}^{k} \binom{k}{l} \int_{-1}^{\infty} \frac{|\lambda^{(k-l)}(t)|}{\kappa_{c,d}(t)} \, dt \sup_{t \in R} \|\kappa_{c,d}(t)\phi^{(l)}(t)\|_A < \infty.$$

Hence, $f \in [\mathscr{L}_{c,d}(A); B]$. Since this is so for every c, d such that $0 < c < d < \infty$, we have $f \in [\mathscr{L}(0, \infty; A); B]$. ◇

Lemma 7.2-2. *Let $f \in [\mathscr{D}(A); B]$ and $p \in R$ with $1 < p < \infty$. If*

$$\int_{-\infty}^{\infty} \|(f * \psi)(t)\|_B{}^p \, dt < \infty \tag{2}$$

for every $\psi \in \mathscr{D}(A)$, then $f \in [\mathscr{D}_{L_q}(A); B]$, where $q = p/(p-1)$.

PROOF. Set

$$\beta \triangleq \left\{ \phi \in \mathscr{D}: \int_{-\infty}^{\infty} |\phi(t)|^q \, dt \le 1 \right\}.$$

Let $\psi \in \mathscr{D}(A)$ and $\phi \in \beta$. In the following, $\check{\psi}(t) \triangleq \psi(-t)$. Then, $f * \check{\phi} \in \mathscr{E}([A; B])$ and $f * \psi \in \mathscr{E}(B)$ according to Theorems 3.5-1 and 5.5-1. Moreover,

$$\langle \check{\phi}(x), \check{\psi}(t + x) \rangle = \langle \psi(x), \phi(t + x) \rangle,$$

and so

$$\langle f * \check{\phi}, \check{\psi} \rangle = \langle f * \psi, \phi \rangle \in B.$$

Hence,

$$\|\langle f * \check{\phi}, \check{\psi} \rangle\|_B \leq \int \|f * \psi\|_B |\phi| \, dt$$

$$\leq \left[\int \|f * \psi\|_B{}^p \, dt \right]^{1/p} \left(\int |\phi|^q \, dt \right)^{1/q}.$$

This inequality is Holder's (Appendix G20). Because of (2), the left-hand side traverses a bounded set of real numbers when ϕ traverses β and ψ and f are held fixed. Thus, $f * \check{\phi}$ traverses a bounded set in $[\mathscr{D}(A); B]^\sigma$.

Now, let K and N be any two compact intervals in R such that $K \subset \mathring{N}$. By Lemma 3.7-1, there exists a constant $M > 0$ and an integer $m \geq 0$ such that, for all $\lambda \in \mathscr{D}_N(A)$,

$$\sup_{\phi \in \beta} \|\langle f * \check{\phi}, \lambda \rangle\|_B < M \max_{0 \leq k \leq m} \sup_{t \in R} \|\lambda^{(k)}(t)\|_A. \tag{3}$$

Also, by Lemma 3.2-2, for any $\theta \in \mathscr{D}_K{}^m(A)$, we can find a sequence $\{\lambda_j\}_{j=1}^\infty \subset \mathscr{D}_N(A)$ such that $\lambda_j \to \theta$ in $\mathscr{D}_N{}^m(A)$. This fact and (3) imply that $f * \check{\phi}$ has a unique extension as a member of the space $[\mathscr{D}_K{}^m(A); B]$. (See Lemma 3.4-5.) This extension, which we also denote by $f * \check{\phi}$, satisfies (3) with λ now allowed to be any member of $\mathscr{D}_K{}^m(A)$. Therefore, as ϕ traverses $\beta, f * \check{\phi}$ traverses a bounded set in $[\mathscr{D}_K{}^m(A); B]^\sigma$. Moreover, for any $\theta \in \mathscr{D}_K{}^m(A)$, we may write

$$\langle f * \theta, \phi \rangle = \langle f * \check{\phi}, \check{\theta} \rangle,$$

where the usual definition of distributional convolution is used. Thus, $\langle f * \theta, \phi \rangle$ traverses a bounded set in B as ϕ traverses β.

Next, set $\theta = \chi a$, where $\chi \in \mathscr{D}_K{}^m$ and $a \in A$. We obtain

$$\langle f * \theta, \phi \rangle = \langle f * \chi, \phi \rangle a,$$

where in the right-hand side, $f \in [\mathscr{D}; [A; B]]$ and $\langle f * \chi, \phi \rangle \in [A; B]$. Since the left-hand side traverses a bounded set in B as ϕ traverses β, we have, from the principle of uniform boundedness (Appendix D12), that $\langle f * \chi, \phi \rangle$ traverses a bounded set in $[A; B]$. Hence $f * \chi$ is a continuous linear mapping of $\mathscr{D}$, supplied with the topology induced by $\mathscr{D}_{L_q}$, into $[A; B]$. [See Appendices C8 and D2(iii).] But $\mathscr{D}$ is dense in $\mathscr{D}_{L_q}$, and therefore $f * \chi$ has a unique extension as a continuous linear mapping of $\mathscr{D}_{L_q}$ into $[A; B]$. Upon denoting this extension by $f * \chi$, we obtain $f * \chi \in [\mathscr{D}_{L_q}; [A; B]]$ for every $\chi \in \mathscr{D}_K{}^m$.

Finally, let $\gamma \in \mathscr{D}$ be such that $\gamma = 1$ on a neighborhood of 0 in R. Let K be a compact interval in R such that $\operatorname{supp} \gamma \subset K$. Set $n = m + 2$, where m is the integer corresponding to $N \supset \mathring{K}$ as above. Set

$$J_m(t) = \frac{t^{m+1}}{(m+1)!} 1_+(t).$$

Then, $\delta = D^n(\gamma J_m) + \xi$, where $\xi \in \mathscr{D}_K$. Thus,

$$f = f * \delta = f * D^n(\gamma J_m) + f * \xi = D^n(f * \gamma J_m) + f * \xi.$$

According to the preceding paragraph, both $f * \gamma J_m$ and $f * \xi$ are members of $[\mathscr{D}_{L_q}; [A; B]]$. But then, so too is f, because $[\mathscr{D}_{L_q}; [A; B]]$ is closed under D^n. By Theorem 3.8-1, $f \in [\mathscr{D}_{L_q}(A); B]$. ◇

It is worth mentioning here that the converse to Lemma 7.2-2 is not true, as can be seen through the following counterexample due to L. Schwartz. Let $p = 2$, $A = C$, and $B = L_2$. Also, let f be the identity operator on L_2. Then, clearly, the restriction of f to $\mathscr{D}_{L_2}$ is a member of $[\mathscr{D}_{L_2}; L_2]$. But, for any $\psi \in \mathscr{D}$,

$$(f * \psi)(t) = \langle f(x), \psi(t - x)\rangle = \psi(t - \cdot).$$

The right-hand side is a mapping of the real line into L_2. However, it does not satisfy (2) because $\|\psi(t - \cdot)\|_{L_2} = \|\psi\|_{L_2}$, so that the integrand of (2) is in this case a constant with respect to t.

This counterexample has still another implication. It is a fact that any $f \in [\mathscr{D}_{L_2}; C]$ has a representation as a finite sum of derivatives of functions in L_2; i.e.,

$$f = \sum_{k=1}^{m} h_k^{(k)}, \qquad h_k \in L_2.$$

(See Schwartz, 1966, p. 201.) This is no longer true for every $f \in [\mathscr{D}_{L_2}; B]$, where B is an arbitrary Banach space. Indeed, if it were true, we could choose $B = L_2$ as above and then write, for any $\psi \in \mathscr{D}$,

$$f * \psi = \sum h_k^{(k)} * \psi = \sum h_k * \psi^{(k)},$$

where $h_k \in L_2(B)$. It can be shown that $h_k * \psi^{(k)} \in L_2(B)$ so that $f * \psi \in L_2(B)$. But we have already noted in the preceding paragraph that $f * \psi \notin L_2(B) = L_2(L_2)$ for a properly chosen f.

Problem 7.2-1. Let $h \in L_2(B)$, where B is a Banach space, and let $\psi \in \mathscr{D}$. Show that $h * \psi \in \mathscr{D}_{L_2}(B)$.

7.3. SCATTER-PASSIVITY

Let q be the input signal on a Hilbert port under the scattering formulism. If q is an ordinary function at some instant of time t, then $\|q(t)\|_H{}^2$ is the instantaneous power injected by the input signal into the Hilbert port.

Similarly, if $\mathfrak{W}$ is the scattering operator and if $\mathfrak{W}q$ is also an ordinary function at the instant t, then $\|(\mathfrak{W}q)(t)\|_H{}^2$ is the instantaneous power carried out of the Hilbert port by the output signal. Therefore, the net power absorbed by the Hilbert port at time t is

$$\|q(t)\|^2 - \|(\mathfrak{W}q)(t)\|^2. \tag{1}$$

In general, neither q nor $\mathfrak{W}q$ need be an ordinary function at t and indeed they may be singular distributions throughout some neighborhood of t. In this case, (1) will not possess a sense as the net power absorption. Nevertheless, our intent is to allow distributional inputs and outputs and at the same time make use of a passivity assumption. This is accomplished by first assuming that $\mathfrak{W}$ is passive on a domain of ordinary functions [namely $\mathscr{D}(H)$], then developing certain representations for $\mathfrak{W}$, and finally extending $\mathfrak{W}$ onto wider domains by means of those representations. (In this regard, see also the discussion at the beginning of Section 4.6.)

***Definition** 7.3-1.* Let $\mathfrak{W}$ be an operator whose domain contains a set $\mathscr{X} \subset L_2(H)$. $\mathfrak{W}$ is said to be *scatter-semipassive on* $\mathscr{X}$ or, alternatively, *contractive from* $\mathscr{X}$ *into* $L_2(H)$ if, for all $q \in \mathscr{X}$ and for $r \triangleq \mathfrak{W}q$, we have that $r \in L_2(H)$ and

$$\int_{-\infty}^{\infty} [\|q(t)\|^2 - \|r(t)\|^2]\, dt \geq 0. \tag{2}$$

If, in addition, $\mathscr{X} = L_2(H)$, then we simply say that $\mathfrak{W}$ is *contractive on* $L_2(H)$.

A stronger condition than scatter-semipassivity is stated by the next definition, where the following notation is used. Given any $T \in R$, we define the function 1_T on R by $1_T(t) = 1$ for $t \leq T$ and $1_T(t) = 0$ for $t > T$. If h is any function on R, the function $t \mapsto 1_T(t)h(t)$ is denoted by $1_T h$.

***Definition** 7.3-2.* Let $\mathfrak{W}$ be an operator whose domain contains a set $\mathscr{Y}$ with the property that $1_T q \in L_2(H)$ for every $q \in \mathscr{Y}$ and every $T \in R$. $\mathfrak{W}$ is said to be *scatter-passive on* $\mathscr{Y}$ if, for all $q \in \mathscr{Y}$ and all $T \in R$ and for $r \triangleq \mathfrak{W}q$, we have that $1_T r \in L_2(H)$ and

$$\int_{-\infty}^{T} [\|q(t)\|^2 - \|r(t)\|^2]\, dt \geq 0. \tag{3}$$

If $\mathfrak{W}$ is scatter-passive on $\mathscr{X} \subset L_2(H)$, then it is also scatter-semipassive on $\mathscr{X}$, as can be seen by letting $T \to \infty$. However, the converse is not true in general. For example, a pure predictor defined by $(\mathfrak{W}q)(t) \triangleq q(t + x)$, where $x > 0$, is scatter-semipassive but not scatter-passive on $\mathscr{D}(H)$.

Lemma 7.3-1. *Let $\mathfrak{W}$ be a linear scatter-semipassive operator on $\mathscr{D}(H)$. Then, $\mathfrak{W}$ is continuous from $\mathscr{D}(H)$ into $L_2(H)$ and therefore into $[\mathscr{D}; H]^\sigma$ as well. Moreover, $\mathfrak{W}$ has a unique linear contractive extension onto $L_2(H)$.*

PROOF. The inequality (2) states in effect that, when $\mathscr{D}(H)$ is equipped with the topology induced by $L_2(H)$, $\mathfrak{W}$ is continuous from $\mathscr{D}(H)$ into $L_2(H)$. But, the canonical injections of $\mathscr{D}(H)$ into $L_2(H)$ and of $L_2(H)$ into $[\mathscr{D}; H]^\sigma$ are both continuous, and this implies the first sentence. The second sentence follows from Appendix D5 and the fact that $\mathscr{D}(H)$ is dense in $L_2(H)$. ◇

Not only does the scatter-passivity of a linear operator imply continuity, it also implies causality. In fact, we have the following result originally pointed out by Wohlers and Beltrami (1965).

Theorem 7.3-1. *Let $\mathfrak{W}$ be a linear operator on $\mathscr{D}(H)$. Then, $\mathfrak{W}$ is causal and scatter-semipassive on $\mathscr{D}(H)$ if and only if $\mathfrak{W}$ is scatter-passive on $\mathscr{D}(H)$.*

PROOF. Let $\mathfrak{W}$ be scatter-passive on $\mathscr{D}(H)$ and, as before, set $r \triangleq \mathfrak{W}q$, where $q \in \mathscr{D}(H)$. We have already noted that (2) can be obtained from (3) by taking $T \to \infty$, and therefore $\mathfrak{W}$ is scatter-semipassive.

Next, assume that $\mathfrak{W}$ is not causal. This means that, for some $T \in R$, we have $q(t) = 0$ for $-\infty < t < T$ and $r(t) \neq 0$ for all t in some set of positive Lebesgue measure contained in $(-\infty, T)$. Then, (3) cannot hold, and this contradicts the scatter-passivity of $\mathfrak{W}$. Hence, $\mathfrak{W}$ must be causal on $\mathscr{D}(H)$.

Conversely, assume that $\mathfrak{W}$ is causal and scatter-semipassive on $\mathscr{D}(H)$. Given any $T \in R$, choose an $X \in R$ such that $T < X$. Let $\theta \in \mathscr{E}$ be real-valued and such that $\theta(t) = 1$ for $-\infty < t \leq 0$, $\theta(t) = 0$ for $1 \leq t < \infty$, and θ is monotonic decreasing for $0 < t < 1$. Set

$$\lambda(t) \triangleq \lambda_{T,X}(t) \triangleq \theta\left(\frac{t-T}{X-T}\right).$$

Given $q \in \mathscr{D}(H)$, set $g = \mathfrak{W}(\lambda q)$ and $r = \mathfrak{W}q$. By scatter-semipassivity, both g and r are members of $L_2(H)$ and by causality, $g = r$ almost everywhere on $(-\infty, T)$. Thus, we may write

$$\int_{-\infty}^{T} \|r\|^2\, dt \leq \int_{-\infty}^{\infty} \|g\|^2\, dt \leq \int_{-\infty}^{T} \|q\|^2\, dt + \int_{T}^{X} \|\lambda q\|^2\, dt.$$

The second inequality is due to the scatter-semipassivity of $\mathfrak{W}$. By choosing X sufficiently close to T, $\int_T^X \|\lambda q\|^2\, dt$ can be made arbitrarily small because

$$\int_{T}^{X} \|\lambda q\|^2\, dt \leq \int_{T}^{X} \|q\|^2\, dt \to 0, \qquad X \to T.$$

This establishes (3) and thereby the scatter-passivity of $\mathfrak{W}$. ◇

Theorem 7.3-1 shows that the following two assumptions on an operator $\mathfrak{W}$ are equivalent:

(i) $\mathfrak{W}$ is linear, causal, and scatter-semipassive on $\mathscr{D}(H)$.
(ii) $\mathfrak{W}$ is linear and scatter-passive on $\mathscr{D}(H)$.

Under a postulational approach to realizability theory, (i) may be considered a better form for a hypothesis because causality and scatter-semipassivity are independent conditions (see Wohlers and Beltrami, 1965 or Zemanian, 1968b). However, statement (ii) recommends itself by virtue of its conciseness. Later on, the reader should bear in mind that causality is a consequence of (ii).

7.4. BOUNDED* SCATTERING TRANSFORMS

Definition 7.4-1. A function S of the complex variable ζ is said to be a *bounded* mapping of H into H* (or simply *bounded**) if, on the half-plane $C_+ \triangleq \{\zeta : \operatorname{Re} \zeta > 0\}$, S is an $[H; H]$-valued analytic function such that $\|S(\zeta)\|_{[H;H]} \leq 1$.

Our aim in this section is to show that the Laplace transform of the unit-impulse response of a linear translation-invariant scatter-passive operator on $\mathscr{D}(H)$ is bounded*. The converse assertion (namely, every bounded* mapping is such a Laplace transform) will be established in the next section.

Theorem 7.4-1. *If the operator $\mathfrak{W}$ is linear, translation-invariant, and scatter-semipassive on $\mathscr{D}(H)$, then $\mathfrak{W} = s\, *$ on $\mathscr{D}(H)$, where $s \in [\mathscr{D}_{L_2}(H); H]$.*

PROOF. By Lemma 7.3-1, $\mathfrak{W}$ is continuous from $\mathscr{D}(H)$ into $[\mathscr{D}; H]^\sigma$. Therefore, $\mathfrak{W} = s\, *$, where $s \in [\mathscr{D}(H); H]$, according to Theorem 5.10-1. Let $\|\cdot\|_{L_2}$ denote the norm for $L_2(H)$. The scatter-semipassivity of $\mathfrak{W}$ implies that, for all $\psi \in \mathscr{D}(H)$,

$$\int_{-\infty}^{\infty} \|(s * \psi)(t)\|_H{}^2 \, dt = \|\mathfrak{W}\psi\|_{L_2}^2 \leq \|\psi\|_{L_2}^2 < \infty.$$

Lemma 7.2-2 now shows that $s \in [\mathscr{D}_{L_2}(H); H]$. ◇

Theorem 7.4-2. *If the operator $\mathfrak{W}$ is linear, translation-invariant, and scatter-passive on $\mathscr{D}(H)$, then its unit-impulse response s possesses a Laplace transform whose strip of definition contains the half-plane $C_+ \triangleq \{\zeta : \operatorname{Re} \zeta > 0\}$.*

PROOF. By Theorems 7.3-1 and 7.4-1, $\mathfrak{W} = s\,*$, where $s \in [\mathscr{D}_{L_2}(H); H]$, and $\mathfrak{W}$ is causal. Hence, $\operatorname{supp} s \subset [0, \infty)$ according to Theorem 5.11-1. We now invoke Lemma 7.2-1 to conclude that $s \in [\mathscr{L}(0, \infty; A); B]$ so that $\mathfrak{L}s$ exists and has a strip of definition containing C_+. ◇

Lemma 7.4-1. *Let $a, b \in R$ be such that $b < 0 < a$. Assume that $s \in [\mathscr{D}_{L_2}(H); H]$ and $\operatorname{supp} s \subset [0, \infty)$. Then, for each $\phi \in \mathscr{L}_{a,b}(H)$, $s * \phi$ exists and is a smooth H-valued function. Moreover, there exists a constant $L > 0$ and an integer $l \geq 0$ such that*

$$\|(s * \phi)(t)\| \leq Le^{-bt} \max_{0 \leq k \leq l} \sup_{t \in R} \|e^{bt}\phi^{(k)}(t)\|. \tag{1}$$

PROOF. Since $[\mathscr{D}_{L_2}(H); H] \subset [\mathscr{D}_{L_1}(H); H]$, we have from Theorem 5.5-2 that $s * \phi$ is a smooth H-valued function and

$$(s * \phi)(t) = \langle s(x), \phi(t - x)\rangle.$$

Let $\lambda \in \mathscr{E}$ be such that $\lambda(x) = 1$ for $-\frac{1}{2} < x < \infty$ and $\lambda(x) = 0$ for $-\infty < x < -1$. Let $B_p \triangleq \sup_x |\lambda^{(p)}(x)|$. Then, there exists a constant $M > 0$ and an integer $l \geq 0$ such that

$$\begin{aligned}
\|(s * \phi)(t)\| &= \|\langle s(x), \lambda(x)\phi(t - x)\rangle\| \\
&\leq M \max_{0 \leq k \leq l} \int_{-1}^{\infty} \|D_x^{\,k}[\lambda(x)\phi(t - x)]\|\, dx \\
&\leq M \max_{0 \leq k \leq l} \sum_{p=0}^{k} \binom{k}{p} B_{k-p} \int_{-1}^{\infty} e^{-b(t-x)} e^{b(t-x)} \|\phi^{(p)}(t - x)\|\, dx \\
&\leq M \max_{0 \leq k \leq l} \sum_{p=0}^{k} \binom{k}{p} B_{k-p} e^{-bt} \int_{-1}^{\infty} e^{bx}\, dx \sup_{t \in R} \|e^{bt}\phi^{(p)}(t)\|.
\end{aligned}$$

This implies (1). ◇

Lemma 7.4-2. *Let $\mathfrak{W}$ be a linear translation-invariant scatter-passive operator on $\mathscr{D}(H)$ and let $q \in \mathscr{L}_{c,d}(H)$, where $d < 0 < c$. Then, for $r \triangleq \mathfrak{W}q$ and for every $T \in R$, we have that*

$$\int_{-\infty}^{T} [\|q(t)\|^2 - \|r(t)\|^2]\, dt \geq 0. \tag{2}$$

PROOF. Let $a, b \in R$ be such that $d < b < 0 < a < c$. Given any $q \in \mathscr{L}_{c,d}(H)$, we can find a sequence $\{\phi_j\}_{j=1}^{\infty} \subset \mathscr{D}(H)$ that tends to q in $\mathscr{L}_{a,b}(H)$. Indeed, let $\lambda \in \mathscr{D}$ be such that $\lambda(t) = 1$ for $|t| < 1$ and $\lambda(t) = 0$ for $|t| > 2$. We may write

$$\kappa_{a,b}(t)D^k\left[\lambda\left(\frac{t}{j}\right)q(t) - q(t)\right] = \sum_{p=0}^{k} \binom{k}{p} \left\{D^{k-p}\left[\lambda\left(\frac{t}{j}\right) - 1\right]\right\} \kappa_{a,b}(t)D^p q(t). \tag{3}$$

Now, $D^{k-p}[\lambda(t/j) - 1]$ is equal to zero for $|t| \leq j$; and for $|t| > j$, it is bounded by a constant that does not depend on j. Moreover, for each p,

$$\sup_{|t|>j} \|\kappa_{a,b}(t)D^p q(t)\| \leq \sup_{t\in R}\|\kappa_{c,d}(t)D^p q(t)\| \sup_{|t|>j} \frac{\kappa_{a,b}(t)}{\kappa_{c,d}(t)} \to 0 \tag{4}$$

as $j \to \infty$. Consequently, the right-hand side of (3) converges uniformly to 0 on R. Thus, upon setting $\phi_j(t) \triangleq \lambda(t/j)q(t)$, we obtain the sequence we seek.

Next, note that $\mathfrak{W} = s\,*$, where s satisfies the hypothesis of Lemma 7.4-1 according to Theorems 5.11-1, 7.3-1, and 7.4-1. Set $\psi_j \triangleq \mathfrak{W}\phi_j = s * \phi_j$. Since $\operatorname{supp} s \subset [0, \infty)$, $\psi_j \in \mathscr{D}_+(H)$. Also, set $r \triangleq \mathfrak{W}q = s * q$. By Theorem 5.5-2, $r \in \mathscr{E}(H)$. So, $\|q\|^2$, $\|r\|^2$, $\|\phi_j\|^2$, and $\|\psi_j\|^2$ are all continuous functions on R. Furthermore, we may invoke Lemma 7.4-1 and the fact that $|e^{bt}/\kappa_{a,b}(t)| \leq 1$ to write

$$e^{bt}\|r(t) - \psi_j(t)\| \leq L \max_{0\leq k\leq l} \sup_{t\in R}\|e^{bt}D^k[q(t) - \phi_j(t)]\| \to 0, \qquad j \to \infty. \tag{5}$$

Now, given any $T \in R$, consider

$$\left|\int_{-\infty}^{T} [\|q\|^2 - \|r\|^2]\,dt - \int_{-\infty}^{T} [\|\phi_j\|^2 - \|\psi_j\|^2]\,dt\right|$$

$$\leq \int_{-\infty}^{T} \|q\|\,\|q - \phi_j\|\,dt + \int_{-\infty}^{T} \|q - \phi_j\|\,\|\phi_j\|\,dt$$

$$+ \int_{-\infty}^{T} \|r\|\,\|r - \psi_j\|\,dt + \int_{-\infty}^{T} \|r - \psi_j\|\,\|\psi_j\|\,dt. \tag{6}$$

Since $b < 0 < a$ and since $\phi_j \to q$ in $\mathscr{L}_{a,b}(H)$, there exists a constant K not depending on j such that $\|q(t)\| \leq Ke^{-bt}$ and $\|\phi_j(t)\| \leq Ke^{-bt}$. Also,

$$\sup_{t\in R} e^{bt}\|q(t) - \phi_j(t)\| \to 0, \qquad j \to \infty.$$

These facts imply that the first and second integrals on the right-hand side of (6) tend to zero as $j \to \infty$. Similarly, by virtue of Lemma 7.4-1 and (5), $\|r(t)\|$, $\|\psi_j(t)\|$, and $e^{bt}\|r(t) - \psi_j(t)\|$ satisfy similar conditions, and, as a result, the third and fourth integrals on the right-hand side of (6) tend to zero as $j \to \infty$. Condition (3) now follows from the scatter-passivity of $\mathfrak{W}$ on $\mathscr{D}(H)$ and the consequent nonnegativity of the second integral on the left-hand side of (6). ◇

The main theorem of this section is the following.

Theorem 7.4-3. *If $\mathfrak{W}$ is a linear translation-invariant scatter-passive operator on $\mathscr{D}(H)$, then $\mathfrak{W}$ has a convolution representation $\mathfrak{W} = s\,*$, where $s \in [\mathscr{D}_{L_2}(H); H]$ and* $\operatorname{supp} s \subset [0, \infty)$. *Moreover, the Laplace transform S of s exists with a strip of definition containing $C_+ \triangleq \{\zeta \in C\colon \operatorname{Re}\zeta > 0\}$ and is bounded*.*

PROOF. We have already established everything except for the condition $\|S(\zeta)\|_{[H;H]} \le 1$ for all $\zeta \in C_+$. (See Theorems 5.11-1, 7.3-1, 7.4-1, and 7.4-2.) Let $a \in H$ and $T \in R$. Choose $\tau \in R$ such that $\tau > T$. Also, let $\lambda \in \mathscr{E}$ be such that $\lambda(t) = 1$ for $-\infty < t < \tau$ and $\lambda(t) = 0$ for $\tau + 1 < t < \infty$. For any $\zeta \in C_+$, set $q(t) = ae^{\zeta t}\lambda(t)$. Thus, with $-\operatorname{Re}\zeta = d < 0$, we have that $q \in \mathscr{L}_{c,d}(H)$ for every $c \in R$. So, we may use Lemma 7.4-2 and (2) for $r = \mathfrak{W}q$.

We now invoke Theorem 5.5-2 to write

$$r(t) = (s * q)(t) = \langle s(x), q(t-x)\rangle.$$

For any fixed $t < T, q(t-x) = ae^{\zeta(t-x)}$ when x is restricted to a sufficiently small neighborhood of $[0, \infty)$. But $\operatorname{supp} s \subset [0, \infty)$, and therefore

$$r(t) = \langle s(x), ae^{\zeta(t-x)}\rangle = e^{\zeta t}S(\zeta)a.$$

Thus,

$$0 \le \int_{-\infty}^{T} [\|q\|^2 - \|r\|^2]\, dt = [\|a\|^2 - \|S(\zeta)a\|^2] \int_{-\infty}^{T} e^{2\operatorname{Re}\zeta t}\, dt.$$

Since this holds for all $a \in H$ and since the integral on the right-hand side is positive, this implies that $\|S(\zeta)\|_{[H;H]} \le 1$ for any $\zeta \in C_+$. ◇

7.5. THE REALIZABILITY OF BOUNDED* SCATTERING TRANSFORMS

We will now show that every bounded* mapping of H into H is the Laplace transform of the unit-impulse response of a linear translation-invariant scatter-passive operator on $\mathscr{D}(H)$. We start with two lemmas.

Lemma 7.5-1. *Let $\phi \in \mathscr{D}(H)$ with $\operatorname{supp}\phi \subset [0, \infty)$. Then, $\Phi \triangleq \mathfrak{L}\phi$ is an analytic H-valued function everywhere on C such that, for each nonnegative integer k, there exists a positive constant C_k for which*

$$\|\Phi(\zeta)\|_H \le C_k|\zeta|^{-k}, \qquad \operatorname{Re}\zeta \ge 0. \tag{1}$$

PROOF. ϕ generates a regular distribution in $[\mathscr{L}(-\infty, \infty); H]$, and consequently

$$\Phi(\zeta) = \int_0^\infty \phi(t)e^{-\zeta t}\, dt, \qquad \zeta \in C.$$

Through successive integrations by parts, this becomes

$$\Phi(\zeta) = \int_0^\infty \phi^{(k)}(t)\zeta^{-k}e^{-\zeta t}\, dt,$$

and (1) follows by estimating the last integral. ◇

Lemma 7.5-2. *Let g be a continuous H-valued function on R such that $\|g(\cdot)\|_H \in L_1$. Also, let $a \in H$. Then,*

$$\int_{-\infty}^{\infty} (a, g(x))\, dx = \left(a, \int_{-\infty}^{\infty} g(x)\, dx\right).$$

This lemma is an immediate consequence of Appendix D15 and Note II of Section 1.4, which continues to hold for improper integrals.

The following is the realizability theorem for bounded* scattering transforms.

Theorem 7.5-1. *Corresponding to each bounded* mapping S of H into H, there exists a unique convolution operator $\mathfrak{W} = s *$ on $[\mathscr{L}(0, \infty); H]$ such that $\mathfrak{L}s = S$ on C_+. Moreover, $s \in [\mathscr{D}_{L_2}(H); H]$ and $\operatorname{supp} s \subset [0, \infty)$. Furthermore, $\mathfrak{W}$ is a linear translation-invariant scatter-passive operator on $\mathscr{D}(H)$.*

PROOF. We may invoke Theorem 6.5-1 to conclude that S is the Laplace transform of an $s \in [\mathscr{L}(0, \infty; H); H]$, that the strip of definition for $\mathfrak{L}s$ contains C_+, and that $\operatorname{supp} s \subset [0, \infty)$. Hence, the convolution operator $\mathfrak{W} \triangleq s *$, which by the uniqueness theorem of the Laplace transform is uniquely determined by S, is a linear translation-invariant causal mapping on $\mathscr{D}(H)$. Moreover, the domain of $s *$ contains $[\mathscr{L}(0, \infty); H]$ according to Problem 5.3-3. It remains for us to prove that $s \in [\mathscr{D}_{L_2}(H); H]$ and that $s *$ is scatter-passive on $\mathscr{D}(H)$.

Let $\phi \in \mathscr{D}(H)$ with $\operatorname{supp} \phi \subset [0, \infty)$. Then, $s * \phi$ is Laplace-transformable, and by Theorem 6.3-2,

$$[\mathfrak{L}(s * f)](\zeta) = S(\zeta)F(\zeta), \qquad \zeta \in C_+.$$

Moreover, for all $\zeta \in C_+$ and for $\omega \triangleq \operatorname{Im} \zeta$,

$$\|S(\zeta)F(\zeta)\|_H \leq \|\Phi(\zeta)\|_H \leq Q(\omega) \triangleq \min(C_0, C_2 \omega^{-2}), \tag{2}$$

where C_0 and C_2 are the constants indicated in (1).

Set $h = s * \phi$. Then, $h \in \mathscr{E}(H)$, and $\operatorname{supp} h \subset [0, \infty)$. We now invoke Lemma 6.5-1 and the uniqueness theorem of the Laplace transformation (Theorem 6.4-2) to obtain the following result. For any fixed $\sigma \triangleq \operatorname{Re} \zeta > 0$ and for each fixed $t \in R$,

$$h(t)e^{-\sigma t} = (1/2\pi) \int_{-\infty}^{\infty} S(\sigma + i\omega)\Phi(\sigma + i\omega)e^{i\omega t}\, d\omega.$$

In view of (2),

$$\|h(t)e^{-\sigma t}\| \leq (1/2\pi) \int_{-}^{\infty} Q(\omega)\, d\omega < \infty.$$

Since this holds for all $\sigma > 0$ and since supp $h \subset [0, \infty)$, it follows that, for any fixed $\sigma > 0$, $\|h(t)e^{-\sigma t}\| \in L_1$.

Now, set $H(\zeta) \triangleq S(\zeta)\Phi(\zeta)$ for $\zeta \in C_+$ and consider

$$\int_{-\infty}^{\infty} \|h(t)e^{-\sigma t}\|^2 \, dt = \int_{-\infty}^{\infty} \left(h(t)e^{-\sigma t}, (1/2\pi) \int_{-\infty}^{\infty} H(\sigma + i\omega)e^{i\omega t} \, d\omega \right) dt \quad (3)$$

where again $\sigma > 0$. By a judicious use of Lemma 7.5-2 and Fubini's theorem, we may first bring the integration on ω outside the inner product in the right-hand side of (3), then reverse the order of integration on ω and t, and finally bring the integration on t inside the inner product to obtain Parseval's equation:

$$\int_{-\infty}^{\infty} \|h(t)e^{-\sigma t}\|^2 \, dt = (1/2\pi) \int_{-\infty}^{\infty} \|H(\sigma + i\omega)\|^2 \, d\omega. \quad (4)$$

Since $\|S(\zeta)\| \leq 1$ for $\zeta \in C_+$, the right-hand side of (4) is bounded by

$$(1/2\pi) \int_{-\infty}^{\infty} \|\Phi(\sigma + i\omega)\|^2 \, d\omega. \quad (5)$$

By (2) and Lesbesgue's theorem of dominated convergence, (5) tends to

$$(1/2\pi) \int_{-\infty}^{\infty} \|\Phi(i\omega)\|^2 \, d\omega. \quad (6)$$

On the other hand, as $\sigma \to 0+$, $\|h(t)e^{-\sigma t}\|^2$ increases monotonically to the limit $\|h(t)\|^2$ at each t. So, by the theorem of B. Levi (Williamson, 1962, p. 62), the left-hand side of (4) tends to $\int_{-\infty}^{\infty} \|h(t)\|^2 \, dt$. We conclude that

$$\int_{-\infty}^{\infty} \|(s * \phi)(t)\|^2 \, dt \leq (1/2\pi) \int_{-\infty}^{\infty} \|\Phi(i\omega)\|^2 \, d\omega < \infty \quad (7)$$

for all $\phi \in \mathscr{D}(H)$ with supp $\phi \subset [0, \infty)$. But, since $s\,*$ commutes with the shifting operator σ_τ,

$$\int_{-\infty}^{\infty} \|(s * \phi)(t)\|^2 \, dt < \infty$$

for all $\phi \in \mathscr{D}(H)$. It now follows from Lemma 7.2-2 that $s \in [\mathscr{D}_{L_2}(H); H]$.

To show that $s\,*$ is scatter-passive on $\mathscr{D}(H)$, we first observe that the middle term in (7) is equal to $\int_{-\infty}^{\infty} \|\phi(t)\|^2 \, dt$. [This is established just as was (4).] Thus, $s\,*$ is scatter-semipassive on every $\phi \in \mathscr{D}(H)$ with supp $\phi \subset [0, \infty)$. But the same is true for all $\phi \in \mathscr{D}(H)$ because again $s\,*$ commutes with σ_τ. The causality of $s\,*$ in conjunction with Theorem 7.3-1 establishes the scatter-passivity of $s\,*$. ◇

Theorems 7.4-3 and 7-5.1 summarize the basic realizability theory under the scattering formulism for arbitrary bounded* mappings of H into H.

Actually, the hypothesis of Theorem 7.4-3 may be weakened somewhat by merely assuming that $\mathfrak{W}$ is linear, translation-invariant, and scatter-passive on the dense subset $\mathscr{D} \odot H$ of $\mathscr{D}(H)$ (see Zemanian, 1970b). Another modification is to impose the passivity assumption on the admittance form of the energy integral and make use of the concepts of augmentation and solvability. This approach was devised by Youla *et al.* (1959) for the scalar case; see also Newcomb (1966). That this procedure can be extended to the Hilbert-port setting and is entirely equivalent to the method discussed in this chapter is shown by Zemanian (1970b, Section 7). An extension of the present theory to Hilbert ports that need be neither translation-invariant nor causal is given by Zemanian (1972b).

Problem 7.5-1. Let H and J be complex Hilbert spaces and let S and T be $[H; J]$-valued bounded analytic functions on C_+. Assume that, for almost all ω and as $\sigma \to 0+$, $S(\sigma + i\omega)$ and $T(\sigma + i\omega)$ converge in the strong operator topology to $S(i\omega)$ and $T(i\omega)$, respectively. Show that, if $S(i\omega) = T(i\omega)$ for almost all ω, then $S = T$ on C_+.

7.6. BOUNDED*-REAL SCATTERING TRANSFORMS

A property of physical systems which we have not as yet considered is reality. That is, they map real signals into real signals. The meaning of a real signal in the context of Hilbert-space-valued functions has yet to be explained, and this is our first objective. The reality of a physical system is reflected in certain symmetries in its scattering operator and scattering transform, and an exposition of this fact is our second objective.

We start with a real Hilbert space H_r and generate a complex Hilbert space H by *complexification*. That is, the elements of H are defined to be $a_1 + ia_2$, where a_1 and a_2 are arbitrary members of H_r. Addition is defined as

$$(a_1 + ia_2) + (b_1 + ib_2) \triangleq (a_1 + b_1) + i(a_2 + b_2),$$

and, for any complex number $\alpha = \alpha_1 + i\alpha_2$, where $\alpha_1, \alpha_2 \in R$, multiplication by α is defined by

$$\alpha(a_1 + ia_2) \triangleq \alpha_1 a_1 - \alpha_2 a_2 + i\alpha_1 a_2 + i\alpha_2 a_1.$$

H_r becomes a subset of H when we identify $a_1 + i0 \in H$ with the element $a_1 \in H_r$. Furthermore, the inner product $(\cdot, \cdot)$ on H_r is extended onto H through the definition

$$(a_1 + ia_2, b_1 + ib_2) \triangleq (a_1, b_1) + (a_2, b_2) + i(a_2, b_1) - i(a_1, b_2), \tag{1}$$

and as a result H becomes a complex Hilbert space. Throughout this section, it is understood that H is the complexification of H_r.

A linear mapping of H into H is called *real* if it maps H_r into H_r. An arbitrary linear mapping Z of H into H has a unique decomposition $Z = Z_1 + iZ_2$, where Z_1 and Z_2 are real linear mappings. Z_1 is called the *real part of* Z, and Z_2 the *imaginary part of* Z. Given Z, we can obtain Z_1 and Z_2 as follows. Z maps any $a_1 \in H_r$ into an element $b_1 + ib_2$ of H. We set $Z_1 a_1 \triangleq b_1$ and $Z_2 a_1 \triangleq b_2$. This defines Z_1 as a linear mapping of H_r into H_r, and it is extended onto H linearly:

$$Z_1(a_1 + ia_2) \triangleq Z_1 a_1 + iZ_1 a_2, \qquad a_1, a_2 \in H_r.$$

We proceed similarly for Z_2. For no other decomposition of Z into the form $Z = F_1 + iF_2$ will both F_1 and F_2 be real mappings. The *complex conjugate of* Z is by definition $\bar{Z} \triangleq Z_1 - iZ_2$. Clearly, Z is a real member of $[H; H]$ if and only if $Z \in [H_r; H_r]$.

We now make a sweeping assertion. All the results we have so far developed in this book can be obtained in the context of real spaces. For example, instead of the basic testing-function space $\mathscr{D}(H)$, we can use the space $\mathscr{D}(H_r)$ of smooth H_r-valued functions on R with compact supports. The members of $[\mathscr{D}(H_r); H_r]$ are *real distributions*, and $f *$ with $f \in [\mathscr{D}(H_r); H_r]$ is a *real convolution operator*. Similarly, a *real signal* on a Hilbert port is an element of $[\mathscr{D}(R); H_r]$. Any $f \in [\mathscr{D}(H_r); H_r]$ has a natural extension onto $\mathscr{D}(H)$ defined by $\langle f, \phi \rangle = \langle f, \phi_1 \rangle + i\langle f, \phi_2 \rangle$, where $\phi = \phi_1 + i\phi_2$ and $\phi_1, \phi_2 \in \mathscr{D}(H_r)$. Upon denoting this extension also by f, we have $f \in [\mathscr{D}(H); H]$. In this way, any one of our standard spaces of real distributions is a subset of the corresponding space of distributions. For example, $[\mathscr{D}_{L_2}(H_r); H_r] \subset [\mathscr{D}_{L_2}(H); H]$.

Reality also has a meaning in the context of bounded* functions as follows.

Definition *7.6-1.* A function S of the complex variable ζ is called *bounded*-real* if S is bounded* and, for every σ on the real positive axis (i.e., for $\sigma \in R$ and $\sigma > 0$), the restriction of $S(\sigma)$ to H_r is a member of $[H_r, H_r]$.

The next theorem is actually a corollary to Theorem 7.4-3.

Theorem 7.6-1. *If $\mathfrak{W}$ is a linear translation-invariant scatter-passive operator on $\mathscr{D}(H)$ and if $\mathfrak{W}$ maps $\mathscr{D}(H_r)$ into $[\mathscr{D}(R); H_r]$, then $\mathfrak{W} = s *$, where $s \in [\mathscr{D}_{L_2}(H_r); H_r]$, and $S \triangleq \mathfrak{L}s$ is bounded*-real.*

PROOF. By Theorem 7.4-3, $\mathfrak{W} = s *$, where $s \in [\mathscr{D}_{L_2}(H); H]$, and S is bounded*. Choose a sequence $\{\psi_j\}_{j=0}^{\infty} \subset \mathscr{D}(R)$ which tends to δ in $[\mathscr{E}(R); R]$. Then, for any $a \in H_r$ and any $\phi \in \mathscr{D}(R)$, $\langle \mathfrak{W}(\psi_j a), \phi \rangle \in H_r$. But

$$\langle \mathfrak{W}(\psi_j a), \phi \rangle = \big\langle s(t), \langle \psi_j(x)a, \phi(t+x) \rangle \big\rangle \to \langle s, \phi a \rangle.$$

Therefore, $\langle s, \phi a \rangle \in H_r$. But $\mathscr{D}(R) \odot H_r$ is total in $\mathscr{D}(H_r)$, and so $s \in [\mathscr{D}(H_r); H_r]$. Hence, by the density of $\mathscr{D}(H_r)$ in $\mathscr{D}_{L_2}(H_r)$, we have $s \in [\mathscr{D}_{L_2}(H_r); H_r]$.

It now follows as in the complex case that $s \in [\mathscr{L}(0, \infty); [H_r; H_r]]$ since $\operatorname{supp} s \subset [0, \infty)$ according to Theorem 7.4-3. Thus, for any $\sigma > 0$, $S(\sigma) = \langle s(t), e^{-\sigma t}\rangle \in [H_r; H_r]$. ◇

As a corollary to Theorem 7.5-1, we have the following result.

Theorem 7.6-2. *If S is bounded*-real, then, in addition to the conclusion of Theorem* 7.5-1, *we have that* $s \in [\mathscr{D}_{L_2}(H_r); H_r]$ *and* $s\,*$ *maps* $\mathscr{D}(H_r)$ *into* $[\mathscr{D}(R); H_r]$.

PROOF. We know from Theorem 7.5-1 that $s \in [\mathscr{D}_{L_2}(H); H]$. The conclusion $s \in [\mathscr{D}_{L_2}(H_r); H_r]$ will follow as in the preceding proof once we show that s maps $\mathscr{D}(H_r)$ into H_r. This will also imply that $s\,*$ maps $\mathscr{D}(H_r)$ into $[\mathscr{D}(R); H_r]$.

Let $\phi \in \mathscr{D}(H_r)$. According to the inversion formula for the Laplace transformation (Theorem 6.4-1),

$$\langle s, \phi\rangle = \lim_{w\to\infty} \left\langle (1/2\pi)\int_{-w}^{w} S(\zeta)e^{\zeta t}\, d\omega, \phi(t)\right\rangle,$$

where $\operatorname{Re}\zeta > 0$ and $\omega = \operatorname{Im}\zeta$. We have to show that $\langle s, \phi\rangle \in H_r$. But this will be established when we prove that

$$\int_{-w}^{w} S(\zeta)e^{\zeta t}\, d\omega \in [H_r; H_r] \tag{2}$$

for each $w > 0$ and t.

Let $a, b \in H_r$. Then, $(S(\sigma)a, b)$ is real for σ real and positive. By the reflection principle and the decomposition $S(\zeta) = S_1(\zeta) + iS_2(\zeta)$, where S_1, $S_2 \in [H_r; H_r]$, we have $(S_1(\bar{\zeta})a, b) = (S_1(\zeta)a, b)$ and $(S_2(\bar{\zeta})a, b) = -(S_2(\zeta)a, b)$ for $\zeta \in C_+$. Since this is so for every $a, b \in H_r$, $S_1(\sigma + i\omega)$ and $S_2(\sigma + i\omega)$ are respectively even and odd functions of ω. We now obtain (2) by noting that the imaginary part of the integral there is the zero member of $[H_r; H_r]$. ◇

Problem 7.6-1. Verify that the complexification of a real Hilbert space is a complex Hilbert space.

Problem 7.6-2. Assume that $\mathfrak{W}$ is a linear translation-invariant scatter-passive operator on $\mathscr{D}(H_r)$ with range in $[\mathscr{D}(R); H_r]$. Show that $\mathfrak{W}$ has a unique extension as a linear translation-invariant scatter-passive operator on $\mathscr{D}(H)$.

7.7. LOSSLESS HILBERT PORTS

There are physical systems with the property that their net energy absorptions for all time from signals of compact support are always zero. Any electrical network consisting exclusively of a finite number of inductors and capacitors has this property. Passive systems of this sort are usually called lossless. In regard to Hilbert ports, we shall use the following definition.

Definition 7.7-1. Let $\mathfrak{W}$ be an operator whose domain contains a set $\mathscr{X} \subset L_2(H)$. $\mathfrak{W}$ is said to be *lossless on* $\mathscr{X}$ if $\mathfrak{W}$ is scatter-passive on $\mathscr{X}$ and, for every $q \in \mathscr{X}$ and $r \triangleq \mathfrak{W}q$,

$$\int_{-\infty}^{\infty} [\|q(t)\|^2 - \|r(t)\|^2]\, dt = 0. \tag{1}$$

When $\mathscr{X}$ supplied with the topology induced by $L_2(H)$ is a normed linear space and when $\mathfrak{W}$ is linear, condition (1) can be restated by saying that $\mathfrak{W}$ is *isometric from* $\mathscr{X}$ *into* $L_2(H)$. It should be borne in mind, however, that losslessness requires in addition that $\mathfrak{W}$ be scatter-passive on $\mathscr{D}(H)$. The pure predictor, defined by $\phi \mapsto \sigma_\tau \phi$, where $\tau < 0$, is isometric from $\mathscr{D}(H)$ into $L_2(H)$ but not scatter-passive on $\mathscr{D}(H)$ and hence not lossless on $\mathscr{D}(H)$ according to our definition. It should also be noted that a linear operator $\mathfrak{W}$ is lossless on $\mathscr{D}(H)$ if and only if (1) is satisfied and $\mathfrak{W}$ is causal on $\mathscr{D}(H)$; this is a consequence of Theorem 7.3-1.

When the Hilbert space H is separable, a lossless convolution operator can be characterized by the fact that its scattering transform $S(\sigma + i\omega)$ is bounded* and, as $\sigma \to 0+$, $S(\sigma + i\omega)$ converges for almost all ω to an operator $S(i\omega)$ such that $\|S(i\omega)a\| = \|a\|$ for all $a \in H$. Any linear operator $T \in [H; H]$ having this property (namely $\|Ta\| = \|a\|$ for all $a \in H$) is said to be *isometric on* H. As our first result along these lines, we have the following assertion.

Theorem 7.7-1. *Let H be a (not necessarily separable) complex Hilbert space. Assume that S is an $[H; H]$-valued analytic function on C_+ such that, for all $\zeta \in C_+$, $\|S(\zeta)\| \le P(|\zeta|)$, where P is a polynomial. Also, assume that, as $\sigma \to 0+$ and for almost all ω, $S(\sigma + i\omega) \to S(i\omega)$ in the strong operator topology, where $S(i\omega)$ is a linear isometric operator on H. Set $s = \mathfrak{L}^{-1}S$* (i.e., *s is the unique member of $[\mathscr{D}(H); H]$ whose Laplace transform coincides with S on C_+). Then, $s\,*$ is lossless on $\mathscr{D}(H)$, and therefore S is bounded*.*

Proof. We first observe that, by Theorem 6.5-1, $s \in [\mathscr{D}(H); H]$ and supp $s \subset [0, \infty)$. Next, we let $\phi \in \mathscr{D}(H)$ with supp $\phi \subset [0, \infty)$ and let $\Phi(\zeta) = \int \phi(t) e^{-\zeta t}\, dt$ as before. Φ is analytic on C. By the estimate of Lemma 7.5-1, we have, for $0 < \sigma < 1$,

$$\|S(\sigma + i\omega)\Phi(\sigma + i\omega)\|^2 \le Q(\omega),$$

where $Q \in L_1$. Moreover, for any fixed ω where $S(\sigma + i\omega)$ converges strongly as $\sigma \to 0+$, we may write

$$\begin{aligned}\|S(\sigma + i\omega)\Phi(\sigma + i\omega) - S(i\omega)\Phi(i\omega)\| &\le \|S(\sigma + i\omega)\|\, \|\Phi(\sigma + i\omega) - \Phi(i\omega)\| \\ &\quad + \|[S(\sigma + i\omega) - S(i\omega)]\Phi(i\omega)\|.\end{aligned}$$

By the principle of uniform boundedness, $\|S(\sigma + i\omega)\|$ is bounded for $0 < \sigma < 1$. We can conclude that the left-hand side tends to zero as $\sigma \to 0+$. Thus, by Lebesgue's theorem of dominated convergence and Parseval's equation,

$$\lim_{\sigma \to 0+} \frac{1}{2\pi} \int_{-\infty}^{\infty} \|S(\sigma + i\omega)\Phi(\sigma + i\omega)\|^2 \, d\omega = \frac{1}{2\pi} \int \|S(i\omega)\Phi(i\omega)\|^2 \, d\omega$$

$$= \frac{1}{2\pi} \int \|\Phi(i\omega)\|^2 \, d\omega = \int \|\phi(t)\|^2 \, dt.$$

Moreover, Parseval's equation [see (4) of Section 7.5] and the theorem of B. Levi show that the left-hand side is equal to

$$\lim_{\sigma \to 0+} \int \|(s * \phi)(t) e^{-\sigma t}\|^2 \, dt = \int \|(s * \phi)(t)\|^2 \, dt.$$

Thus, $s\,*$ satisfies (1) for every $q = \phi \in \mathscr{D}(H)$ with $\operatorname{supp} \phi \subset [0, \infty)$. The translation invariance of $s\,*$ shows that $s\,*$ satisfies (1) for all $q \in \mathscr{D}(H)$. Since $\operatorname{supp} s \subset [0, \infty)$, $s\,*$ is causal and therefore lossless on $\mathscr{D}(H)$. Theorem 7.4-3 now implies that S is bounded*. ◇

Note. A small modification of the foregoing proof shows that, if we weaken the hypothesis of Theorem 7.7-1 by replacing the assumption that $S(i\omega)$ is a linear isometric operator on H by the condition $\|S(i\omega\| \leq 1$ for almost all ω, then we can still conclude that S is bounded*.

The next theorem is a sort of converse to Theorem 7.7-1. Its proof makes use of the following known fact. Let H and J be separable complex Hilbert spaces. If F is an $[\mathrm{H}; J]$-valued function that is analytic and bounded on C_+, then there is a set $\Omega \subset R$ whose complement $R \backslash \Omega$ has measure zero such that, for each $\omega \in \Omega$ and as $\sigma \to 0+$, $F(\sigma + i\omega)$ converges in the strong operator topology. This is proven by Sz.-Nagy and Foias (1970, pp. 185–187) for $[H; J]$-valued bounded analytic functions on the unit disc $\{z \in C\colon |z| < 1\}$ and strong convergence along radial lines. The present version can be obtained by mapping the unit disc into C_+ through $\zeta = (1 + z)(1 - z)^{-1}$. The radial lines become circles centered on the imaginary axis and passing through $\zeta = 1$. That strong convergence for $G(\zeta) \triangleq F(z)$ also occurs along horizontal lines in the ζ plane follows from the inequality

$$\|G(\sigma + i\omega) - G(\sigma + i\eta)\| \leq 4(|\omega - \eta|/\sigma) \sup_{\zeta \in C_+} \|G(\zeta)\|, \qquad 0 < \sigma < \tfrac{1}{2} \tag{2}$$

which is a result of Cauchy's integral formula.

Theorem 7.7-2. *Let H be a separable complex Hilbert space. Assume that the convolution operator $s\,*$, where $s \in [\mathscr{D}(H); H]$, is lossless on $\mathscr{D}(H)$. Then,*

$S \triangleq \mathfrak{L}s$ is bounded, and, for almost all ω and as $\sigma \to 0+$, $S(\sigma + i\omega)$ converges in the strong operator topology to a linear isometric operator $S(i\omega)$ on H.*

PROOF. By the definition of losslessness, $s\,*$ is scatter-passive on $\mathscr{D}(H)$, and consequently S is bounded* according to Theorem 7.4-3 and the linearity and translation invariance of convolution operators. We again invoke Parseval's equation to write, for any $\phi \in \mathscr{D}(H)$ with supp $\phi \subset [0, \infty)$,

$$\int \|(s * \phi)(t)e^{-\sigma t}\|^2 \, dt = (1/2\pi) \int \|S(\sigma + i\omega)\Phi(\sigma + i\omega)\|^2 \, d\omega. \tag{3}$$

As before, the left-hand side tends to $\int \|(s * \phi)(t)\|^2 \, dt$, which, by losslessness and Parseval's equation, is equal to

$$(1/2\pi) \int \|\Phi(i\omega)\|^2 \, d\omega.$$

On the other hand, since $\|S(\zeta)\| \leq 1$ for $\zeta \in C_+$, there exists a set $\Omega \subset R$ with $R \backslash \Omega$ of measure zero such that, as $\sigma \to 0+$, $S(\sigma + i\omega) \to S(i\omega) \in [H; H]$ strongly for every $\omega \in \Omega$. It follows readily that $\|S(i\omega)\| \leq 1$ for all $\omega \in \Omega$. We can now employ the principle of uniform boundedness and Lebesgue's theorem of dominated convergence as in the preceding proof to conclude that the right-hand side of (3) converges as $\sigma \to 0+$. Altogether, then,

$$\int \|\Phi(i\omega)\|^2 \, d\omega = \int \|S(i\omega)\Phi(i\omega)\|^2 \, d\omega. \tag{4}$$

But $(\mathfrak{L}\sigma_\tau \phi)(i\omega) = e^{-i\omega\tau}\Phi(i\omega)$, and consequently (4) holds for all $\phi \in \mathscr{D}(H)$ and not merely for those ϕ with supp $\phi \subset [0, \infty)$.

We shall use (4) to show that $\|S(i\omega)a\| = \|a\|$ for every $a \in H$ and almost all ω, thereby completing the proof. To this end, let $f_{c,d}(\omega) = 1$ for $c < \omega < d$ and $f_{c,d}(\omega) = 0$ otherwise. We use the classical fact that the Fourier transformation $\mathfrak{F}$ is an automorphism on L_2. Choose a sequence $\{\theta_j\}_{j=1}^{\infty} \subset \mathscr{D}$ that converges in L_2 to the inverse Fourier transform of $f_{c,d}$. We have that $\Theta_j(i\omega) \triangleq (\mathfrak{L}\theta_j)(i\omega) = (\mathfrak{F}\theta_j)(\omega)$. Thus, $\Theta_j(i\,\cdot)a \to f_{c,d}(\cdot)a$ in $L_2(H)$. Observe that

$$\int \|S(i\omega)[f_{c,d}(\omega)a - \Theta_j(i\omega)a]\|^2 \, d\omega \leq \int \|f_{c,d}(\omega)a - \Theta_j(i\omega)a\|^2 \, d\omega \to 0.$$

Thus, upon setting $\Phi(i\omega) = \Theta_j(i\omega)a$ in (4) and taking the limit, we get

$$\int_c^d \|a\|^2 \, d\omega = \int_c^d \|S(i\omega)a\|^2 \, d\omega.$$

Since this holds for every c and d, we have $\|a\| = \|S(i\omega)a\|$ for all a and almost all ω. ◇

A *unitary operator on* H is a linear isometric operator that maps H onto H. When H is finite-dimensional, every linear isometric operator on H is unitary.

However, this need not be the case when H is infinite-dimensional. This is reflected in the fact that, when $s*$ is lossless on $\mathscr{D}(H)$, $S(i\omega)$ is isometric almost everywhere but not necessarily unitary. The following is an illustration of this.

Example 7.7-1. Let $\{e_j\}_{j=1}^{\infty}$ be an orthonormal basis in the separable, infinite-dimensional Hilbert space H. Every $a \in H$ has the unique expansion $a = \sum \alpha_j e_j$, where $\alpha_j \in C$. Moreover, $\|a\|^2 = \sum \|\alpha_j\|^2$. Define an operator F on a by

$$Fa = \sum_{j=1}^{\infty} \alpha_j e_{j+1}.$$

F is linear and isometric on H but not unitary. We now define an operator $\mathfrak{W}$ on $\mathscr{D}(H)$ by $(\mathfrak{W}\phi)(t) \triangleq F[\phi(t)]$, where $\phi \in \mathscr{D}(H)$. Clearly, $\mathfrak{W} = s*$, where $s = F\delta$. Therefore, $S(\zeta) \triangleq (\mathfrak{L}s)(\zeta) = F$ for every ζ. Thus, $S(i\omega)$ is isometric but not unitary for every ω. ◇

Problem 7.7-1. Derive (2) and prove the assertion about the strong convergence of $G(\zeta)$ along horizontal lines being a consequence of strong convergence along certain circles.

7.8. THE LOSSLESS HILBERT n-PORT

The Hilbert n-port can be viewed as a mathematical model for a system having n places of access through which energy can be injected or extracted. A cavity resonator having n waveguides connected to it is such a system. We define the Hilbert n-port as follows.

Let H_j, where $j = 1, \ldots, n$, be complex Hilbert spaces with the inner products $(\cdot, \cdot)_j$ and norms $\|\cdot\|_j$. We shall drop the subscription j on the inner product and norm notations when it is clear which Hilbert space H_j is meant. Let $H \triangleq H_1 \times \cdots \times H_n$ be the Cartesian product of the H_j and assign to H the product topology. H is also a Hilbert space whose inner product $(\cdot, \cdot)$ is defined on any $a = \{a_j\}$ and $b = \{b_j\}$ in H by

$$(a, b) \triangleq \sum_{j=1}^{n} (a_j, b_j).$$

Thus, the norm for H satisfies

$$\|a\|^2 = \sum_{j=1}^{n} \|a_j\|^2.$$

Every $F \in [H; H]$ has a unique $n \times n$ matrix representation $[F_{kj}]$, where $F_{kj} \in [H_j; H_k]$, and conversely every such $n \times n$ matrix defines a member of

$[H; H]$ when the customary rules for matrix multiplication are followed. Upon denoting the adjoint of an operator with a prime, we have $F' = [F_{kj}]' = [F'_{jk}]$; that is, we obtain the adjoint of $[F_{kj}]$ by interchanging the rows and columns and then taking the adjoint of each element.

Moreover, if $F(\cdot)$ is an $[H; H]$-valued analytic function on some open set $\Omega \subset C$, then we may write, for any $a, b \in H$,

$$(F(\cdot)a, b) = \sum_k \sum_j (F_{kj}(\cdot)a_j, b_k)_k.$$

Since the weak analyticity of an operator-valued function is equivalent to its analyticity (Theorem 1.7-1), $F(\cdot)$ is analytic on Ω if and only if $F_{kj}(\cdot)$ is analytic on Ω for every k and j.

With $H = H_1 \times \cdots \times H_n$ given, we define a *Hilbert n-port* as a Banach system that relates two H-valued distributions q and r that are complementary in the following sense. Whenever q and r are ordinary functions, $\|q(t)\|^2 - \|r(t)\|^2$ is the instantaneous net real power entering the Hilbert n-port at the instant t. We obtain the ordinary n-port of electrical network theory upon setting each $H_j = C$. The mapping $q \mapsto r$ is the *scattering operator* $\mathfrak{W}$. According to our prior results, $\mathfrak{W}$ is a linear translation-invariant scatter-passive operator on $\mathscr{D}(H)$ if and only if $\mathfrak{W} = s\,*$, where $S \triangleq \mathfrak{L}s$ is bounded*. Upon setting $v = q + r$ and $u = q - r$, we define the *admittance operator* $\mathfrak{N}$ as the mapping $v \mapsto u$.

Let H and J be complex Hilbert spaces and let T be an $[H; J]$-valued function on some set Ξ. We shall say that T *is bounded by* 1 *on* Ξ if $\|T(\zeta)\| \leq 1$ for all $\zeta \in \Xi$. The matrix representation $[S_{kj}]$ of any bounded* mapping S of H into H has the following property: Every S_{kj} is bounded by 1 on C_+, and therefore each S_{kk} is a bounded* mapping of H_k into H_k. To demonstrate this, we write

$$0 \leq \|S(\zeta)a\|^2 = \sum_k \Big\|\sum_j S_{kj}(\zeta)a_j\Big\|^2 \leq \|a\|^2 = \sum_j \|a_j\|^2.$$

Upon setting all $a_j = 0$ except when $j = l$, we obtain

$$\sum_k \|S_{kl}(\zeta)a_l\|^2 \leq \|a_l\|^2,$$

which implies that $\|S_{kl}(\zeta)\| \leq 1$.

Our purpose in this section is to examine the properties of lossless Hilbert n-ports that have some additional special properties. The results to be given here are taken from the thesis of D'Amato (1971). They are similar to certain known facts concerning the ordinary lossless n-port (Carlin and Giordano, 1964), but there are also some notable differences.

Throughout the rest of this section, we shall assume that the complex Hilbert spaces H, J, and H_0 are separable. Let T be an $[H; J]$-valued bounded analytic function on C_+. We have already noted in Section 7.7 that, as $\sigma \to 0+$ and for almost all ω, $T(\sigma + i\omega)$ tends to a limit $T(i\omega)$ in the strong operator topology. In the following, whenever we write $T(i\omega)$, it is understood that $T(i\omega)$ is the strong limit of $T(\sigma + i\omega)$ as $\sigma \to 0+$ for those ω at which the limit exists. As was indicated in Problem 7.5-1, the boundary values $T(i\omega)$ uniquely determine T on C_+.

We shall call a scattering transform S *lossless* if $s\,*$ is lossless on $\mathscr{D}(H)$, where $s \in [\mathscr{D}(H); H]$ and $S = \mathfrak{L}s$. By Theorems 7.7-1 and 7.7-2, S is lossless if and only if S is bounded* and $S(i\omega)$ is isometric for almost all ω. The isometry of $S(i\omega)$ is equivalent to the condition $S'(i\omega)S(i\omega) = I$, where I is the identity operator on H and $S'(i\omega) \triangleq [S(i\omega)]'$. In terms of the matrix representation $S = [S_{kj}]$, this condition reads $[S'_{jk}][S_{kj}] = [I_{kj}]$, where $I_{kj} = 0$ if $k \neq j$ and $I_{jj} \triangleq I_j$ is the identity operator on H_j. Upon expansion, we get

$$S'_{1k}(i\omega)S_{1j}(i\omega) + \cdots + S'_{nk}(i\omega)S_{nj}(i\omega) = 0, \qquad k \neq j \tag{1}$$

and

$$S'_{1j}(i\omega)S_{1j}(i\omega) + \cdots + S'_{nj}(i\omega)S_{nj}(i\omega) = I_j \tag{2}$$

for almost all ω. Thus, $[S_{kj}]$ is lossless if and only if it is bounded* and satisfies (1) and (2) for all k and j.

The scattering transform of a Hilbert n-port is called *reciprocal* if its matrix representation $[S_{kj}]$ satisfies $S_{kj} = S_{jk}$ for all k and j. The physical significance of this is that the transmission between any two ports does not depend upon the direction of the transmission. Many systems possess this property. It is implicit in the definition of reciprocality that the Hilbert spaces H_j comprising $H = H_1 \times \cdots \times H_n$ are all the same; that is, $H_j = H_0$ for some Hilbert space H_0.

If $S = [S_{kj}]$ is a reciprocal lossless scattering transform of a Hilbert 2-port, then (2) yields

$$S'_{11}(i\omega)S_{11}(i\omega) + S'_{21}(i\omega)S_{21}(i\omega) = I_0$$

and

$$S'_{21}(i\omega)S_{21}(i\omega) + S'_{22}(i\omega)S_{22}(i\omega) = I_0,$$

where I_0 is the identity operator on H_0. Consequently,

$$S'_{11}(i\omega)S_{11}(i\omega) = S'_{22}(i\omega)S_{22}(i\omega) \tag{3}$$

or equivalently $\|S_{11}(i\omega)a\| = \|S_{22}(i\omega)a\|$ for all $a \in H_0$ and almost all ω.

To get a physical interpretation of (3), let $\phi = \{\phi_1, 0\} \in \mathscr{D}(H_0 \times H_0)$ and expand the lossless condition

$$\int \|\phi(t)\|^2\, dt = \int \|(s * \phi)(t)\|^2\, dt,$$

where $s = [s_{kj}]$ and $s_{12} = s_{21}$, through the usual matrix manipulations. We obtain

$$\int \|\phi_1(t)\|^2 \, dt = \int \|(s_{11} * \phi_1)(t)\|^2 \, dt + \int \|(s_{21} * \phi_1)(t)\|^2 \, dt.$$

As we have seen in the preceding section, this is equivalent to

$$\int \|\Phi_1(i\omega)\|^2 \, d\omega = \int \|S_{11}(i\omega)\Phi_1(i\omega)\|^2 \, d\omega + \int \|S_{21}(i\omega)\Phi_1(i\omega)\|^2 \, d\omega.$$

The left-hand side represents the energy impressed upon the first port for all time. The first term on the right-hand side is the reflected energy at that port, and the last term is the energy transmitted from the first port to the second port, again for all time. A similar equation holds for the second port, and so we conclude that (3) has the following significance. Both ports of a Hilbert 2-port, whose scattering transform is reciprocal and lossless, behave the same way so far as total energy reflection is concerned. This is a classical result for ordinary 2-ports (Carlin and Giordano, 1964, p. 276).

Another (albeit less common) property of Hilbert n-ports is the following. The scattering transform $[S_{kj}]$ of a Hilbert n-port is said to be *matched at the ith port* if $S_{jj} = 0$. This means that the jth port does not reflect any energy. As an immediate consequence of our foregoing equations, we have the following extension of another classical result. If the scattering transform $[S_{kj}]$ of a Hilbert 2-port is reciprocal, lossless, and matched at the first port, then it is also matched at the second port, and in addition S_{21} is lossless. For ordinary 2-ports, reciprocality is not needed in order for $S_{11} = 0$ to imply $S_{22} = 0$ (Carlin and Giordano, 1964, p. 379). However, when H_0 is infinite-dimensional, reciprocality is needed, as is shown by the following example.

Example 7.8-1. We shall construct a lossless scattering transform $[S_{kj}]$ for a Hilbert 2-port such that $S_{11} = 0$, $S_{22} \neq 0$, and $S_{12} \neq S_{21}$. Let $H = H_0 \times H_0$ and let $\{e_i\}_{i=1}^{\infty}$ be an orthonormal basis for H_0. An arbitrary $a \in H_0$ has the unique expansion

$$a = \sum_{i=1}^{\infty} a_i e_i, \qquad a_i \in C.$$

We define S_{12}, S_{21}, and S_{22} as constant operators not depending on ζ, through the following equations:

$$S_{12} a \triangleq \sum_{i=2}^{\infty} a_i e_i,$$

$$S_{21} a \triangleq \sum_{i=1}^{\infty} a_i e_{i+1},$$

$$S_{22} a = a_1 e_1.$$

It follows readily that $S'_{12} = S_{12}$, $S'_{12} = S_{22}$, whereas

$$S'_{21}a = \sum_{i=1}^{\infty} a_{i+1}e_i.$$

Consequently, $S'_{21}S_{21} = I_0$, $S'_{12}S_{12} + S'_{22}S_{22} = I_0$, and $S'_{22}S_{21} = S'_{21}S_{22} = 0$. Thus, (1) and (2) are satisfied, and, for

$$S = \begin{bmatrix} 0 & S_{12} \\ S_{21} & S_{22} \end{bmatrix},$$

we have that $S'S = I$, where I is the identity operator on H. Thus, S is isometric, and $\|S\| = 1$. Since S does not depend on ζ, we conclude that S is bounded* and therefore lossless. ◇

We now turn to the Hilbert 3-port. Another standard result is that there does not exist an ordinary 3-port whose scattering transform is reciprocal, lossless, and matched at all three ports (Carlin and Giordano, 1964, p. 278). This restriction is no longer in force when $H = H_0 \times H_0 \times H_0$ and H_0 is infinite-dimensional. However, we can say the following. If S is a reciprocal lossless scattering transform for a Hilbert 3-port and is matched at all three ports, then H_0 is infinite-dimensional and the following conditions are satisfied for almost all ω:

$$S'_{12}(i\omega)S_{12}(i\omega) = S'_{13}(i\omega)S_{13}(i\omega) = S'_{23}(i\omega)S_{23}(i\omega) = \tfrac{1}{2}I_0, \tag{4}$$

$$S'_{12}(i\omega)S_{13}(i\omega) = S'_{12}(i\omega)S_{23}(i\omega) = S'_{13}(i\omega)S_{23}(i\omega) = 0. \tag{5}$$

These equations are obtained by substituting the conditions $S_{11} = S_{22} = S_{23} = 0$ and $I_j = I_0$ in (1) and (2) and solving. To show that H_0 is infinite-dimensional, we first observe from (4) that both $\sqrt{2}\,S_{12}(i\omega)$ and $\sqrt{2}\,S_{13}(i\omega)$ are isometric on H_0 for almost all ω. Now, let $\mathscr{R}[T]$ denote the range of an operator T. Also, let $\mathscr{N}[T]$ be the null set of T, that is, the set of all points a for which $Ta = 0$. The isometry of $\sqrt{2}\,S_{13}(i\omega)$ implies that $\mathscr{R}[S_{13}(i\omega)] \neq \{0\}$, and (5) indicates that $\mathscr{N}[S'_{12}(i\omega)] \supset \mathscr{R}[S_{13}(i\omega)]$, again for almost all ω. But $\mathscr{N}[S'_{12}(i\omega)]$ is equal to the orthogonal complement of $\mathscr{R}[S_{12}(i\omega)]$, that is, to the set of all $a \in H_0$ for which $(a, b) = 0$ for all $b \in \mathscr{R}[S_{12}(i\omega)]$ (see Berberian [1], p. 133). We conclude that there exists at least one ω for which $\mathscr{R}[S_{12}(i\omega)] \neq H_0$, whereas $\sqrt{2}\,S_{12}(i\omega)$ is an isometry. This can happen only when H_0 is infinite-dimensional (Berberian, 1961, p. 146).

Example 7.8-2. Here is an example of a reciprocal lossless scattering transform $S = [S_{kj}]$ for a Hilbert 3-port that is matched at all three ports. As before, let $H = H_0 \times H_0 \times H_0$, where H_0 is infinite-dimensional, and let $\{e_i\}_{i=0}^{\infty}$ be an orthonormal basis for H_0. We take S to be constant on the

ζ plane and set $S_{11} = S_{22} = S_{33} = 0$. The other elements of $S = [S_{kj}]$ are defined on any $a = \sum_{i=0}^{\infty} a_i e_i$ by

$$S_{12}a = \frac{1}{\sqrt{2}} \sum_{i=0}^{\infty} a_i e_{3i},$$

$$S_{13}a = \frac{1}{\sqrt{2}} \sum_{i=0}^{\infty} a_i e_{3i+1},$$

$$S_{23}a = \frac{1}{\sqrt{2}} \sum_{i=0}^{\infty} a_i e_{3i+2},$$

$$S_{21} = S_{12}, \qquad S_{31} = S_{13}, \qquad S_{32} = S_{23}.$$

It follows that

$$S'_{12}a = \frac{1}{\sqrt{2}} \sum_{i=0}^{\infty} a_{3i} e_i,$$

$$S'_{13}a = \frac{1}{\sqrt{2}} \sum_{i=0}^{\infty} a_{3i+1} e_i,$$

$$S'_{23}a = \frac{1}{\sqrt{2}} \sum_{i=0}^{\infty} a_{3i+2} e_i.$$

Therefore, $S'S = I$. Since S is constant on the ζ plane, S is reciprocal, lossless, and matched at all three ports. ◇

Problem 7.8-1. Let $H = H_0 \times \cdots \times H_0$, where H_0 appears n times. Assume that $S = [S_{kj}]$ is a bounded* mapping of H into H. Also, assume that S_{il} is a lossless scattering transform for one fixed pair i and l. Show that $S_{kl} = 0$ for all $k \neq i$.

Chapter 8

The Admittance Formulism

8.1. INTRODUCTION

In this final chapter, we discuss the admittance formulism for the Hilbert port. Its distinguishing characteristic is that the net real energy absorbed by the Hilbert port over the time interval I is

$$e(I) = \operatorname{Re} \int_I (\mathfrak{N}v(t), v(t))\, dt,$$

where $\mathfrak{N}$ is the admittance operator. The passivity of $\mathfrak{N}$ is reflected by the condition $e(I_T) \geq 0$ for every time interval of the form $\{t\colon -\infty < t \leq T\}$, where $T < \infty$.

The admittance formulism is inherently more complicated than the scattering formulism. For example, as was indicated in Lemma 7.3-1, a linear scatter-passive operator on $\mathscr{D}(H)$ has a continuous extension onto $L_2(H)$, which is contractive. The use of this fact in conjunction with certain forms of Parseval's equation led to the realizability theorems of Sections 7.4

and 7.5. On the other hand, under the admittance formulism, a linear passive operator on $\mathscr{D}(H)$ need not have a continuous extension onto $L_2(H)$. An illustration of this is given in the next section. As another example, recall that the causality and scatter-semipassivity of a linear scattering operator is equivalent to its scatter-passivity. In contrast to this, the connection between semipassivity and passivity for a linear admittance operator is more complicated and involves a certain restriction on the singular behavior of the admittance operator; see Theorem 8.13-1.

Most of this chapter is based on the work of Hackenbroch (1968, 1969), which is an extension in several ways of the classical paper of König and Meixner (1958) dealing with systems having complex-valued signals. In addition, some of these results are reworked in a distributional context to obtain the conclusions of Zemanian (1963, 1970a).

In this chapter, the integration theory of Chapter 2 is used in an essential way. Moreover, we assume throughout that $t \in R$ and $\mathscr{D} = \mathscr{D}_{R^1}$.

8.2. PASSIVITY

We start with the definition of semipassivity, which is a weaker condition than passivity. In the following, $L_1^{\text{loc}}(H)$ denotes the set of all H-valued functions on R which, for every compact interval $I \subset R$, are members of $L_1(I, \mathfrak{C}; \mu; H)$, where $\mathfrak{C}$ is the σ-algebra of Borel subsets of I and μ is Lebesgue measure. L_1 is the space obtained by setting $I = (-\infty, \infty)$ and $H = C$.

Definition 8.2-1. Let $\mathfrak{N}$ be an operator whose domain contains a set $\mathscr{X} \subset L_1^{\text{loc}}(H)$. $\mathfrak{N}$ is said to be *semipassive on* $\mathscr{X}$ if $\mathfrak{N}$ maps $\mathscr{X}$ into $L_1^{\text{loc}}(H)$ and if, for every $v \in \mathscr{X}$ and for $u \triangleq \mathfrak{N}v$, we have that

$$(u(\cdot), v(\cdot)) \in L_1 \tag{1}$$

and

$$\operatorname{Re} \int_{-\infty}^{\infty} (u(t), v(t))\, dt \geq 0. \tag{2}$$

In the next definition, 1_T is the function defined in Section 7.3; namely $1_T(t) = 1$ for $t \leq T$ and $1_T = 0$ for $t > T$.

Definition 8.2-2. Let $\mathfrak{N}$ be an operator whose domain contains a set $\mathscr{X} \subset L_1^{\text{loc}}(H)$. $\mathfrak{N}$ is said to be *passive on* $\mathscr{X}$ if $\mathfrak{N}$ maps $\mathscr{X}$ into $L_1^{\text{loc}}(H)$ and if, for every $v \in \mathscr{X}$ and every $T \in R$ and for $u \triangleq \mathfrak{N}v$, we have that

$$(u(\cdot), v(\cdot))1_T(\cdot) \in L_1 \tag{3}$$

and

$$\operatorname{Re} \int_{-\infty}^{T} (u(t), v(t))\, dt \geq 0. \tag{4}$$

Whenever $\mathscr{X}$ is a set of continuous H-valued functions on R with compact supports [in particular, when $\mathscr{X} = \mathscr{D}(H)$], conditions (1) and (3) are automatically satisfied by virtue of the assumption that $\mathfrak{N}$ maps $\mathscr{X}$ into $L_1^{\text{loc}}(H)$. For, $(u(\cdot), v(\cdot))$ is clearly measurable (see Appendix G9). Moreover, if K is a compact interval containing supp v, then

$$\int_{-\infty}^{\infty} |(u(t), v(t))|\, dt \leq \int_K \|u(t)\|\, dt \sup_{t \in K} \|v(t)\| < \infty.$$

Obviously, the passivity of $\mathfrak{N}$ on $\mathscr{D}(H)$ implies its semipassivity on $\mathscr{D}(H)$. However, the converse is not true in general even when $\mathfrak{N}$ is causal. Indeed, let $u \triangleq \mathfrak{N}v \triangleq -v^{(1)}$ for all $v \in \mathscr{D}(H)$. Then, upon integrating by parts, we see that

$$\operatorname{Re} \int_{-\infty}^{T} (u(t), v(t))\, dt = -\tfrac{1}{2}\|v(T)\|^2.$$

As $T \to \infty$, the right-hand side tends to zero, and so $\mathfrak{N}$ is semipassive on $\mathscr{D}(H)$. However, $\mathfrak{N}$ is certainly not passive on $\mathscr{D}(H)$. In Section 8.14, we will determine the precise conditions under which passivity and semipassivity are equivalent for a convolution operator.

On the other hand, $\mathfrak{N}: v \mapsto v^{(1)}$ is clearly passive on $\mathscr{D}(H)$. This is an example of a linear passive operator on $\mathscr{D}(H)$ that does not have a continuous extension onto $L_2(H)$.

Another point should be made here. The admittance variables v and u are related to the scattering variables q and r by

$$v = q + r, \qquad u = q - r.$$

Upon substituting these equations into (4), we obtain the energy condition for scatter-passive operators:

$$\int_{-\infty}^{T} [\|q(t)\|^2 - \|r(t)\|^2]\, dt \geq 0.$$

However, the assumption that the admittance operator $v \mapsto u$ is passive on $\mathscr{D}(H)$ is not quite the same as the assumption that the scattering operator $q \mapsto r$ is scatter-passive on $\mathscr{D}(H)$. This is because q is not in general a member of $\mathscr{D}(H)$ when v is a member of $\mathscr{D}(H)$, and conversely.

We now present a result of Youla *et al.* (1959). Linearity and passivity imply causality. The analog to this under the scattering formulism is contained in Theorem 7.3-1. Under the admittance formulism, we have the following.

Theorem 8.2-1. *Let $\mathfrak{N}$ be a linear passive operator on $\mathscr{D}(H)$. Then, $\mathfrak{N}$ is causal on $\mathscr{D}(H)$.*

PROOF. Let v, $v_1 \in \mathscr{D}(H)$. Set $u = \mathfrak{N}v$ and $u_1 = \mathfrak{N}v_1$. By the definition of passivity, u, $u_1 \in L_1^{\text{loc}}(H)$. Given an arbitrary $T \in R$, assume that $v(t) = 0$ for $t < T$. We shall show that $u(t) = 0$ for almost all $t < T$, proving thereby that $\mathfrak{N}$ is causal.

Let α be any real number and set $v_2 \triangleq v_1 + \alpha v$ and $u_2 \triangleq u_1 + \alpha u$. By the linearity of $\mathfrak{N}$, $u_2 = \mathfrak{N}v_2$. Therefore, by passivity,

$$\operatorname{Re} \int_{-\infty}^{T} (u_2, v_2)\, dt \geq 0.$$

Since $v_2(t) = v_1(t)$ for $t < T$, this can be rewritten as

$$\operatorname{Re} \int_{-\infty}^{T} (u_1, v_1)\, dt + \alpha \operatorname{Re} \cdot \int_{-\infty}^{T} (u, v_1)\, dt \geq 0.$$

But α is arbitrary, and hence

$$\operatorname{Re} \int_{-\infty}^{T} (u, v_1)\, dt = 0.$$

Upon replacing v_1 by iv_1, we see that the imaginary part of the integral is also equal to zero. Therefore,

$$\int_{-\infty}^{T} (u, v_1)\, dt = 0. \tag{5}$$

Now, let K be any compact interval contained in $(-\infty, T]$. By Schwarz's inequality,

$$\left| \int_K \|u\|^2\, dt - \int_K (u, v_1)\, dt \right|^2 \leq \int_K \|u\|^2\, dt \int_K \|u - v_1\|^2\, dt.$$

Since $\mathscr{D}(H)$ is dense in $L_2(H)$, it follows easily that $\int_K \|u - v_1\|^2\, dt$ can be made arbitrarily small by choosing an appropriate $v_1 \in \mathscr{D}_K(H)$. In view of (5), this implies that $\int_K \|u\|^2\, dt = 0$. Since K is arbitrary, we conclude that $u(t) = 0$ for almost all $t \in (-\infty, T]$. ◇

Problem 8.2-1. Let H be the complexification of a real Hilbert space H_r. Assume that $\mathfrak{N}$ is a real linear mapping on $\mathscr{D}(H)$. Thus, $\mathfrak{N}$ maps $\mathscr{D}(H_r)$ into $[\mathscr{D}(R); H_r]$. Show that, if $\mathfrak{N}$ is passive on $\mathscr{D}(H_r)$, then it is also passive on $\mathscr{D}(H)$.

Problem 8.2-2. Let $\mathfrak{N} = f\cdot$, where $f(t, x) = h(t)h(x)$, $h \in L_1^{\text{loc}}(R)$, and h is not a function that equals zero almost everywhere. Show that $\mathfrak{N}$ is a linear semipassive operator on $\mathscr{D}$ but is not passive on $\mathscr{D}$. Also, show that $\mathfrak{N}$ is not causal on $\mathscr{D}$ if supp h intersects the open interval $(-\infty, 0)$. What must h be if $\mathfrak{N}$ is to be translation-invariant?

8.3. LINEARITY AND SEMIPASSIVITY IMPLY CONTINUITY

A remarkable result of König (1959) is that linearity and passivity imply continuity. [See also Hackenbroch, 1969 and Hackenbroch (to be published).] In the present context, we can state the following.

Theorem 8.3-1. *Let $\mathfrak{N}$ be a linear semipassive mapping on $\mathscr{D}(H)$. Then, $\mathfrak{N}$ is continuous from $\mathscr{D}(H)$ into $[\mathscr{D}; H]$.*

PROOF. Let

$$\mathfrak{B}(\phi, \psi) \triangleq \mathfrak{B}_{\mathfrak{N}}(\phi, \psi) \triangleq \int_{-\infty}^{\infty} (\mathfrak{N}\phi(t), \psi(t))dt, \qquad \phi, \psi \in \mathscr{D}(H). \tag{1}$$

Then, $\mathfrak{B}$ is a sesquilinear form on $\mathscr{D}(H)$, and $\operatorname{Re} \mathfrak{B}(\phi, \phi) \geq 0$ by semipassivity. We first prove that, for any compact interval $K \subset R$, $\mathfrak{B}$ is continuous on $\mathscr{D}_K(H) \times \mathscr{D}_K(H)$ supplied with the product topology.

Upon applying the Schwarz inequality (Appendix C15) to (1), we see that

$$|\mathfrak{B}(\phi, \psi)| \leq q(\phi)\|\psi\|_{L_2}, \qquad \phi, \psi \in \mathscr{D}_K(H), \tag{2}$$

where $\|\cdot\|_{L_2}$ is the norm for $L_2(H)$ and q is a function from $\mathscr{D}_K(H)$ into R such that $q(\phi) > 0$ for all $\phi \in \mathscr{D}_K(H)$. Now, consider

$$\hat{\mathfrak{B}}(\phi, \psi) \triangleq \hat{\mathfrak{B}}_{\mathfrak{N}}(\phi, \psi) \triangleq \tfrac{1}{2}[\mathfrak{B}(\phi, \psi) + \overline{\mathfrak{B}(\psi, \phi)}].$$

By virtue of the semipassivity of $\mathfrak{N}$, $\hat{\mathfrak{B}}$ is a positive sesquilinear form on $\mathscr{D}_K(H) \times \mathscr{D}_K(H)$. The Schwarz inequality (Appendix A6) yields

$$|\hat{\mathfrak{B}}(\phi, \psi)| \leq [\hat{\mathfrak{B}}(\phi, \phi)\hat{\mathfrak{B}}(\psi, \psi)]^{1/2}.$$

Since $\hat{\mathfrak{B}}(\phi, \phi) \leq |\mathfrak{B}(\phi, \phi)|$, we get

$$|\mathfrak{B}(\phi, \psi)| \leq |\mathfrak{B}(\psi, \phi)| + 2[\hat{\mathfrak{B}}(\phi, \phi)\hat{\mathfrak{B}}(\psi, \psi)]^{1/2} \leq W(\psi)Z(\phi), \tag{3}$$

where

$$W(\psi) = q(\psi) + 2|\mathfrak{B}(\psi, \psi)|^{1/2} \tag{4}$$

and

$$Z(\phi) = \|\phi\|_{L_2} + |\mathfrak{B}(\phi, \phi)|^{1/2}. \tag{5}$$

Now, for every $\phi \in \mathscr{D}_K(H)$ other than the zero function, the mapping

$$f_\phi: \quad \psi \mapsto \overline{\mathfrak{B}(\phi, \psi)}/Z(\phi)$$

is continuous and linear from $\mathscr{D}_K(H)$ into C. Thus, $f_\phi \in [\mathscr{D}_K(H); C]$. In view of (3), the collection of all such f_ϕ is a bounded set in $[\mathscr{D}_K(H); C]^\sigma$. Therefore, that collection is equicontinuous on $\mathscr{D}_K(H)$ (Appendix D9). As a result of this and the continuity of $\|\cdot\|_{L_2}$ on $\mathscr{D}_K(H)$, we can find a continuous seminorm ρ on $\mathscr{D}_K(H)$ such that

$$|\mathfrak{B}(\phi, \psi)| \leq \rho(\psi)Z(\phi) \tag{6}$$

and simultaneously $\|\psi\|_{L_2} \leq \rho(\psi)$ for all $\phi, \psi \in \mathscr{D}_K(H)$. Combining this with (5), we now obtain

$$Z(\phi)/\rho(\phi) \leq 1 + [Z(\phi)/\rho(\phi)]^{1/2}.$$

Consequently, there exists a constant $M > 0$ such that $Z(\phi) \leq M\rho(\phi)$. By (6) again,

$$|\mathfrak{B}(\phi, \psi)| \leq M\rho(\phi)\rho(\psi), \tag{7}$$

which verifies the continuity of $\mathfrak{B}$ on $\mathscr{D}_K(H) \times \mathscr{D}_K(H)$.

Next, set $\psi = \bar{\theta}a$, where $\theta \in \mathscr{D}_K$ and $a \in H$. By (7),

$$\begin{aligned} M\rho(\phi)\rho(\psi) &\geq \left| \int_{-\infty}^{\infty} (\mathfrak{N}\phi, \psi)\, dt \right| = \left| \int (\theta\mathfrak{N}\phi, a)\, dt \right| \\ &= \left| \left(\int \theta\mathfrak{N}\phi\, dt, a \right) \right| = |(\langle \mathfrak{N}\phi, \theta \rangle, a)|. \end{aligned}$$

So,

$$\|\langle \mathfrak{N}\phi, \theta\rangle\|_H = \sup_{\|a\|=1} |(\langle \mathfrak{N}\phi, \theta\rangle, a)| \leq \sup_{\|a\|=1} M\rho(\phi)\rho(a\theta).$$

Let Θ be a bounded set in $\mathscr{D}$. Then, Θ is a bounded set in $\mathscr{D}_N$ for some compact interval N. Moreover, given any other compact interval J, we can choose K to contain N and J. Then, ρ will be a continuous seminorm on $\mathscr{D}_J(H)$, and Θ will be a bounded set in $\mathscr{D}_K$. Consequently, for every $\phi \in \mathscr{D}_J(H)$,

$$\sup_{\theta \in \Theta} \|\langle \mathfrak{N}\phi, \theta\rangle\|_H \leq P\rho(\phi),$$

where P is a constant. This proves that $\mathfrak{N}$ is continuous from $\mathscr{D}_J(H)$ into $[\mathscr{D}; H]$. Since J can be chosen arbitrarily, $\mathfrak{N}$ is continuous from $\mathscr{D}(H)$ into $[\mathscr{D}; H]$. ◇

Problem 8.3-1. Assign to $L_1^{\text{loc}}(H)$ the topology generated by the collection $\{\omega_k\}_{k=1}^\infty$ of seminorms defined by

$$\omega_k(f) = \int_{K_k} \|f(t)\|_H\, dt, \qquad f \in L_1^{\text{loc}}(H),$$

where $\{K_k\}_{k=1}^{\infty}$ is a nested closed cover of R. Then, $L_1^{\text{loc}}(H)$ becomes a Fréchet space. Use the closed-graph theorem (Appendix D10) to prove that every linear semipassive mapping on $\mathscr{D}(H)$ is continuous from $\mathscr{D}(H)$ into $L_1^{\text{loc}}(H)$.

8.4. THE FOURIER TRANSFORMATION ON $\mathscr{S}(H)$

Subsequently, we shall need a variety of results concerning the *Fourier transformation* $\mathfrak{F}$. We gather them in this section. The *Fourier transform* of any $\phi \in \mathscr{S}(H)$ is denoted by $\tilde{\phi} \triangleq \mathfrak{F}\phi$ and defined by

$$\tilde{\phi}(\omega) \triangleq (\mathfrak{F}\phi)(\omega) \triangleq \int_{-\infty}^{\infty} \phi(t)e^{-i\omega t}\, dt. \tag{1}$$

Some differentiations under the integral sign and integrations by parts show that

$$(i\omega)^p \tilde{\phi}^{(k)}(\omega) = \int_{-\infty}^{\infty} e^{-i\omega t} D_t^p[(-it)^k \phi(t)]\, dt,$$

and this implies that $\mathfrak{F}$ is a continuous linear mapping of $\mathscr{S}(H)$ into $\mathscr{S}(H)$. The same argument as that which holds for complex-valued functions (Zemanian, 1965, Section 7.2) establishes the following inversion formula for $\mathfrak{F}$:

$$\phi(t) = (\mathfrak{F}^{-1}\tilde{\phi})(t) = (1/2\pi)\int_{-\infty}^{\infty} \tilde{\phi}(\omega)e^{i\omega t}\, d\omega. \tag{2}$$

Here, $\mathfrak{F}^{-1}$ denotes the *inverse Fourier transformation*. By the same manipulations as above, we see that $\mathfrak{F}^{-1}$, too, is a continuous linear mapping of $\mathscr{S}(H)$ into $\mathscr{S}(H)$. It also follows that $\mathfrak{F}$ must be a bijection. Thus, $\mathfrak{F}$ is an automorphism on $\mathscr{S}(H)$.

We take note of the following version of Parseval's equation:

$$2\pi \int_{-\infty}^{\infty} (\phi(t), \psi(t))\, dt = \int_{-\infty}^{\infty} (\tilde{\phi}(\omega), \tilde{\psi}(\omega))\, d\omega, \qquad \phi, \psi \in \mathscr{S}(H). \tag{3}$$

This can be obtained by using Fubini's theorem and Lemma 7.5-2.

We now present three lemmas concerning complex-valued functions.

Lemma 8.4-1. *Let $\mathscr{D} * \mathscr{D}$ denote the set of all functions ϕ with the representation $\phi = \theta * \psi$, where $\theta, \psi \in \mathscr{D}$. Then, $\mathscr{D} * \mathscr{D}$ is dense in $\mathscr{D}$.*

This is established by choosing a sequence $\{\theta_k\}_{k=1}^{\infty} \subset \mathscr{D}$ such that $\theta_k \geq 0$, $\int \theta_k \, dt = 1$, and

$$\operatorname{supp} \theta_k \subset \{t: |t| \leq k^{-1}\}$$

and then showing that $\theta_k * \phi \to \phi$ in $\mathscr{D}$.

Lemma 8.4-2. *Let* $\mathscr{Z} \triangleq \mathfrak{F}(\mathscr{D})$. *Also, let* $\mathscr{Z} \cdot \mathscr{Z}$ *denote the set of all functions* χ *with the representation* $\chi = \tilde{\theta}\tilde{\psi}$, *where* $\theta, \psi \in \mathscr{D}$. *Then,* $\mathscr{Z} \cdot \mathscr{Z}$ *is dense in* $\mathscr{S}$.

Proof. Choose any $\tilde{\phi} \in \mathscr{S}$ and any $\mathscr{S}$-neighborhood $\tilde{\Omega}$ of $\tilde{\phi}$. Then, $\phi \triangleq \mathfrak{F}^{-1}\tilde{\phi} \in \mathscr{S}$, and $\Omega \triangleq \mathfrak{F}^{-1}(\tilde{\Omega})$ is an $\mathscr{S}$-neighborhood of ϕ. By the density of $\mathscr{D}$ in $\mathscr{S}$, there exists a $\zeta \in \mathscr{D} \cap \mathring{\Omega}$. But, since the topology of $\mathscr{D}$ is stronger than that induced on it by $\mathscr{S}$ and since $\mathring{\Omega}$ is also an $\mathscr{S}$-neighborhood of ζ, there exists a $\mathscr{D}$-neighborhood Ξ of ζ such that $\Xi \subset \mathring{\Omega}$. By the preceding lemma, we can find a $\theta * \psi \in \mathscr{D} * \mathscr{D}$ in Ξ. Thus, $\theta * \psi \in \Omega$. But $\mathfrak{F}(\theta * \psi) = \tilde{\theta}\tilde{\psi}$. Therefore, $\tilde{\theta}\tilde{\psi} \in \tilde{\Omega}$. ◇

Lemma 8.4-3. *Given any bounded Borel set* $E \subset R$, *there exists a function* $\theta \in \mathscr{D}$ *such that* $|\tilde{\theta}(\eta)| \geq 1$ *for all* $\eta \in E$.

Proof. We can choose a compact interval $K \supset E$ and a function $\tilde{\psi} \in \mathscr{D}$ such that $\tilde{\psi}(\eta) > 2$ for all $\eta \in K$. Therefore, $\psi \in \mathscr{S}$. We can also choose a sequence $\{\phi_k\}_{k=1}^{\infty} \subset \mathscr{D}$ that converges in $\mathscr{S}$ to ψ. Therefore, $\tilde{\phi}_k \to \tilde{\psi}$ in $\mathscr{S}$. This implies that $\tilde{\phi}_k$ tends to $\tilde{\psi}$ uniformly on K. Thus, the sequence $\{\phi_k\}$ contains an element θ with the required properties. ◇

We finally take note of a standard result concerning complex-valued distributions, namely the Bochner–Schwartz theorem. In the following, $\mathfrak{C}$ denotes the collection of all Borel subsets of R. A *tempered positive measure* μ is a positive measure on $\mathfrak{C}$ such that $(1 + t^2)^{-p} \in L_1(R, \mathfrak{C}; \mu; C)$ for some integer p. (See Appendix G, Sections G4, G10, and G13.) On the other hand, $f \in [\mathscr{D}; C]$ is called *positive-definite* if, for every $\phi \in \mathscr{D}$,

$$\int_{-\infty}^{\infty} (f * \phi)(t)\overline{\phi(t)} \, dt \geq 0.$$

Theorem 8.4-1. $f \in [\mathscr{D}; C]$ *is positive-definite if and only if there exists a tempered positive measure* μ *such that, for every* $\phi \in \mathscr{D}$,

$$\langle f, \phi \rangle = \int_R d\mu \, \tilde{\phi}. \tag{4}$$

Proofs of this theorem are given by Schwartz (1966, pp. 276–277) and Gelfand and Vilenkin (1964, Section 3.3). According to the standard definition of the Fourier transformation on $[\mathscr{D}; C]$ (see, for example, Zemanian,

1965, Section 7.8), (4) means that f is the Fourier transform of the distribution generated by μ.

Later on, we shall meet a *tempered complex measure*. This is a σ-finite complex measure ν (see Appendix G12) for which

$$\int_{[-k,k]} d|\nu| \, (1+t^2)^{-p}$$

converges as $k \to \infty$ if p is chosen sufficiently large.

8.5. LOCAL MAPPINGS

Let $\mathfrak{N}$ be a linear translation-invariant semipassive mapping of $\mathscr{D}(H)$ into $[\mathscr{D}; H]$. By Theorem 8.3-1, $\mathfrak{N}$ is also continuous. Therefore, $\mathfrak{N} = y\,*$, where $y \in [\mathscr{D}(H); H]$, according to Theorem 5.10-1. It now follows from Theorem 5.5-1 that $\mathfrak{N}$ is a continuous linear mapping of $\mathscr{D}(H)$ into $\mathscr{E}(H)$. We define the sesquilinear mappings $\mathfrak{B}$ and $\hat{\mathfrak{B}}$ on $\mathscr{D}(H) \times \mathscr{D}(H)$ as before. For instance,

$$\mathfrak{B}(\phi, \psi) \triangleq \mathfrak{B}_{\mathfrak{N}}(\phi, \psi) \triangleq \int_{-\infty}^{\infty} (\mathfrak{N}\phi(t), \psi(t))\, dt, \qquad \phi, \psi \in \mathscr{D}(H). \tag{1}$$

By the proof of Theorem 8.3-1, $\mathfrak{B}$ is a continuous sesquilinear form on $\mathscr{D}_K(H) \times \mathscr{D}_K(H)$ for every compact interval K, and therefore so too is $\hat{\mathfrak{B}}$ because

$$\hat{\mathfrak{B}}(\phi, \psi) \triangleq \hat{\mathfrak{B}}_{\mathfrak{N}}(\phi, \psi) \triangleq \tfrac{1}{2} \int_{-\infty}^{\infty} [(\mathfrak{N}\phi, \psi) + (\phi, \mathfrak{N}\psi)]\, dt. \tag{2}$$

Another result from Hackenbroch (1969) is a characterization of those causal operators $\mathfrak{N}$ for which $\hat{\mathfrak{B}} = 0$. One property such operators have is that they respond only to present values of input signals, but not to past values nor, by causality, to future values. A precise definition of this property is the following.

Definition 8.5-1. A mapping $\mathfrak{N}$ of $\mathscr{D}(H)$ into $[\mathscr{D}; H]$ is said to be *local* if $\operatorname{supp} \mathfrak{N}\phi \subset \operatorname{supp} \phi$ for all $\phi \in \mathscr{D}(H)$.

For linear mappings, local operators are the same thing as the so-called memoryless operators of systems theory (Willems, 1971, p. 14).

Theorem 8.5-1. *Let $\mathfrak{N}$ be a linear translation-invariant causal mapping of $\mathscr{D}(H)$ into $[\mathscr{D}; H]$ and such that $\hat{\mathfrak{B}} = 0$. Then, $\mathfrak{N}$ is local.*

Note. The condition $\hat{\mathfrak{B}} = 0$ implies that $\mathfrak{N}$ is semipassive on $\mathscr{D}(H)$.

PROOF. That $\mathfrak{N}$ is local is equivalent to the following assertion. If $\phi \in \mathscr{D}(H)$ vanishes on a neighborhood Λ of $T \in R$, then $\mathfrak{N}\phi$ also vanishes on Λ. Now, ϕ can be written in the form $\phi = \phi_1 + \phi_2$, where $\phi_1, \phi_2 \in \mathscr{D}(H)$, supp $\phi_1 \subset [t_1, t_2]$, and supp $\phi_2 \subset [t_3, t_4]$ for certain $t_1, t_2, t_3, t_4 \in R$ with $t_1 < t_2 < T < t_3 < t_4$. So we need merely prove that, if $\phi \in \mathscr{D}(H)$ and supp $\phi \subset [c, d]$, then supp $\mathfrak{N}\phi \subset [c, d]$.

Since $\mathfrak{N}$ is causal by hypothesis, $\mathfrak{N}\phi$ vanishes for $t < c$. Again by hypothesis, $\mathfrak{B}(\phi, \psi) = 0$ for all $\psi \in \mathscr{D}(H)$, so that

$$\int (\mathfrak{N}\phi, \psi)\, dt = -\int (\phi, \mathfrak{N}\psi)\, dt. \tag{3}$$

Choose ψ such that $\psi(t) = 0$ for $t < d$. Therefore, $(\mathfrak{N}\psi)(t) = 0$ for $t < d$ by causality. Hence, the right-hand side of (3) is equal to zero. But the behavior of ψ for $t > d$ is unrestricted, and we know that $\mathfrak{N}\phi \in \mathscr{E}(H)$. Therefore, $(\mathfrak{N}\phi)(t) = 0$ for $t > d$ as well. ◇

In order to prove Hackenbroch's characterization of $\mathfrak{N}$ (see Theorem 8.5-2), we shall need the following result.

Lemma 8.5-1. *Let A and B be complex Banach spaces. Assume that $\{L_k\}_{k=1}^{\infty} \subset [A; B]$, where $L_k \neq 0$ for every k. Then, there exists an $a \in A$ such that $L_k a \neq 0$ for every k.*

PROOF. Let $F_k \triangleq \{a \in A\colon L_k a = 0\}$. Our lemma will be proved when we show that $A \neq \bigcup_{k=1}^{\infty} F_k$. First, note that each F_k is a closed linear subspace of A and that $F_k \neq A$ because $L_k \neq 0$. We now show that F_k is nowhere dense; that is, its closure $\bar{F}_k$ has no interior points. Suppose that b is an interior point of $\bar{F}_k$. Since $\bar{F}_k = F_k$, this means that there exists an $\varepsilon \in R_+$ such that $O(b, \varepsilon) \subset F_k$, where

$$O(b, \varepsilon) \triangleq \{a \in A\colon \|a - b\| < \varepsilon\}.$$

Also, since $L_k \neq 0$, there exists a nonzero $c \in A$ such that $L_k c \neq 0$. Then,

$$d = b + \tfrac{1}{2}\varepsilon\|c\|^{-1}c \in O(b, \varepsilon).$$

Upon applying L_k to d and using the fact that $L_k b = 0$, we obtain

$$L_k d = \tfrac{1}{2}\varepsilon\|c\|^{-1}L_k c \neq 0.$$

This is a contradiction, and therefore F_k is truly nowhere dense. Baire's category theorem (Appendix C13) now implies that $A \neq \bigcup F_k$. ◇

Theorem 8.5-2. *$\mathfrak{N}$ is a linear translation-invariant causal mapping of $\mathscr{D}(H)$ into $[\mathscr{D}; H]$ such that $\mathfrak{B} = 0$ if and only if, for all $\phi \in \mathscr{D}(H)$,*

$$\mathfrak{N}\phi = \sum_{k=0}^{n} P_k \phi^{(k)}, \tag{4}$$

where $P_0, P_1, \ldots, P_n \in [H; H]$ *and*

$$P_k' = (-1)^{k+1} P_k, \qquad k = 0, 1, \ldots, n. \tag{5}$$

Note. As before, the prime denotes the adjoint operator.

PROOF. If $\mathfrak{N}$ is defined by (4), then it is clearly a linear translation-invariant causal mapping on $\mathscr{D}(H)$ into $\mathscr{D}(H)$. Some integrations by parts show that $\hat{\mathfrak{B}} = 0$.

Conversely, assume that $\mathfrak{N}$ is a linear translation-invariant causal mapping of $\mathscr{D}(H)$ into $[\mathscr{D}; H]$ and such that $\hat{\mathfrak{B}} = 0$. For any fixed $T \in R$ and $a, b \in H$, consider the mapping

$$\theta \mapsto ((\mathfrak{N}\theta a)(T), b), \qquad \theta \in D. \tag{6}$$

Since $\mathfrak{N}$ is a continuous linear mapping of $\mathscr{D}(H)$ into $\mathscr{E}(H)$, (6) is a complex-valued distribution. Its support is $\{T\}$ because $\mathfrak{N}$ is local. Therefore, by a standard result for complex-valued distributions (Zemanian, 1965, Theorem 3.5-2),

$$((\mathfrak{N}\theta a)(T), b) = \sum_{k=0}^{\infty} \alpha_k(a, b)\theta^{(k)}(T), \tag{7}$$

where the $\alpha_k(a, b)$ denote complex numbers and satisfy the following condition. For each fixed choice of the pair $a, b \in H$, all but a finite number of the $\alpha_k(a, b)$ are equal to zero. This finite set will change in general for different choices of a and b. However, the $\alpha_k(a, b)$ do not depend on T, because of the translation invariance of $\mathfrak{N}$. Moreover, it is not difficult to show that each α_k is continuous and sesquilinear on $H \times H$ supplied with the product topology. Therefore, according to Appendix D15, there exists a $P_k \in [H; H]$ such that $\alpha_k(a, b) = (P_k a, b)$. Thus, (7) becomes

$$((\mathfrak{N}\theta a)(T), b) = \sum_{k=0}^{\infty} (P_k a, b)\theta^{(k)}(T). \tag{8}$$

We now show that there exists a finite integer n such that $(P_k a, b) = 0$ for all $k > n$ and all $a, b \in H$. If this is not true, there exists a subsequence $\{Q_j\}_{j=1}^{\infty}$ of $\{P_k\}_{k=1}^{\infty}$ such that $Q_j \neq 0$ for all j. By Lemma 8.5-1, there is an $a \in H$ such that $Q_j a \neq 0$ for all j. But $(Q_j a, \cdot) \in [H; C]$, and so, by Lemma 8.5-1 again, there is a $b \in H$ such that $(Q_j a, b) \neq 0$ for all j. This contradicts the fact that only a finite number of coefficients in the right-hand side of (8) can be nonzero.

It now follows that, for any $\phi \in \mathscr{D} \odot H$,

$$((\mathfrak{N}\phi)(T), b) = \sum_{k=0}^{n} (P_k \phi^{(k)}(T), b). \tag{9}$$

Since $\mathscr{D} \odot H$ is dense in $\mathscr{D}(H)$, (9) remains true for all $\phi \in \mathscr{D}(H)$. This establishes (4) since b and T are arbitrary.

We turn now to the proof of (5). Set $Q_k \triangleq P_k + (-1)^k P_k'$. Some integrations by parts show that, for any $\phi, \psi \in \mathscr{D}(H)$,

$$\int_{-\infty}^{\infty} \left(\sum_{k=0}^{n} Q_k \phi^{(k)}(t), \psi(t) \right) dt = 2\hat{\mathfrak{B}}(\phi, \psi). \tag{10}$$

The right-hand side is equal to zero by hypothesis. Thus, for any $a, b \in H$ and any $\theta, \lambda \in \mathscr{D}$,

$$\int \sum (Q_k a, b) \theta^{(k)}(t) \overline{\lambda(t)} \, dt = 0.$$

We may apply Parseval's equation [see (3) of Section 8.4] to this and employ the fact that $(\mathfrak{F}\theta^{(k)})(\eta) = (i\eta)^k \tilde{\theta}(\eta)$ to get

$$\int \sum (Q_k a, b)(i\eta)^k \tilde{\theta}(\eta) \overline{\tilde{\lambda}(\eta)} \, d\eta = 0 \tag{11}$$

for all $\tilde{\theta}, \tilde{\lambda} \in \mathscr{Z}$. According to Lemma 8.4-2, $\mathscr{Z} \cdot \mathscr{Z}$ is dense in $\mathscr{S}$. Therefore, (11) continues to hold with $\tilde{\theta}\tilde{\lambda}$ replaced by any $\chi \in \mathscr{S}$. Consequently,

$$\sum_{k=0}^{n} (Q_k a, b)(i\eta)^k = 0$$

for all η. Whence, $(Q_k a, b) = 0$ for all $a, b \in H$, which implies that $Q_k = 0$. This proves (5). ◇

Problem 8.5-1. Verify that the quantities α_k occurring in the proof of Theorem 8.5-2 are continuous sesquilinear forms on $H \times H$ and do not depend on T.

Problem 8.5-2. Let $\mathfrak{N}$ be a continuous linear translation-invariant local mapping of $\mathscr{D}(H)$ into $[\mathscr{D}; H]$. Show that

$$\mathfrak{N} = \sum_{k=0}^{n} Q_k \delta^{(k)} *,$$

where $Q_k \in [H; H]$.

8.6. POSITIVE SESQUILINEAR FORMS ON $\mathscr{D} \times \mathscr{D}$

We shall use the Bochner–Schwartz theorem to obtain a representation for positive, separately continuous, translation-invariant sesquilinear forms on $\mathscr{D} \times \mathscr{D}$. See Appendices A6 and F1 for the definition of a positive, separately continuous, sesquilinear form on $\mathscr{D} \times \mathscr{D}$. As for translation invariance, we have the following.

Definition 8.6-1. Let $\mathscr{X}$ be a space of functions from R into H that is closed under translations (i.e., if $\phi \in \mathscr{X}$, then $\sigma_\tau \phi \in \mathscr{X}$ for every $\tau \in R$). A sesquilinear form $\mathfrak{B}$ on $\mathscr{X} \times \mathscr{X}$ is said to be *translation invariant* if

$$\mathfrak{B}(\sigma_\tau \phi, \sigma_\tau \psi) = \mathfrak{B}(\phi, \psi)$$

for all $\phi, \psi \in \mathscr{X}$.

Given any $y \in [\mathscr{D}; C]$, define the mapping $\mathfrak{B}$ on $\mathscr{D} \times \mathscr{D}$ by

$$\mathfrak{B}(\theta, \lambda) \triangleq \langle y, \theta * \lambda^\dagger \rangle, \qquad \theta, \lambda \in \mathscr{D}, \tag{1}$$

where $\lambda^\dagger(t) \triangleq \overline{\lambda(-t)}$. Then, $\mathfrak{B}$ is a separately continuous translation-invariant sesquilinear form on $\mathscr{D} \times \mathscr{D}$. Indeed, the sesquilinearity is clear. The separate continuity follows from the fact that both $\theta \mapsto \theta * \lambda^\dagger$ and $\lambda \mapsto \theta * \lambda^\dagger$ are continuous mappings of $\mathscr{D}$ into $\mathscr{D}$. Finally, the translation invariance follows from the easily established identity

$$\theta * \lambda^\dagger = (\sigma_\tau \theta) * (\sigma_\tau \lambda)^\dagger.$$

Conversely, every separately continuous translation-invariant sesquilinear form on $\mathscr{D} \times \mathscr{D}$ has the representation (1). Indeed, by Theorem 4.5-1, there exists a unique $f \in [\mathscr{D}_{R^2}; C]$ such that

$$\mathfrak{B}(\theta, \lambda) = \langle f(t, x), \theta(t)\overline{\lambda(x)} \rangle$$

for all $\theta, \lambda \in \mathscr{D}$. From the translation invariance of $\mathfrak{B}$ and the totality in $\mathscr{D}_{R^2}$ of the set of functions of the form $\theta(t)\overline{\lambda(x)}$, we see that

$$f(t, x) = f(t - \tau, x - \tau)$$

for every $\tau \in R$. We may now proceed exactly as in Section 5.10 to show that

$$\mathfrak{B}(\theta, \lambda) = \langle y * \bar{\lambda}, \theta \rangle, \tag{2}$$

where $y \in [\mathscr{D}; C]$ is uniquely determined by $\mathfrak{B}$. A simple manipulation converts the right-hand side into $\langle y, \theta * \lambda^\dagger \rangle$. Thus, we have established the following.

Theorem 8.6-1. *$\mathfrak{B}$ is a separately continuous translation-invariant sesquilinear form on $\mathscr{D} \times \mathscr{D}$ if and only if $\mathfrak{B}$ has the representation* (1), *where $y \in [\mathscr{D}; C]$. y is uniquely determined by $\mathfrak{B}$, and conversely.*

If, in addition, $\mathfrak{B}$ is positive, we have the next result.

Theorem 8.6-2. $\mathfrak{B}$ *is a positive, separately continuous, translation-invariant sesquilinear form on* $\mathscr{D} \times \mathscr{D}$ *if and only if* $\mathfrak{B}$ *has the representation*

$$\mathfrak{B}(\theta, \lambda) = \int_R d\mu_t \, \tilde{\theta}(t)\overline{\tilde{\lambda}(t)}, \qquad \theta, \lambda \in \mathscr{D}, \tag{3}$$

where μ *is a tempered positive measure on the Borel subsets of* R.

PROOF. The "if" part of this theorem follows quite readily. Indeed, the sesquilinearity is again clear. The separate continuity follows from the continuity of the Fourier transformation from $\mathscr{D}$ into $\mathscr{S}$ and the fact that μ is a tempered positive measure. The translation invariance is an immediate result of the identity $(\mathfrak{F}\sigma_\tau \theta)(\omega) = e^{-i\omega\tau}\tilde{\theta}(\omega)$. Finally, the positivity of $\mathfrak{B}$ follows from the positivity of μ because

$$\mathfrak{B}(\theta, \theta) = \int d\mu_t |\tilde{\theta}(t)|^2 \geq 0.$$

For the "only if" part of this theorem, we first invoke Theorem 8.6-1 to obtain (2). Upon setting $\theta = \bar{\lambda} \in \mathscr{D}$, we may write

$$\int (y * \theta)\bar{\theta} \, dt = \langle y * \bar{\lambda}, \lambda \rangle = \mathfrak{B}(\lambda, \lambda) \geq 0.$$

Thus, $y \in [\mathscr{D}; C]$ is positive-definite. By the Bochner–Schwartz theorem (Theorem 8.4-1), there exists a tempered positive measure μ on the Borel subsets of R such that

$$\mathfrak{B}(\theta, \lambda) = \langle y, \theta * \lambda^\dagger \rangle = \int d\mu \, \mathfrak{F}(\theta * \lambda^\dagger).$$

But $\mathfrak{F}(\theta * \lambda^\dagger) = \tilde{\theta}\mathfrak{F}(\lambda^\dagger) = \tilde{\theta}\bar{\tilde{\lambda}}$. This completes the proof. ◇

8.7. POSITIVE SESQUILINEAR FORMS ON $\mathscr{D}(H) \times \mathscr{D}(H)$

Our purpose in this section is to extend Theorem 8.6-2 to certain sesquilinear forms on $\mathscr{D}(H) \times \mathscr{D}(H)$. $\mathfrak{C}$ will denote the collection of all Borel subsets of R, and $\mathfrak{C}_\infty$ will denote the set of all bounded Borel subsets of R. In Definition 2.3-1, we choose $\mathfrak{C}_k$ as the collection of all Borel subsets of $\{t: |t| \leq k\}$, where $k = 1, 2, \ldots$.

Definition 8.7-1. A *tempered PO measure* P is a σ-finite PO measure on $\mathfrak{C}_\infty$ such that

$$g_q(t) \triangleq (1 + |t|^{2q})^{-1}$$

is integrable with respect to P for some nonnegative integer q.

As in Section 2.6 [see Equation (10) there], we set

$$\mathscr{G}_{g_q}(H) \triangleq \{f \in \mathscr{G}(H) : g_q^{-1} f \in \mathscr{G} \mathbin{\hat{\odot}} H\}.$$

Lemma 8.7-1. *For each nonnegative integer q, $\mathscr{S}(H) \subset \mathscr{G}_{g_q}(H)$. Moreover, $\phi \mapsto g_q^{-1}\phi$ is a continuous linear mapping of $\mathscr{S}(H)$ into $\mathscr{G} \mathbin{\hat{\odot}} H$.*

PROOF. Let $\phi \in \mathscr{S}(H)$ and consider the Fourier inversion formula:

$$\phi(t) = (1/2\pi) \int \tilde{\phi}(\omega) e^{i\omega t}\, d\omega.$$

We know that $\tilde{\phi} \in \mathscr{S}(H) \subset L_1(H)$. By Theorem 2.5-1, $\phi \in \mathscr{G} \mathbin{\hat{\odot}} H$. Also, by (6) of Section 2.5, $\|\phi\|_1 \leq \|\tilde{\phi}\|_{L_1}$, where $\|\cdot\|_1$ is the norm for $\mathscr{G} \mathbin{\hat{\odot}} H$ (see Corollary 2.4-1a). Hence,

$$\begin{aligned} \|\phi\|_1 &\leq \int_R \|\tilde{\phi}(\omega)\|_H \, d\omega \\ &\leq \sup_{\omega \in R} \|\tilde{\phi}(\omega)[1 + |\omega|^2]\| \int_R \frac{d\omega}{1 + |\omega|^2} < \infty. \end{aligned}$$

Since the Fourier transformation is an automorphism on $\mathscr{S}(H)$, this shows that the canonical injection of $\mathscr{S}(H)$ into $\mathscr{G} \mathbin{\hat{\odot}} H$ is continuous. Our lemma now follows from the fact that multiplication by g_q^{-1} is a continuous linear mapping of $\mathscr{S}(H)$ into $\mathscr{S}(H)$. ◇

By virtue of this lemma and Definitions 2.6-2 and 2.6-3, the integrals

$$\int_R dP_t\, \phi(t) \qquad \text{and} \qquad \int_R d(P_t\, \phi(t), \psi(t))$$

have meaning for any ϕ, $\psi \in \mathscr{S}(H)$ and any tempered PO measure P.

Theorem 8.7-1. *If P is a tempered PO measure, then*

$$\phi \mapsto \int dP_t\, \phi(t) \tag{1}$$

is a continuous linear mapping of $\mathscr{S}(H)$ into H. Moreover,

$$\{\phi, \psi\} \mapsto \int d(P_t\, \phi(t), \psi(t)) \tag{2}$$

is a positive continuous sesquilinear form on $\mathscr{S}(H) \times \mathscr{S}(H)$.

Note. The mapping (1) is called the *tempered $[H; H]$-valued distribution generated by P.*

PROOF. As in Section 2.6, we set

$$Q(E) \triangleq \int_E dP_t \, g_q(t),$$

where E is any Borel subset of R and q is chosen large enough to ensure that $\|Q(R)\|_{[H;H]} < \infty$. We may then write, for any $\phi \in \mathscr{S}(H)$,

$$\left\| \int dP_t \, \phi(t) \right\|_H = \left\| \int dQ_t \, [g_q(t)]^{-1} \phi(t) \right\|_H$$
$$\leq \|Q(R)\|_{[H;H]} \|[g_q(t)]^{-1} \phi(t)\|_1 .$$

[See (2) of Section 2.5.] This result combined with Lemma 8.7-1 implies the first conclusion.

With respect to the second conclusion, we already know that (2) is a positive sesquilinear form on $\mathscr{G}_{g_q}(H) \times \mathscr{G}_{g_q}(H)$ (see the paragraph before Definition 2.6-1) and therefore on $\mathscr{S}(H) \times \mathscr{S}(H)$. To show its continuity, we invoke (2) of Section 2.6 to write, for any $\phi, \psi \in \mathscr{S}(H)$,

$$\left| \int d(P_t \phi(t), \psi(t)) \right| = \left| \int d(Q_t [g_q(t)]^{-1/2} \phi(t), [g_q(t)]^{-1/2} \psi(t)) \right|$$
$$\leq \|Q(R)\| \, \|(1 + t^{2q})^{1/2} \phi(t)\|_1 \, \|(1 + t^{2q})^{1/2} \psi(t)\|_1 . \qquad (3)$$

We now appeal again to Lemma 8.7-1 to complete the proof. ◇

Theorem 8.7-2. *With P again being a tempered PO measure, set*

$$\mathfrak{A}(\phi, \psi) \triangleq \int_R d(P_t \tilde{\phi}(t), \tilde{\psi}(t)), \qquad \phi, \psi \in \mathscr{S}(H). \qquad (4)$$

Then, $\mathfrak{A}$ is a positive, continuous, translation-invariant, sesquilinear form on $\mathscr{S}(H) \times \mathscr{S}(H)$.

PROOF. Sesquilinearity is again clear. The translation invariance follows readily from the identity

$$(\mathfrak{F}\sigma_\tau \phi)(t) = e^{-i\tau t} \tilde{\phi}(t).$$

For the continuity, combine an estimate like (3) with the facts that $\phi \mapsto g_q^{-1} \phi$ is a continuous linear mapping of $\mathscr{S}(H)$ into $\mathscr{G} \hat{\odot} H$ and the Fourier transformation is an automorphism on $\mathscr{S}(H)$. The positivity is asserted by Theorem 8.7-1 since $\tilde{\phi}, \tilde{\psi} \in \mathscr{S}(H)$. ◇

The principal theorem of this section is the following converse to Theorem 8.7-2 due to Hackenbroch.

Theorem 8.7-3. Every positive, separately continuous, translation-invariant, sesquilinear form $\mathfrak{A}$ on $\mathscr{D}(H) \times \mathscr{D}(H)$ has the representation

$$\mathfrak{A}(\phi, \psi) = \int_R d(P_t \tilde{\phi}(t), \tilde{\psi}(t)), \qquad \phi, \psi \in \mathscr{D}(H), \tag{5}$$

where P is a tempered PO measure. P is uniquely determined by $\mathfrak{A}$.

PROOF. (i) Assuming for the moment that the representation (5) holds true, we first show that the tempered PO measure P is uniquely determined by $\mathfrak{A}$. Let $\phi = \theta a$ and $\psi = \lambda b$, where $\theta, \lambda \in \mathscr{D}$ and $a, b \in H$. By virtue of (1) of Section 2.6, Definition 2.6-3, and Theorem 2.3-4, we may write

$$\begin{aligned}\mathfrak{A}(\theta a, \lambda b) &= \int d(P_t \tilde{\theta}(t) a, \tilde{\lambda}(t) b) \\ &= \left(\left[\int dP_t \tilde{\theta}(t) \overline{\tilde{\lambda}(t)}\right] a, b\right).\end{aligned}$$

By (3) of Section 2.3, this becomes

$$\mathfrak{A}(\theta a, \lambda b) = \int d(P_t a, b) \tilde{\theta}(t) \overline{\tilde{\lambda}(t)}, \tag{6}$$

where $\tilde{\theta}, \bar{\tilde{\lambda}} \in \mathscr{Z}$. Now, $\mathscr{Z} \cdot \mathscr{Z}$ is dense in $\mathscr{S}$ according to Lemma 8.4-2. Moreover, it follows from Theorem 8.7-1 that

$$\zeta \mapsto \int d(P_t a, b) \zeta(t)$$

is a continuous linear mapping of $\mathscr{S}$ into C. Hence, given any $a, b \in H$, $\mathfrak{A}$ determines $\int d(P_t a, b)\zeta(t)$ for all $\zeta \in \mathscr{S}$. This means that $(P_t a, b)$ is uniquely determined as a σ-finite complex measure on the bounded Borel sets in R (Appendix G12). Since this is true for every $a, b \in H$, $\mathfrak{A}$ uniquely determines P on those sets.

(ii) Let $a, b \in H$ as before. Set $\mathfrak{A}_{a,b}(\theta, \lambda) \triangleq \mathfrak{A}(\theta a, \lambda b)$, where $\theta, \lambda \in \mathscr{D}$. Then, $\mathfrak{A}_{a,b}$ is a separately continuous translation-invariant sesquilinear form on $\mathscr{D} \times \mathscr{D}$. Upon setting $\mathfrak{A}_a = \mathfrak{A}_{a,a}$ and using the polarization equation (Appendix A7), we may write

$$\mathfrak{A}_{a,b} = \tfrac{1}{4}[\mathfrak{A}_{a+b} - \mathfrak{A}_{a-b} + i\mathfrak{A}_{a+ib} - i\mathfrak{A}_{a-ib}].$$

Note that $\mathfrak{A}_a$ is positive on $\mathscr{D} \times \mathscr{D}$. So, by Theorem 8.6-2, for each $a \in H$, there exists a tempered positive measure μ_a such that

$$\mathfrak{A}_a(\theta, \lambda) = \int d\mu_{a,t} \tilde{\theta}(t) \overline{\tilde{\lambda}(t)}.$$

Upon setting

$$\mu_{a,b} = \tfrac{1}{4}[\mu_{a+b} - \mu_{a-b} + i\mu_{a+ib} - i\mu_{a-ib}],$$

we have that $\mu_{a,b}$ is a tempered complex measure and

$$\mathfrak{A}_{a,b}(\theta, \lambda) = \int d\mu_{a,b,t}\,\tilde{\theta}(t)\overline{\tilde{\lambda}(t)}, \qquad \theta, \lambda \in \mathscr{D}. \tag{7}$$

Note that $\mathfrak{A}$ uniquely determines $\mathfrak{A}_{a,b}$ for any given $a, b \in H$. Hence, by the argument in part (i) of this proof, $\mu_{a,b}$ is uniquely determined by the given $\mathfrak{A}$, a, and b.

(iii) We now show that, for each $a, b \in H$, $\mu_{a,b}$ has the representation

$$\mu_{a,b}(E) = \big(P(E)a, b\big), \tag{8}$$

where E is any bounded Borel set in R and $P(E) \in [H; H]_+$. First, note that, for any fixed $\theta, \lambda \in \mathscr{D}$, $\mathfrak{A}_{a,b}(\theta, \lambda)$ is sesquilinear in its dependence on a and b. It follows from (7) that $\{a, b\} \mapsto \mu_{a,b}(E)$ is a positive sesquilinear form on $H \times H$ because μ_a is a tempered positive measure. Furthermore, by Lemma 8.4-3, there exists a $\theta \in \mathscr{D}$ such that $|\tilde{\theta}(t)| \geq 1$ for all $t \in E$. So,

$$\mu_a(E) \leq \int_E d\mu_{a,t}|\tilde{\theta}(t)|^2 = \mathfrak{A}_a(\theta, \theta) = \mathfrak{A}(\theta a, \theta a).$$

Now, $\mathfrak{A}$ is separately continuous on $\mathscr{D}(H) \times \mathscr{D}(H)$ and therefore, according to Appendix F2, continuous. Hence, $\mu_a(E) \leq \mathfrak{A}(\theta a, \theta a) \leq M\|a\|^2$, where M does not depend on a. Therefore, by Appendix D15, (8) holds, and $P(E) \in [H; H]_+$.

(iv) Our next objective is to demonstrate that P is tempered. Since $\mu_{a,b}$ is a σ-finite complex measure, it follows from Theorem 2.2-1 that P is a σ-finite PO measure on $\mathfrak{C}_\infty$. We have to prove that g_q is integrable with respect to P for some q.

We start by showing that the mapping

$$\{\theta, a, b\} \mapsto \int d\mu_{a,b,t}\,\theta(t) \tag{9}$$

is continuous from $\mathscr{S} \times H \times H$ into C, as well as linear with respect to θ and a and antilinear with respect to b. (It is understood that $\mathscr{S} \times H \times H$ is supplied with the product topology.) Since both $\mathscr{S}$ and H are Fréchet spaces, we need merely establish that the mapping (9) is separately continuous (Appendix F2). Since $\mu_{a,b}$ is a tempered complex measure, the mapping is continuous and linear on $\mathscr{S}$ for fixed a and b.

Now, assume that θ and b are fixed and define the mapping $F_{\theta,b}\colon H \rightarrow C$ by

$$F_{\theta,b}(a) = \int d\mu_{a,b,t}\,\theta(t).$$

For the moment, assume in addition that

$$\theta = \zeta\bar{\lambda}, \qquad \zeta, \lambda \in \mathscr{D}. \tag{10}$$

Then, by (7),

$$F_{\theta, b}(a) = \int d\mu_{a, b, t}\, \tilde{\zeta}(t)\overline{\tilde{\lambda}(t)} = \mathfrak{A}_{a, b}(\zeta, \lambda) = \mathfrak{A}(\zeta a, \lambda b).$$

This implies that $F_{\theta, b}$ is continuous and linear because $\mathfrak{A}$ is separately continuous and sesquilinear. We now invoke Lemma 8.4-2; thus, for arbitrary $\theta \in \mathscr{S}$, we can choose a sequence $\{\theta_k\}_{k=1}^{\infty}$ such that each θ_k has the form (10) and $\theta_k \to \theta$ in $\mathscr{S}$. By what we have already shown, $F_{\theta_k, b} \in [H; C]$ and $F_{\theta_k, b}(a) \to F_{\theta, b}(a)$ for each $a \in H$. It follows that $F_{\theta, b} \in [H; C]$ also (Appendix D11). A similar argument shows that the mapping (9) is continuous and antilinear with respect to b.

We can summarize our results so far as follows. There exist two nonnegative integers q and k and a constant $M > 0$ such that, for all $a, b \in H$ with $\|a\| \leq 1$ and $\|b\| \leq 1$ and for all $\theta \in \mathscr{S}$ with

$$\max_{0 \leq p \leq k} \sup_{t \in R} |(1 + |t|^{2q})\theta^{(p)}(t)| \leq 1,$$

we have that

$$\left| \int d\mu_{a, b, t}\theta(t) \right| \leq M. \tag{11}$$

(See Appendix F2.)

Now, let $\theta \in \mathscr{D}$ be nonnegative and such that $\theta(t) = 1$ for $|t| \leq 1$. Set

$$\theta_j(t) = (1 + |t|^{2q})^{-1}\theta(j^{-1}t), \qquad j = 1, 2, \ldots .$$

Then,

$$\sup\{|(1 + |t|^{2q})\theta_j^{(p)}(t): t \in R, \quad 0 \leq p \leq k, \quad j = 1, 2, \ldots\} \triangleq c < \infty.$$

So, by (11), for all $a \in H$ with $\|a\| \leq 1$,

$$0 \leq \int d\mu_{a, t}\, \theta_j(t) \leq cM, \qquad j = 1, 2, \ldots,$$

where again $\mu_{a, t} \triangleq \mu_{a, a, t}$. So,

$$\left(\int_{|t| \leq j} dP_t(1 + |t|^{2q})^{-1}a, a \right) = \int_{|t| \leq j} d\mu_{a, t}(1 + |t|^{2q})^{-1}$$

$$\leq \int_R d\mu_{a, t}\theta_j(t) \leq cM.$$

This proves that, for all j,

$$\left\| \int_{|t| \leq j} dP_t(1 + |t|^{2q})^{-1} \right\|_{[H; H]} \leq cM.$$

(See Appendix D15.) By Theorem 2.3-1, g_q is integrable with respect to P. Thus, P is tempered.

(v) So far, we have shown that

$$\mathfrak{A}(\theta a, \lambda b) = \int d(P_t a, b)\tilde{\theta}(t)\overline{\tilde{\lambda}(t)}$$
$$= \left(\left[\int dP_t \tilde{\theta}(t)\overline{\tilde{\lambda}(t)}\right]a, b\right)$$

for all $\theta, \lambda \in \mathscr{D}$ and all $a, b \in H$. From the sesquilinearity of $\mathfrak{A}$, (1) of Section 2.6, and Definition 2.6-1, we see that (5) holds for all $\phi, \psi \in \mathscr{D} \odot H$. By the hypothesis and Appendix F2, $\mathfrak{A}$ is continuous on $\mathscr{D}(H) \times \mathscr{D}(H)$. Also, by Theorem 8.7-2, the right-hand side of (5) defines a continuous mapping on $\mathscr{S}(H) \times \mathscr{S}(H)$ and therefore on $\mathscr{D}(H) \times \mathscr{D}(H)$. But, since $\mathscr{D} \odot H$ is dense in $\mathscr{D}(H)$, $\mathscr{D} \odot H \times \mathscr{D} \odot H$ is dense in $\mathscr{D}(H) \times \mathscr{D}(H)$. Therefore, (5) holds for all $\phi, \psi \in \mathscr{D}(H)$. ◇

8.8. CERTAIN SEMIPASSIVE MAPPINGS OF $\mathscr{D}(H)$ INTO $\mathscr{E}(H)$

In this section, we study the properties of certain mappings of $\mathscr{D}(H)$ into $\mathscr{E}(H)$, which occur in subsequent realizability theorems. For any H-valued function ϕ on R, we set

$$\phi_t(x) \triangleq \phi(t - x)$$

and

$$\phi_{[t]}(x) \triangleq \begin{cases} \phi(t - x), & x \geq 0 \\ 0, & x < 0. \end{cases}$$

Also, $\tilde{\phi}_t \triangleq \mathfrak{F}(\phi_t)$, and $\tilde{\phi}_{[t]} \triangleq \mathfrak{F}(\phi_{[t]})$.

Lemma 8.8-1. *For any $\phi \in \mathscr{D}(H)$ and any positive p,*

$$\eta^p \tilde{\phi}_t(\eta) = i^p \int_{-\infty}^{\infty} \phi^{(p)}(t - x)(e^{-ix\eta} - 1)\, dx,$$

$$\eta^p \tilde{\phi}_{[t]}(\eta) = i^p \int_0^{\infty} \phi^{(p)}(t - x)(e^{-i\eta x} - 1)\, dx - \sum_{r=1}^{p-1} i^r \eta^{p-r} \phi^{(r-1)}(t).$$

This lemma can be established by integrating by parts.

Lemma 8.8-2. *Let P be a PO measure and set*

$$j(x) \triangleq \int_{-\infty}^{\infty} dP_\eta e^{ix\eta}.$$

Also, for $\phi \in \mathscr{D}(H)$, set

$$(\mathfrak{M}_1 \phi)(t) \triangleq \int_{-\infty}^{\infty} j(x)\phi(t-x)\, dx \tag{1}$$

and

$$(\mathfrak{M}_2 \phi)(t) \triangleq \int_{0}^{\infty} j(x)\phi(t-x)\, dx. \tag{2}$$

These equations are equivalent to

$$(\mathfrak{M}_1 \phi)(t) = \int_{-\infty}^{\infty} dP_\eta \tilde{\phi}_t(-\eta) \tag{3}$$

and

$$(\mathfrak{M}_2 \phi)(t) = \int_{-\infty}^{\infty} dP_\eta \tilde{\phi}_{[t]}(-\eta). \tag{4}$$

Moreover,

$$\int_{-\infty}^{\infty} (\mathfrak{M}_1 \phi(t), \psi(t))\, dt = \int_{-\infty}^{\infty} d(P_\eta \tilde{\phi}(\eta), \tilde{\psi}(\eta)), \qquad \phi, \psi \in \mathscr{D}(H), \tag{5}$$

and

$$\operatorname{Re} \int_{-\infty}^{\tau} (\mathfrak{M}_2 \phi(t), \phi(t))\, dt = \tfrac{1}{2} \int_{-\infty}^{\infty} d(P_\eta \tilde{\psi}(\eta), \tilde{\psi}(\eta)), \qquad \phi \in \mathscr{D}(H), \quad \tau \in R, \tag{6}$$

where in the last equation,

$$\psi(t) \triangleq \begin{cases} \phi(t), & t \le \tau, \\ 0, & t > 0. \end{cases} \tag{7}$$

PROOF: Equations (3) and (4) follow directly from Theorem 2.5-2, which also asserts that the integrals in (1) and (2) exist as Bochner integrals. It can be shown that $x \mapsto j(x)\phi(t-x)$, $\mathfrak{M}_1\phi$, and $\mathfrak{M}_2 \phi$ are all continuous H-valued functions. (See Problem 8.8-1.) Consequently, the left-hand sides of (5) and (6) exist. The right-hand sides of (5) and (6) also exist by virtue of Theorem 2.6-1. Moreover,

$$\begin{aligned} \int_R (\mathfrak{M}_1\phi(t), \psi(t))\, dt &= \int_R \left(\int_R j(x)\phi(t-x)\, dx, \psi(t) \right) dt \\ &= \int dt \int (j(t-x)\phi(x), \psi(t))\, dx \\ &= \int dt \int \left(\int dP_\eta e^{i(t-x)\eta} \phi(x), \psi(t) \right) dx. \end{aligned}$$

By Theorem 2.6-1 again, the right-hand side is equal to

$$\int d(P_\eta \tilde{\phi}(\eta), \tilde{\psi}(\eta)).$$

This verifies (5).

Next, consider

$$I_\tau(\phi) \triangleq \int_{-\infty}^{\tau} (\mathfrak{M}_2 \phi(t), \phi(t))\, dt$$
$$= \int_{-\infty}^{\tau} dt \int_{-\infty}^{t} (j(t-x)\phi(x), \phi(t))\, dx.$$

Since the values of P_η are positive operators and therefore self-adjoint, it follows that the adjoint of $j(x)$ is $j(-x)$. By using this fact, we see that

$$\overline{I_\tau(\phi)} = \int_{-\infty}^{\tau} dt \int_{-\infty}^{t} (\phi(t), j(t-x)\phi(x))\, dx$$
$$= \int_{-\infty}^{\tau} dx \int_{x}^{\tau} (j(x-t)\phi(t), \phi(x))\, dt.$$

Upon reversing the roles of t and x in the right-hand side, adding $I_\tau(\phi)$ to $\overline{I_\tau(\phi)}$, and using (7), we get

$$2 \operatorname{Re} I_\tau(\phi) = \int_{-\infty}^{\tau} dt \int_{-\infty}^{\tau} (j(t-x)\phi(x), \phi(t))\, dx$$
$$= \int_R dt \int_R (j(t-x)\psi(x), \psi(t))\, dx.$$

An application of Theorem 2.6-1 as before yields (6). ◇

In the rest of this section, Q is a tempered PO measure. Also, $p = 2q$, where $q = 1, 3, 5, \ldots$; hence, $i^p = -1$. Set

$$P(E) = \int_E dQ_\eta (1 + \eta^p)^{-1},$$

where E is any Borel set in R and p is chosen so large that $(1+\eta^p)^{-1}$ is integrable with respect to Q. Thus, P is a PO measure according to Theorem 2.3-3. We define j as in the preceding lemma. Also, $\mathfrak{B}$ and $\hat{\mathfrak{B}}$ denote the sesquilinear forms defined in Section 8.3; namely, for any operator $\mathfrak{N}$ from $\mathscr{D}(H)$ into $L_1^{\text{loc}}(H)$,

$$\mathfrak{B}(\phi, \psi) \triangleq \mathfrak{B}_{\mathfrak{N}}(\phi, \psi) \triangleq \int_R (\mathfrak{N}\phi(t), \psi(t))\, dt, \qquad \phi, \psi \in \mathscr{D}(H)$$

and

$$\hat{\mathfrak{B}}(\phi, \psi) \triangleq \hat{\mathfrak{B}}_{\mathfrak{N}}(\phi, \psi) \triangleq \tfrac{1}{2}[\mathfrak{B}(\phi, \psi) + \overline{\mathfrak{B}(\psi, \phi)}].$$

Theorem 8.8-1. *For any $\phi \in \mathscr{D}(H)$, set*

$$\mathfrak{N}_1 \phi(t) \triangleq \int_{-\infty}^{\infty} j(x)\phi(t-x)\,dx + \int_{-\infty}^{\infty} [j(0) - j(x)]\phi^{(p)}(t-x)\,dx \tag{8}$$

and

$$\mathfrak{N}_2 \phi(t) \triangleq 2\int_0^{\infty} j(x)\phi(t-x)\,dx + 2\int_0^{\infty} [j(0) - j(x)]\phi^{(p)}(t-x)\,dx. \tag{9}$$

Then, $\mathfrak{N}_1$ and $\mathfrak{N}_2$ are linear translation-invariant semipassive mappings of $\mathscr{D}(H)$ into $\mathscr{E}(H)$. In addition, $\mathfrak{N}_2$ is causal on $\mathscr{D}(H)$. Finally, for $\mathfrak{N} = \mathfrak{N}_1$ or $\mathfrak{N} = \mathfrak{N}_2$,

$$\hat{\mathfrak{B}}(\phi, \psi) \triangleq \hat{\mathfrak{B}}_{\mathfrak{N}}(\phi, \psi) = \int_R d(Q_\eta \tilde{\phi}(\eta), \tilde{\psi}(\eta)) \tag{10}$$

for all $\phi, \psi \in \mathscr{D}(H)$.

PROOF. As before, the integrals in (8) and (9) exist as Bochner integrals, and $x \mapsto j(x)\phi(t-x)$, $\mathfrak{N}_1\phi$, and $\mathfrak{N}_2\phi$ are continuous H-valued functions. Clearly, $\mathfrak{N}_1$ and $\mathfrak{N}_2$ are linear and translation invariant, and in addition $\mathfrak{N}_2$ is causal. If we can show that $\mathfrak{N}_1$ and $\mathfrak{N}_2$ are semipassive on $\mathscr{D}(H)$, then it will follow from Theorem 8.3-1 that $\mathfrak{N}_1$ and $\mathfrak{N}_2$ are continuous from $\mathscr{D}(H)$ into $[\mathscr{D}; H]$. This will imply in turn that $\mathfrak{N}_1$ and $\mathfrak{N}_2$ are convolution operators (Theorem 5.10-1) and therefore map $\mathscr{D}(H)$ into $\mathscr{E}(H)$ (Theorem 5.5-1).

First, consider the case of $\mathfrak{N} = \mathfrak{N}_1$. An application of Theorem 2.5-2 yields

$$\begin{aligned}(\mathfrak{N}_1\phi)(t) = &\int_R dP_\eta \int_R e^{i\eta x}\phi(t-x)\,dx \\ &+ \int_R dP_\eta \int_R (1 - e^{i\eta x})\phi^{(p)}(t-x)\,dx.\end{aligned} \tag{11}$$

For each $k = 1, 2, \ldots$, set

$$\begin{aligned}(\mathfrak{N}_{1,k}\phi)(t) \triangleq &\int_{-k}^{k} dP_\eta \int_R e^{i\eta x}\phi(t-x)\,dx \\ &+ \int_{-k}^{k} dP_\eta \int_R (1 - e^{i\eta x})\phi^{(p)}(t-x)\,dx. \qquad (12\end{aligned}$$

By using Theorems 2.2-3 and 2.5-1 and the extension of (2) of Section 2.5 onto $\mathscr{G} \hat{\odot} H \supset \mathscr{D}(H)$, we obtain the estimate

$$\begin{aligned}|(\mathfrak{N}_{1,k}\phi(t), \psi(t))| \leq \|P(R)\|_{[H;H]} &\int_R [\|\phi(x)\|_H + 2\|\phi^{(p)}(x)\|_H] \\ &\times dx \|\psi(t)\|_H.\end{aligned} \tag{13}$$

where $\phi, \psi \in \mathscr{D}(H)$.

By Theorem 2.5-2 again, we may also write

$$(\mathfrak{N}_{1,k}\phi)(t) = \int_R \left[\int_{-k}^{k} dP_\eta e^{i\eta x}\right]\phi(t-x)\,dx$$
$$+ \int_R \left[\int_{-k}^{k} dP_\eta(1 - e^{i\eta x})\right]\phi^{(p)}(t-x)\,dx. \tag{14}$$

It is not difficult to show that the quantities in both pairs of brackets on the right-hand side converge in the strong operator topology as $k \to \infty$ uniformly for all $x \in R$. We can infer from this that $(\mathfrak{N}_{1,k}\phi)(t) \to (\mathfrak{N}_1\phi)(t)$ in H for each $t \in R$. This result in conjunction with (13) allows us to invoke Lebesgue's theorem of dominated convergence (Appendix G18) to conclude that, as $k \to \infty$,

$$\int_R (\mathfrak{N}_{1,k}\phi(t), \psi(t))\,dt \to \int_R (\mathfrak{N}_1\phi(t), \psi(t))\,dt. \tag{15}$$

On the other hand, by Lemma 8.8-1, Definition 2.6-2, and Theorem 2.5-2 again,

$$(\mathfrak{N}_{1,k}\phi)(t) = \int_{-k}^{k} dP_\eta(1 + \eta^p)\tilde{\phi}_t(-\eta) = \int_{-k}^{k} dQ_\eta \tilde{\phi}_t(-\eta).$$

Therefore, by Lemma 8.8-2,

$$\int_{-\infty}^{\infty} (\mathfrak{N}_{1,k}\phi(t), \psi(t))\,dt = \int_{-k}^{k} d(Q_\eta \tilde{\phi}(\eta), \tilde{\psi}(\eta)), \qquad \phi, \psi \in \mathscr{D}(H). \tag{16}$$

According to Problem 2.6-4, as $k \to \infty$,

$$\int_{-k}^{k} d(Q_\eta \tilde{\phi}(\eta), \tilde{\psi}(\eta)) \to \int_{-\infty}^{\infty} d(Q_\eta \tilde{\phi}(\eta), \tilde{\psi}(\eta)). \tag{17}$$

Upon combining (15)–(17), we get

$$\mathfrak{B}_{\mathfrak{N}_1}(\phi, \psi) \triangleq \int_R (\mathfrak{N}_1\phi(t), \psi(t))\,dt = \int_{-\infty}^{\infty} d(Q_\eta \tilde{\phi}(\eta), \tilde{\psi}(\eta)).$$

This implies (10). Indeed, the values of Q are positive operators and therefore self-adjoint. Hence, in view of Definition 2.6-3 and (1) of Section 2.6, we may write in effect

$$\overline{\mathfrak{B}_{\mathfrak{N}_1}(\psi, \phi)} = \int \overline{(\mathfrak{N}_1\psi, \phi)}\,dt = \int d\overline{(Q\tilde{\psi}, \tilde{\phi})} = \int d(\tilde{\phi}, Q\tilde{\psi})$$
$$= \int d(Q\tilde{\phi}, \tilde{\psi}) = \int (\mathfrak{N}_1\phi, \psi)\,dt = \mathfrak{B}_{\mathfrak{N}_1}(\phi, \psi).$$

But, for $\phi = \psi$, the right-hand side of (10) is nonnegative. Hence, $\mathfrak{N}_1$ is semipassive on $\mathscr{D}(H)$.

We now take up the case where $\mathfrak{N} = \mathfrak{N}_2$. Again by Theorem 2.5-2,

$$(\mathfrak{N}_2 \phi)(t) = 2 \int_R dP_\eta \int_0^\infty e^{i\eta x}\phi(t-x)\,dx + 2\int_R dP_\eta \int_0^\infty (1 - e^{i\eta x})\phi^{(p)}(t-x)\,dx.$$

Following our previous procedure, we set

$$(\mathfrak{N}_{2,k}\phi)(t) \triangleq 2\int_{-k}^{k} dP_\eta \int_0^\infty e^{i\eta x}\phi(t-x)\,dx + 2\int_{-k}^{k} dP_\eta \int_0^\infty (1-e^{i\eta x})\phi^{(p)}(t-x)\,dx$$

and conclude again that

$$\int_R (\mathfrak{N}_{2,k}\phi(t), \psi(t))\,dt \to \int_R (\mathfrak{N}_2\phi(t), \psi(t))\,dt \tag{18}$$

as $k \to \infty$.

As the next step, we shall show that

$$\operatorname{Re}\int_R (\mathfrak{N}_{2,k}\phi(t), \phi(t))\,dt = \int_{-k}^{k} d(Q_\eta \tilde{\phi}(\eta), \tilde{\phi}(\eta)), \qquad \phi \in \mathscr{D}(H). \tag{19}$$

From Lemma 8.8-1, we see that

$$\tfrac{1}{2}\mathfrak{N}_{2,k}\phi(t) = \int_{-k}^{k} dP_\eta(1+\eta^p)\tilde{\phi}_{[t]}(-\eta) + \sum_{r=1}^{p-1} i^r \int_{-k}^{k} dP_\eta(-\eta)^{p-r}\phi^{(r-1)}(t). \tag{20}$$

We now set $\mathfrak{T}_r \triangleq i^r \int_{-k}^{k} dP_\eta(-\eta)^{p-r}$. Since the values of P are positive operators and therefore self-adjoint, it follows that $\mathfrak{T}_r' = (-1)^r\mathfrak{T}_r$, where $r = 1, \ldots, p-1$. Upon integrating by parts, we see that

$$\operatorname{Re}\int_R (\mathfrak{T}_r\phi^{(r-1)}(t), \phi(t))\,dt = 0.$$

We infer from this result and (20) that

$$\tfrac{1}{2}\operatorname{Re}\int_R (\mathfrak{N}_{2,k}\phi(t), \phi(t))\,dt = \operatorname{Re}\int_R \left(\int_{-k}^{k} dQ_\eta \tilde{\phi}_{[t]}(-\eta), \phi(t)\right)dt.$$

Since Q is a PO measure on $[-k, k]$, we may invoke (6) and take τ larger than any support point of ϕ to equate the right-hand side to

$$\tfrac{1}{2}\int_{-k}^{k} d(Q_\eta\tilde{\phi}(\eta), \tilde{\phi}(\eta)).$$

This establishes (19).

Observe that (17) still holds in the present case. The combination of (17)–(19) establishes

$$\mathfrak{B}_{\mathfrak{N}_2}(\phi, \phi) \triangleq \operatorname{Re}\int_R (\mathfrak{N}_2\phi(t), \phi(t))\,dt = \int_R d(Q_\eta\tilde{\phi}(\eta), \tilde{\phi}(\eta)). \tag{21}$$

Since the right-hand side is nonnegative, we have hereby proved the semi-passivity of $\mathfrak{N}_2$ on $\mathscr{D}(H)$. Finally, we need merely apply the polarization equation (Appendix A7) to both sides of (21) in order to get (10). ◇

Problem 8.8-1. With P being a PO measure, prove that

$$\int_{-k}^{k} dP_\eta e^{i\eta x} \to \int_{-\infty}^{\infty} dP_\eta e^{i\eta x}$$

in the strong operator topology uniformly for all $x \in R$. Then, prove the assertion made in the proof of Lemma 8.8-2 that $x \mapsto j(x)\phi(t-x)$, $\mathfrak{M}_1\phi$, and $\mathfrak{M}_2\phi$ are all continuous H-valued functions. Also, prove the assertion made in the proof of Theorem 8.8-1 that $(\mathfrak{N}_{1,k}\phi)(t) \to (\mathfrak{N}_1\phi)(t)$ in H for every $t \in R$.

8.9. AN EXTENSION OF THE BOCHNER–SCHWARTZ THEOREM

The Bochner–Schwartz theorem (Theorem 8.4-1) gives a representation of any complex-valued positive-definite distribution as the Fourier transform of a distribution generated by a tempered positive measure. Hackenbroch's extension (Hackenbroch, 1969, Corollary 3.5) of this representation to operator-valued distributions is the subject of the present section. A similar extension is given by Kritt (1968).

Definition 8.9-1. An $f \in [\mathscr{D}(H); H]$ is said to be *positive-definite* if, for all $\phi \in \mathscr{D}(H)$,

$$\int_{-\infty}^{\infty} (f * \phi(t), \phi(t))\, dt \geq 0. \tag{1}$$

Theorem 8.9-1. *Corresponding to each positive-definite* $f \in [\mathscr{D}(H); H]$, *there exists a tempered* PO *measure* M *such that, for every* $\phi \in \mathscr{D}(H)$,

$$\langle f, \phi \rangle = \int_R dM\tilde{\phi}. \tag{2}$$

Note. The customary definition of the Fourier transform of a complex-valued distribution (see, for example, Zemanian, 1965, p. 203) can be extended to the members of $[\mathscr{D}(H); H]$. By virtue of this, this theorem can be restated by saying that every positive-definite distribution in $[\mathscr{D}(H); H]$ is the Fourier transform of a tempered $[H; H]$-valued distribution generated by a tempered PO measure. In this regard, see also Theorem 8.7-1.

PROOF. Set $\mathfrak{N} = f*$. Then, $\mathfrak{N}$ is a continuous, linear, translation-invariant mapping of $\mathscr{D}(H)$ into $\mathscr{E}(H)$. As usual, set

$$\mathfrak{B}(\phi, \psi) \triangleq \mathfrak{B}_{\mathfrak{N}}(\phi, \psi) \triangleq \int_R (\mathfrak{N}\phi(t), \psi(t))\, dt, \qquad \phi, \psi \in \mathscr{D}(H).$$

Then, $\mathfrak{B}$ is a positive, separately continuous, translation-invariant, sesquilinear form on $\mathscr{D}(H) \times \mathscr{D}(H)$. So, by Theorem 8.7-3, there exists a unique tempered PO measure Q such that

$$\mathfrak{B}(\phi, \psi) = \int_R d(Q_\eta \tilde{\phi}(\eta), \tilde{\psi}(\eta)), \qquad \phi, \psi \in \mathscr{D}(H). \tag{3}$$

Now, starting with Q, we proceed as in Section 8.8 to define P and j. We then define $\mathfrak{N}_1$ by (8) of Section 8.8. As was shown in the proof of Theorem 8.8-1, $\mathfrak{B}_{\mathfrak{N}_1}(\phi, \psi)$ is equal to the right-hand side of (3). Therefore,

$$\int_R (\mathfrak{N}\phi, \psi)\, dt = \int_R (\mathfrak{N}_1\phi, \psi)\, dt.$$

Upon replacing ψ by $\theta_j a$, where $a \in H$, $j = 1, 2, \ldots$, and $\theta_j \to \sigma_\tau \delta$ for an arbitrarily chosen $\tau \in R$, we see that $\mathfrak{N} = \mathfrak{N}_1$. Furthermore, (11) of Section 8.8 and Lemma 8.8-1 show that

$$(\mathfrak{N}_1\phi)(t) = \int_R dQ_\eta\, \tilde{\phi}_t(-\eta).$$

Since $\phi_t(x) \triangleq \phi(t - x)$ and $\check{\phi}(x) \triangleq \phi(-x)$, we may write $\check{\phi}_0 = \phi$ and

$$\langle f, \phi\rangle = (\mathfrak{N}\check{\phi})(0) = (\mathfrak{N}_1\check{\phi})(0) = \int_R dQ\tilde{\phi}(-\eta).$$

This yields (2) when we set $M(E) \triangleq Q(-E)$ for any bounded Borel set E. ◇

Problem 8.9-1. Show that, if $f \in [\mathscr{D}(H); H]$ possesses the representation (2), then f is positive-definite.

8.10. REPRESENTATIONS FOR CERTAIN CAUSAL SEMIPASSIVE MAPPINGS

We now develop Hackenbroch's representations for linear, translation-invariant, causal, semipassive mappings on $\mathscr{D}(H)$. The first result is a time-domain characterization.

Theorem 8.10-1. *$\mathfrak{N}$ is a linear, translation-invariant, causal, semipassive mapping on $\mathscr{D}(H)$ if and only if, for every $\phi \in \mathscr{D}(H)$,*

$$(\mathfrak{N}\phi)(t) = \sum_{k=0}^{n} P_k \phi^{(k)}(t) + \int_0^\infty j(x)\phi(t-x)\,dx$$

$$+ \int_0^\infty [j(0) - j(x)]\phi^{(p)}(t-x)\,dx, \tag{1}$$

where the following conditions are satisfied: $P_k \in [H; H]$ and $P_k' = (-1)^{k+1}P_k$. (As before, P_k' is the adjoint of P_k.) Also, $p = 2q$, where q is an odd positive integer. Finally,

$$j(x) = \int_R dP_\eta e^{i\eta x}, \tag{2}$$

where P_η is a PO *measure on the Borel subsets of R.*

PROOF. We first derive (1) from the stated properties of $\mathfrak{N}$. As in the proof of Theorem 8.9-1, $\hat{\mathfrak{B}}_{\mathfrak{N}}$ is a positive, separately continuous, translation-invariant, sesquilinear form on $\mathscr{D}(H) \times \mathscr{D}(H)$. By Theorem 8.7-3, there exists a unique tempered PO measure Q such that

$$\hat{\mathfrak{B}}_{\mathfrak{N}}(\phi, \psi) = \tfrac{1}{2}\int_R d(Q_\eta \tilde{\phi}(\eta), \tilde{\psi}(\eta)). \tag{3}$$

We define P from Q as in Section 8.8, j by (2), and $\mathfrak{N}_2$ by (9) of Section 8.8. We also set $\mathfrak{M} \triangleq \frac{1}{2}\mathfrak{N}_2$. Upon comparing (10) of Section 8.8 with (3), we see that $\hat{\mathfrak{B}}_{\mathfrak{M}} = \hat{\mathfrak{B}}_{\mathfrak{N}}$. Theorem 8.8-1 also states that $\mathfrak{M}$ is a linear, translation-invariant, causal, semipassive mapping on $\mathscr{D}(H)$. Therefore, so, too, is $\mathfrak{N} - \mathfrak{M}$. Since $\hat{\mathfrak{B}}_{\mathfrak{N}-\mathfrak{M}} = \hat{\mathfrak{B}}_{\mathfrak{N}} - \hat{\mathfrak{B}}_{\mathfrak{M}} = 0$, it follows from Theorem 8.5-2 that

$$[(\mathfrak{N} - \mathfrak{M})\phi](t) = \sum_{k=0}^{n} P_k \phi^{(k)}(t), \qquad \phi \in \mathscr{D}(H).$$

This combined with the definition of $\mathfrak{M}$ yields (1).

That any operator satisfying (1) possesses the stated properties follows from Theorem 8.8-1 and the identity

$$\operatorname{Re} \int_R (P_k \phi^{(k)}(t), \phi(t))\,dt = 0,$$

which is a result of the condition $P_k' = (-1)^{k+1}P_k$. ◇

We turn to a frequency-domain characterization for the Laplace transform of the unit-impulse response of the operator $\mathfrak{N}$. With j defined by (2), we have that

$$\| j(x) \|_H \leq \| P(R) \|_{[H;H]}.$$

Moreover, for any $\phi \in \mathscr{E}(H)$, $x \mapsto j(x)\phi(x)$ is a continuous H-valued function according to Problem 8.8-1. As a result, the mapping

$$\phi \mapsto \int_0^\infty j(x)\phi(x)\,dx + \int_0^\infty [j(0) - j(x)]\phi^{(p)}(x)\,dx$$

is a member of $[\mathscr{D}_{L_1}(H); H]$. Indeed,

$$\left\| \int_0^\infty j(x)\phi(x)\,dx \right\|_H \le \|P(R)\|_{[H;\,H]} \int_{-\infty}^\infty \|\phi(x)\|_H\,dx,$$

and a similar estimate holds for the second integral.

With $\mathfrak{N}$ defined by (1), we have from Theorem 5.5-2 that $\mathfrak{N} = y\,*$, where $y \in [\mathscr{D}_{L_1}(H); H]$ is defined on any $\phi \in \mathscr{L}_{a,b}(H)$ with $b < 0 < a$ by

$$\langle y, \phi\rangle \triangleq \sum_{k=0}^n (-1)^k P_k \phi^{(k)}(0) + \int_0^\infty j(x)\phi(x)\,dx + \int_0^\infty [j(0) - j(x)]\phi^{(p)}(x)\,dx. \tag{4}$$

By continuous extension, this equation is seen to hold for all $\phi \in \mathscr{D}_{L_1}(H)$. Clearly, supp $y \subset [0, \infty)$. By Lemma 7.2-1, we now have that $y \in [\mathscr{L}(0, \infty; H); H]$. Consequently, the Laplace transform Y of y exists and, for every $\zeta \in C_+$ and $a \in H$, we have

$$\begin{aligned} Y(\zeta)a &\triangleq [(\mathfrak{L}y)(\zeta)]a = \langle y(x), ae^{-\zeta x}\rangle \\ &= \sum_{k=0}^n P_k a\zeta^k + \int_0^\infty dx \int_R dP_\eta a[e^{i\eta x} + \zeta^p(1 - e^{i\eta x})]e^{-\zeta x}. \end{aligned}$$

Theorem 2.5-2 allows us to reverse the order of integration. We then integrate with respect to x and note that $a \in H$ is arbitrary, to get

$$Y(\zeta) = \sum_{k=0}^n P_k \zeta^k + \int_R dP_\eta \frac{1 - i\eta\zeta^{p-1}}{\zeta - i\eta}, \qquad \zeta \in C_+. \tag{5}$$

These results combined with Theorem 8.10-1 and the uniqueness property of the Laplace transformation yield the following frequency-domain realizability theorem.

Theorem 8.10-2. *If $\mathfrak{N}$ is a linear, translation-invariant, causal, semipassive mapping on $\mathscr{D}(H)$, then $\mathfrak{N}$ is a convolution operator $y\,*$ where $y \in [\mathscr{D}_{L_1}(H); H]$ satisfies* (4). *Moreover, $Y(\zeta) \triangleq (\mathfrak{L}y)(\zeta)$ exists for at least all $\zeta \in C_+$ and satisfies* (5). *Here, P_k, P_η, and p satisfy the conditions stated in Theorem* 8.10-1.

Conversely, for every function Y having the representation (5), *there exists a* *unique convolution operator* $\mathfrak{N} = y\,*$ *such that* $(\mathfrak{L}y)(\zeta) = Y(\zeta)$ *for* $\zeta \in C_+$. *Moreover,* $y \in [\mathscr{D}_{L_1}(H); H]$, *and y is given on any* $\phi \in \mathscr{D}_{L_1}(H)$ *by* (4). *Finally,* $\mathfrak{N}$ *is a linear, translation-invariant, causal, semipassive mapping on* $\mathscr{D}(H)$.

8.11. A REPRESENTATION FOR POSITIVE* TRANSFORMS

If in Theorem 8.10-2 the condition of semipassivity on $\mathscr{D}(H)$ is replaced by the stronger requirement of passivity on $\mathscr{D}(H)$, then the representation (5) in the preceding section takes on a stricter form. In particular, the following additional conditions occur: $n = 1$, $P_1 \in [H; H]_+$, and $p = 2$. The proof of this result, which is given in the next section, is based upon a representation due to Schwindt (1965) for certain $[H; H]$-valued functions, which we call positive*.

Definition 8.11-1. A function Y of the complex variable ζ is said to be a *positive* mapping of H into H* (or simply *positive**) if Y is an $[H; H]$-valued analytic function on C_+ such that $\operatorname{Re}\big(Y(\zeta)a, a\big) \geq 0$ for every $\zeta \in C_+$ and $a \in H$.

When $H = C$, $[C; C]$ may be identified with C, in which case the following classical result is in force.

Theorem 8.11-1. *F is a complex-valued positive* function if and only if, for all* $\zeta \in C_+$, *F admits the representation*

$$F(\zeta) = i \operatorname{Im} F(1) + X\zeta + \int_R dN_\eta \frac{1 - i\zeta\eta}{\zeta - i\eta}, \tag{1}$$

where $X \in R$, $X \geq 0$, *and N is a finite positive measure on the Borel subsets of R. F uniquely determines X among the complex numbers and N among the complex measures.*

A proof of this is given by Akhiezer and Glazman (1963, Section 59). Actually, in their version, N is a bounded nondecreasing function on R and the integral is interpreted in the Stieltjes sense. However, this is entirely equivalent to the present statement; see Zaanen, 1967, pp. 33–34, 63–64.

Schwindt's representation is a generalization of (1) to the case where F is $[H; H]$-valued, as follows.

Theorem 8.11-2. *Y is a positive* mapping of H into H if and only if, for all $\zeta \in C_+$, Y can be represented by*

$$Y(\zeta) = P_1\zeta + P_0 + \int_R dP_\eta \frac{1 - i\eta\zeta}{\zeta - i\eta}, \tag{2}$$

where $P_1 \in [H; H]_+$, P_0 is a skew-adjoint member of $[H; H]$, and P_η is a PO *measure on the Borel subsets of R.*

PROOF. Assume that Y is defined by (2). Since

$$\eta \mapsto (1 - i\eta\zeta)/(\zeta - i\eta)$$

is a member of $\mathscr{G}$, it follows from (4) of Section 2.2 and Rudin (1966, p. 201) that the integral on the right-hand side of (2) is weakly analytic and therefore analytic on C_+. Therefore, so, too, is Y. Furthermore, for all $a \in H$, we have that $(P_1 a, a) \geq 0$, $(P(\cdot)a, a)$ is a finite positive measure, and $(P_0 a, a)$ is imaginary. Also,

$$\operatorname{Re} \frac{1 - i\eta\zeta}{\zeta - i\eta} \geq 0$$

for all $\zeta \in C_+$. It follows that $\operatorname{Re}(Y(\zeta)a, a) \geq 0$ for $\zeta \in C_+$ and $a \in H$. So, truly, Y is positive*.

Conversely, assume that Y is positive* and set

$$F_\zeta(a, b) = (Y(\zeta)a, b), \qquad a, b \in H. \tag{3}$$

Thus, for each fixed $\zeta \in C_+$, F_ζ is a continuous sesquilinear form on $H \times H$. Also, for fixed $a \in H$, $\zeta \mapsto F_\zeta(a, a)$ is a complex-valued positive* function. So, by Theorem 8.11-1,

$$F_\zeta(a, a) = i \operatorname{Im} F_1(a, a) + X(a)\zeta + \int_R dN_\eta(a) \frac{1 - i\zeta\eta}{\zeta - i\eta}, \tag{4}$$

where, for each fixed $a \in H$, $X(a)$ is a real nonnegative number and $N_\eta(a)$ is a finite positive measure. The notation $N_\eta(a)$ denotes the measure $E \mapsto [N_\eta(a)](E)$, where E is any Borel subset of R. Upon setting $P_0 \triangleq \frac{1}{2}[Y(1) - Y(1)']$, we see immediately that P_0 is skew-adjoint and that

$$(P_0 a, a) = i \operatorname{Im} F_1(a, a). \tag{5}$$

Now, from (3) and (4) and the fact that $X(a)\zeta$ is uniquely determined by $F_\zeta(a, a)$, we see that X satisfies the following two identities. For all $a, b \in H$ and $\beta \in C$,

$$X(\beta a) = |\beta|^2 X(a) \tag{6}$$

and

$$X(a+b) + X(a-b) = 2X(a) + 2X(b). \tag{7}$$

Similarly,

$$N_\eta(\beta a) = |\beta|^2 N_\eta(a) \tag{8}$$

and

$$N_\eta(a+b) + N_\eta(a-b) = 2N_\eta(a) + 2N_\eta(b). \tag{9}$$

Next, set

$$x(a, b) \triangleq \tfrac{1}{4}[X(a+b) - X(a-b) + iX(a+ib) - iX(a-ib)]. \tag{10}$$

Any functional on H that takes on only nonnegative values and satisfies (6) and (7) defines through (10) a positive sesquilinear form x on $H \times H$ such that

$$x(a, a) = X(a) \geq 0. \tag{11}$$

(See, for example, Kurepa, 1965.) Similarly, we define the complex measure $Q_\eta(a, b)\colon E \mapsto [Q_\eta(a, b)](E)$ on the Borel subsets E of R by

$$Q_\eta(a, b) \triangleq \tfrac{1}{4}[N_\eta(a+b) - N_\eta(a-b) + iN_\eta(a+ib) - iN_\eta(a-ib)]. \tag{12}$$

Again (see Kurepa, 1965), for every E, $\{a, b\} \mapsto [Q_\eta(a, b)](E)$ is a positive sesquilinear form on $H \times H$ such that

$$[Q_\eta(a, a)](E) = [N_\eta(a)](E) \geq 0. \tag{13}$$

With these results, we see from (4), (11), and (13) that

$$\operatorname{Re} F_1(a, a) = x(a, a) + [Q_\eta(a, a)](R).$$

Thus,

$$x(a, a) \leq |F_1(a, a)| \leq \|Y(1)\| \, \|a\|^2.$$

We may now conclude from Appendix D15 that there exists a unique $P_1 \in [H; H]_+$ such that $(P_1 a, b) = x(a, b)$ for all $a, b \in H$. In the same way, we see that, for every Borel set $E \subset R$, there exists a $P_\eta(E) \in [H; H]_+$ such that $\big(P_\eta(E)a, b\big) = [Q_\eta(a, b)](E)$. In fact, Theorem 2.2-1 asserts that P_η is a PO measure.

Upon combining these results, we see from (4) that, for all $a \in H$,

$$\big(Y(\zeta)a, a\big) = (P_0 a, a) + (P_1 a, a)\zeta + \int_R d(P_\eta a, a)\, \frac{1 - i\zeta\eta}{\zeta - i\eta},$$

where P_0, P_1, and P_η satisfy the conditions stated in the theorem. Upon appealing to the polarization equation and Equation (4) of Section 2.2, we finally obtain (2). ◇

For any $a \in H$ and any $\sigma \in R_+$, (2) yields

$$\frac{(Y(\sigma)a, a)}{\sigma} = (P_1 a, a) + \frac{(P_0 a, a)}{\sigma} + \int_R d(P_\eta a, a)\left[\frac{1 + \eta^2 + i\eta(\sigma^{-1} - \sigma)}{\sigma^2 + \eta^2}\right].$$

Both the real and imaginary parts of the quantity within the brackets are bounded on the domain

$$\{\{\sigma, \eta\}: 1 < \sigma < \infty, \quad -\infty < \eta < \infty\}$$

and tend uniformly to zero as $\sigma \to \infty$ on any compact subset of the η axis. As a consequence, the integral tends to zero and $\sigma^{-1}(Y(\sigma)a, a)$ tends to $(P_1 a, a)$. By the polarization equation, therefore,

$$\lim_{\sigma \to \infty} \sigma^{-1}(Y(\sigma)a, b) = (P_1 a, b) \tag{14}$$

for all $a, b \in H$. We shall make use of this result in a moment.

When studying real passive operators in Section 8.13, we will meet a special type of positive* functions, namely, the positive*-real functions. In anticipation of this, we determine the special form that Schwindt's theorem assumes for these functions. Assume throughout the rest of this section that H is the complexification of a real Hilbert space H_r.

Definition 8.11-2. A function Y is said to be a *positive*-real mapping of H into H* (or simply *positive*-real*) if Y is positive* and, for each real positive number σ, the restriction of $Y(\sigma)$ to H_r is a member of $[H_r; H_r]$.

Lemma 8.11-1. *Let Q be a skew-adjoint member of $[H; H]$. Then, Q is real if and only if $(Qa, a) = 0$ for all $a \in H_r$.*

PROOF. If Q is real, (Qa, a) is real for all $a \in H_r$. But, (Qa, a) is imaginary because Q is skew-adjoint. Therefore, $(Qa, a) = 0$.

Conversely, assume that $(Qa, a) = 0$ for all $a \in H_r$ and set $Q = Q_1 + iQ_2$, where Q_1 and Q_2 are real. Then, $Q' = Q_1' - iQ_2'$. Since Q is skew-adjoint, $Q_1 = -Q_1'$ and $Q_2 = Q_2'$. By the preceding paragraph, $(Q_1 a, a) = 0$ for all $a \in H_r$, so that $(Q_2 a, a) = 0$. Since Q_2 is self-adjoint,

$$\|Q_2\| = \sup\{(Q_2(a + ib), a + ib): a, b \in H_r, \quad \|a + ib\| = 1\},$$

and, moreover, $(Q_2(a + ib), a + ib)$ is real. Therefore,

$$(Q_2(a + ib), a + ib) = (Q_2 a, a) + (Q_2 b, b) = 0.$$

Hence, $Q_2 = 0$, so that Q is real. ◇

Theorem 8.11-3. *Y is a positive*-real mapping of H into H if and only if the representation* (2) *possesses the following additional properties: P_0 and P_1 are real mappings, and $P_\eta = \bar{P}_{-\eta}$.*

Note. The measure $\bar{P}_{-\eta}$ is defined as follows. For any Borel set E, $\bar{P}_{-\eta}(E) \triangleq \overline{P_\eta(-E)}$.

PROOF. Assume that P_0, P_1, and P_η have the stated properties. We may set $P_\eta = L_\eta + iM_\eta$, where the measure L_η and M_η take their values in $[H_r; H_r]$. It follows that $L_\eta = L_{-\eta}$ and $M_\eta = -M_{-\eta}$. For $\zeta = \sigma \in R_+$, the imaginary part of the integral in (2) is

$$\int_R dL_\eta \frac{\eta(1-\sigma^2)}{\sigma^2+\eta^2} + \int_R dM_\eta \frac{\sigma(1+\eta^2)}{\sigma^2+\eta^2},$$

and this is equal to zero by the oddness and evenness of the integrands. Thus, $Y(\sigma)$ is real, so that Y is positive*-real.

On the other hand, when Y is positive*-real, (14) shows that P_1 is real. Moreover, for any $a \in H_r$, we have

$$\big(Y(1)a, a\big) = (P_1 a, a) + (P_0 a, a) + \int_R d(P_\eta a, a).$$

The left-hand side as well as the first and last terms on the right-hand side are real numbers. Therefore, so, too, is $(P_0 a, a)$. But P_0 is skew-adjoint, which requires that $(P_0 a, a)$ be imaginary. Hence, $(P_0 a, a) = 0$. By Lemma 8.11-1, P_0 is real.

We can now conclude that

$$G(\sigma) \triangleq \int_R d(P_\eta a, b) \frac{1 - i\eta\sigma}{\sigma - i\eta} = \big([Y(\sigma) - P_1\sigma - P_0]a, b\big)$$

is a real number for all $a, b \in H_r$ and all $\sigma \in R_+$. It follows from the reflection principle that $G(\zeta) = \overline{G(\bar{\zeta})}$ for all $\zeta \in C_+$. We infer from this that

$$\int_R d((P_\eta - \bar{P}_{-\eta})a, b) \frac{1 - i\eta\zeta}{\zeta - i\eta} = 0, \qquad a, b \in H_r, \quad \zeta \in C_+.$$

By the uniqueness assertion of Theorem 8.11-1, $\big((P_\eta - \bar{P}_{-\eta})a, b\big) = 0$ for all $a, b \in H_r$, and therefore $P_\eta = \bar{P}_{-\eta}$. ◇

Problem 8.11-1. An $[H; H]$-valued analytic function Y is said to have a pole at a point $z \in C$ if, for some neighborhood Ω of z, for all $\zeta \in \Omega \backslash \{z\}$, and for some positive integer n,

$$Y(\zeta) = \sum_{k=-n}^{\infty} F_k(\zeta - z)^k,$$

where $F_k \in [H; H]$ and the series converges in the uniform operator topology. The pole is called simple if $n = 1$. Assume that Y is positive* and has a pole at a point $z = i\eta$ of the imaginary axis. Show that the pole is simple and $F_{-1} \in [H; H]_+$.

8.12. POSITIVE* ADMITTANCE TRANSFORMS

We are at last ready to establish the realizability conditions for a linear translation-invariant passive admittance operator. Consider the distribution $y \in [\mathscr{D}(H); H]$ defined by

$$\langle y, \phi \rangle \triangleq -P_1 \phi^{(1)}(0) + P_0 \phi(0) + \int_0^\infty j(x)\phi(x)\,dx + \int_0^\infty [j(0) - j(x)]\phi^{(2)}(x)\,dx, \tag{1}$$

where $P_1 \in [H; H]_+$, P_0 is a skew-adjoint member of $[H; H]$,

$$j(x) \triangleq \int_R dP_\eta e^{i\eta x},$$

and P_η is a PO measure on the Borel subsets of R. As was shown in Section 8.10, $y \in [\mathscr{D}_{L_1}(H); H]$, and (1) continues to hold for all $\phi \in \mathscr{D}_{L_1}(H)$. Moreover, the Laplace transform of y is precisely Schwindt's representation for a positive* mapping of H into H. One of the things we shall show is that (1) characterizes the unit-impulse response y of a passive convolution operator.

Lemma 8.12-1. *Assume that $y \in [\mathscr{D}_{L_1}(H); H]$ and that $y\,*$ is passive on $\mathscr{D}(H)$. Choose $c, d \in R$ such that $d < 0 < c$. Then, $y\,*$ is passive on $\mathscr{L}_{c,d}(H)$.*

PROOF. Choose $a, b \in R$ such that $d < b < 0 < a < c$. As was shown in the proof of Lemma 7.4-2, given any $\phi \in \mathscr{L}_{c,d}(H)$, we can find a sequence $\{\phi_j\}_{j=1}^\infty \subset \mathscr{D}(H)$ that converges in $\mathscr{L}_{a,b}(H)$ to ϕ. Upon setting $\psi = y * \phi$ and $\psi_j = y * \phi_j$, we obtain from Theorem 5.5-2 that $\psi_j \to \psi$ in $\mathscr{B}(H)$. The lemma now follows from the estimate

$$\left| \int_{-\infty}^T [(\psi_j, \phi_j) - (\psi, \phi)]\,dt \right| \le \int_{-\infty}^T \|\psi_j\|\,\|\phi_j - \phi\|\,dt + \int_{-\infty}^T \|\psi_j - \psi\|\,\|\phi\|\,dt. \quad \diamond$$

Theorem 8.12-1. *If* $\mathfrak{N}$ *is a linear translation-invariant passive operator on* $\mathscr{D}(H)$, *then* $\mathfrak{N}$ *is a convolution operator* $y\,*$, *where* $y \in [\mathscr{D}_{L_1}(H); H]$ *satisfies* (1). *Moreover,* $Y \triangleq \mathfrak{L}y$ *exists and is positive.**

Conversely, for every positive mapping* Y *of* H *into* H, *there exists a unique convolution operator* $\mathfrak{N} = y\,*$ *such that* $(\mathfrak{L}y)(\zeta) = Y(\zeta)$ *for all* $\zeta \in C_+$. *Moreover,* y *is represented by* (1), *so that* $y \in [\mathscr{D}_{L_1}(H); H]$ *and* supp $y \subset [0, \infty)$. *Also,* $\mathfrak{N}$ *is a linear translation-invariant passive mapping on* $\mathscr{D}(H)$.

Proof. Assume that the operator $\mathfrak{N}$ is linear, translation-invariant, and passive on $\mathscr{D}(H)$. By Theorem 8.2-1, $\mathfrak{N}$ is causal on $\mathscr{D}(H)$. So, by Theorem 8.10-2, $\mathfrak{N} = y\,*$, where $y \in [\mathscr{D}_{L_1}(H);\ H]$ and supp $y \subset [0, \infty)$. By Lemma 7.2-1, $y \in [\mathscr{L}(0, \infty; H); H]$ so that $Y \triangleq \mathfrak{L}y$ exists and has a strip of definition containing C_+. We will now show that $\operatorname{Re}(Y(\zeta)a, a) \geq 0$ for all $\zeta \in C_+$ and $a \in H$.

To this end, we first invoke Theorem 5.5-2 to write

$$(\mathfrak{N}\phi)(t) = \langle y(x), \phi(t - x)\rangle$$

for all $\phi \in \mathscr{L}_{c,d}(H)$, where $d < 0 < c$. By Lemma 8.12-1, for any $T \in R$,

$$\operatorname{Re} \int_{-\infty}^{T} (\langle y(x), \phi(t - x)\rangle, \phi(t))\, dt \geq 0. \tag{2}$$

Fix upon a $\zeta \in C_+$ and choose $X > T$. Also, choose $\theta \in \mathscr{E}$ such that $\theta(t) = e^{\zeta t}$ on $-\infty < t < X$ and $\theta(t) = 0$ on $X + 1 < t < \infty$. Finally, fix c, $d \in R$ such that $-\operatorname{Re}\zeta < d < 0 < c$. Then, $\theta \in \mathscr{L}_{c,d}$. Upon putting $\phi = \theta a$ in (2) and noting that $\phi(t - x) = ae^{\zeta(t-x)}$ for $-\infty < t < T$ and for all x in some neighborhood of supp y, we can manipulate (2) into

$$\operatorname{Re}(Y(\zeta)a, a) \int_{-\infty}^{T} e^{2\operatorname{Re}\zeta t}\, dt \geq 0.$$

This implies that $\operatorname{Re}(Y(\zeta)a, a) \geq 0$. Thus, Y is positive*.

We now appeal to Schwindt's representation and the uniqueness property of the Laplace transformation to conclude that y satisfies (1). This establishes the first half of our theorem.

Next, assume that Y is positive*. By Schwindt's representation, $Y = \mathfrak{L}y$, where y has the representation (1). In view of Theorem 8.10-2, the remaining statements of Theorem 8.12-1 are all clear except for the assertion that $\mathfrak{N} \triangleq y\,*$ is passive on $\mathscr{D}(H)$.

To show this, let $T \in R$, $\phi \in \mathscr{D}(H)$, and $a \in H$. Since P_0 is skew-adjoint, $(P_0 a, a)$ is imaginary, so that

$$\operatorname{Re} \int_{-\infty}^{T} (P_0 \phi(t), \phi(t))\, dt = 0. \tag{3}$$

Moreover, P_1 is a positive operator and therefore self-adjoint. Consequently,

$$D(P_1\phi, \phi) = (P_1\phi^{(1)}, \phi) + (P_1\phi, \phi^{(1)})$$
$$= (P_1\phi^{(1)}, \phi) + \overline{(P_1\phi^{(1)}, \phi)} = 2\,\mathrm{Re}(P_1\phi^{(1)}, \phi),$$

so that

$$\mathrm{Re}\int_{-\infty}^{T} (P_1\phi^{(1)}(t), \phi(t))\,dt = \tfrac{1}{2}(P_1\phi(T), \phi(T)) \geq 0. \tag{4}$$

Now, set

$$(\mathfrak{N}_1\phi)(t) \triangleq \int_0^\infty j(x)\phi(t-x)\,dx + \int_0^\infty [j(0) - j(x)]\phi^{(2)}(t-x)\,dx.$$

The substitution of the definition of $j(x)$ and an application of Theorem 2.5-2 converts this into

$$(\mathfrak{N}_1\phi)(t) = \int_R dP_\eta\,\tilde{\phi}_{[t]}(-\eta) + \int_R dP_\eta[-i\eta\phi(t) + \eta^2\tilde{\phi}_{[t]}(-\eta)].$$

We are using here the notation defined at the beginning of Section 8.8. For any positive integer k, we define $\mathfrak{N}_{1,k}$ by

$$(\mathfrak{N}_{1,k}\phi)(t) \triangleq \int_{-k}^{k} dP_\eta\,\tilde{\phi}_{[t]}(-\eta)(1+\eta^2) - iQ_k\phi(t), \tag{5}$$

where $Q_k = \int_{-k}^{k} dP_\eta\,\eta$. Note that $(Q_k a, a) = \int_{-k}^{k} d(P_\eta a, a)\eta$, which is a real number. Hence,

$$\mathrm{Re}\int_{-\infty}^{T} (-iQ_k\phi(t), \phi(t))\,dt = 0. \tag{6}$$

By the argument given in the proof of Theorem 8.8-1 [see, in particular, (18) of Section 8.8],

$$\mathrm{Re}\int_{-\infty}^{T} (\mathfrak{N}_{1,k}\phi(t), \phi(t))\,dt \to \mathrm{Re}\int_{-\infty}^{T} (\mathfrak{N}_1\,\phi(t), \phi(t))\,dt \tag{7}$$

as $k \to \infty$. We define the measure M on any Borel subset E of $[-k, k]$ by

$$M(E) \triangleq \int_E dP_\eta(1+\eta^2)$$

and set $M(J) = 0$ if J is a Borel set that does not intersect $[-k, k]$. Upon identifying M with the PO measure P given in Lemma 8.8-2, we may invoke (6) of Section 8.8 in conjunction with (5) and (6) to write

$$\mathrm{Re}\int_{-\infty}^{T} (\mathfrak{N}_{1,k}\phi(t), \phi(t))\,dt = \tfrac{1}{2}\int_{-k}^{k} d\big(P_\eta(1+\eta^2)\tilde{\psi}(\eta), \tilde{\psi}(\eta)\big) \geq 0.$$

In view of (7),

$$\operatorname{Re} \int_{-\infty}^{T} (\mathfrak{N}_1 \phi(t), \phi(t))\, dt \geq 0. \tag{8}$$

Upon combining (3), (4), and (8), we see that $\mathfrak{N}$ is passive on $\mathscr{D}(H)$. ◇

Theorem 8.12-1 states the fundamental realizability conditions relating to positive* functions. A similar proof for it, which, however, is not based on Theorem 8.10-2, is given by Zemanian (1970a). As is indicated there, we need merely assume that $\mathfrak{N}$ is linear, translation-invariant, and passive on $\mathscr{D} \odot H$ in order to establish the first half of Theorem 8.10-2. It should also be noted that Theorem 8.12-1 possesses an extension to n dimensions (that is, to the case where $\mathscr{D} = \mathscr{D}_{R^1}$ is replaced by $\mathscr{D}_{R^n}$) due to Vladimirov (1969a, b).

8.13. POSITIVE*-REAL ADMITTANCE TRANSFORMS

Here we show how the realizability conditions for an admittance operator are sharpened when that operator is real. Once again, we assume that H is the complexification of a real Hilbert space H_r.

Theorem 8.13-1. *If $\mathfrak{N}$ is a linear translation-invariant passive operator on $\mathscr{D}(H)$ and if $\mathfrak{N}$ maps $\mathscr{D}(H_r)$ into $[\mathscr{D}(R); H_r]$, then $\mathfrak{N} = y\,*$, where $y \in [\mathscr{D}_{L_1}(H_r); H_r]$ satisfies* (1) *of Section* 8.12 *with P_1, P_0, and P_η restricted as in Theorem* 8.11-3. *Moreover, $Y \triangleq \mathfrak{L}y$ is positive*-real.*

Conversely, if Y is a positive-real mapping on H into H, then $Y = \mathfrak{L}y$ on C_+, where y is a member of $[\mathscr{D}_{L_1}(H_r); H_r]$ and satisfies* (1) *of Section* 8.12 *with P_1, P_0, and P_η restricted as in Theorem* 8.11-3. *Moreover, $\mathfrak{N} \triangleq y\,*$ maps $\mathscr{D}(H_r)$ into $[\mathscr{D}(R); H_r]$ and is a linear translation-invariant passive operator on $\mathscr{D}(H)$.*

PROOF. We have already seen that every positive* function is the Laplace transform of a unique $y \in [\mathscr{D}_{L_1}(H); H]$ given by (1) of Section 8.12. We now invoke Theorems 8.11-2 and 8.11-3, which give the necessary and sufficient conditions on P_1, P_0, and P_η in order for Y to be positive*-real. In this case, $y \in [\mathscr{D}_{L_1}(H_r); H_r]$ because the imaginary part of $\int dP_\eta e^{i\eta x}$ equals zero by virtue of the condition $P_\eta = \bar{P}_{-\eta}$. The second half of our theorem now follows from the second half of Theorem 8.12-1.

Under the hypothesis of the first half of the theorem, we can prove that $y \in [\mathscr{D}(H_r); H_r]$ and $Y(\sigma) \in [H_r; H_r]$ for every $\sigma \in R_+$ exactly as in the proof of Theorem 7.6-1. Hence, Y is positive*-real, and, by Theorem 8.12-1 again, $\mathfrak{N} = y\,*$, where y has the stated properties. ◇

Problem 8.13-1. Assume that $\mathfrak{N}$ is a linear translation-invariant passive operator on $\mathscr{D}(H_r)$ with range in $[\mathscr{D}(R); H_r]$. Show that $\mathfrak{N}$ has a unique extension as a linear translation-invariant passive operator on $\mathscr{D}(H)$.

8.14. A CONNECTION BETWEEN PASSIVITY AND SEMIPASSIVITY

Some conditions under which a semipassive convolution operator is passive is given by the next theorem, the scalar version of which was given by König and Zemanian (1965). Its proof is based upon the following observations.

The function j in the representation (1) of Section 8.12 is a strongly continuous $[H; H]$-valued function by virtue of Problem 8.8-1. Therefore, j is strongly measurable, and, by the principle of uniform boundedness, it is bounded in the uniform operator topology on every compact set. Thus, in accordance with Problem 3.3-2, j defines a member of $[\mathscr{D}; [H; H]]$ (which we also denote by j) by means of the equation

$$\langle j, \theta \rangle a \triangleq \int j(x) a \theta(x)\, dx, \qquad a \in H, \quad \theta \in \mathscr{D}.$$

Now,

$$\int_0^x j(t) 1_+(t)\, dt \tag{1}$$

converges in the strong operator topology and defines a strongly continuous function of x. Therefore, (1) is also a member of $[\mathscr{D}; [H; H]]$. Through an integration by parts,

$$\left\langle \int_0^x j(t) 1_+(t)\, dt, \; -\theta^{(1)}(x) \right\rangle a = \int_0^\infty j(x) a \theta(x)\, dx = \langle j, \theta \rangle a.$$

By our usual identification between $[\mathscr{D}(H); H]$ and $[\mathscr{D}; [H; H]]$, $\int_0^\infty j(x)\phi(x)\, dx$ is the value assigned to any $\phi \in \mathscr{D}(H)$ by the generalized derivative of (1).

In the same way, $[j(0) - j]1_+$ is strongly continuous and therefore a member of $[\mathscr{D}; [H; H]]$. Thus

$$\int_0^\infty [j(0) - j(x)] \phi^{(2)}(x)\, dx$$

is the value assigned to any $\phi \in \mathscr{D}(H)$ by the second generalized derivative of $[j(0) - j]1_+$.

However, for $p \geq 4$,

$$\int_0^\infty [j(0) - j(x)] \phi^{(p)}(x)\, dx$$

is not the value assigned to ϕ by the second generalized derivative of a strongly continuous $[H; H]$-valued function. This is because $f \triangleq D^2\{[j(0) - j]1_+\}$ and therefore $D^{p-2}\{[j(0) - j]1_+\}$ are not strongly continuous. To show this, let

$$\theta_k(x) \triangleq e^{-k^2x^2} \bigg/ \int_{-\infty}^{\infty} e^{-k^2x^2}\, dx, \qquad k = 1, 2, \ldots .$$

Since $\theta_k \in \mathscr{S} \subset \mathscr{D}_{L_1}$ and $[j(0) - j]1_+ \in [\mathscr{D}_{L_1}(H); H]$, we may write, for any $a \in H$,

$$\begin{aligned}(\langle fa, \theta_k\rangle, a) &= (\langle D^2\{[j(0) - j]1_+\}, \theta_k\rangle a, a) = (\langle [j(0) - j]1_+, \theta_k^{(2)}\rangle a, a)\\ &= \left(\int_0^{\infty} dx \int_R dP_\eta(1 - e^{i\eta x})a\theta^{(2)}(x), a\right).\end{aligned}$$

An application of Theorem 2.5-2 and two integrations by parts convert the right-hand side into

$$\int_R d(P_\eta a, a)\eta^2 \int_0^{\infty} \theta_k(x)e^{i\eta x}\, dx.$$

This is equal to

$$\tfrac{1}{2}\int d(P_\eta a, a)\eta^2[\exp(-\eta^2/4k^2)] \operatorname{erfc}(-i\eta/2k), \tag{2}$$

where erfc denotes the complementary error function (see Zemanian, 1965, pp. 350, 357). But

$$\operatorname{Re} \operatorname{erfc}(-i\eta/2k) = 1,$$

and thus (2) clearly does not tend to zero as $k \to \infty$. It would have to do so if $f \triangleq D^2\{[j(0) - j]1_+\}$ were strongly continuous at the origin because $\theta_k \to \delta$ and $\operatorname{supp} f \subset [0, \infty)$.

Finally, we note again that $Y \triangleq \mathfrak{L}y$ is positive* if and only if y is represented by (1) of Section 8.12.

In view of these results, we need merely compare Theorems 8.10-1 and 8.12-1 in order to conclude the following.

Theorem 8.14-1. *Let $y \in [\mathscr{D}(H); H]$. Then, $y\,*$ is passive on $\mathscr{D}(H)$ if and only if y satisfies the following conditions.*

(i) *$y = P_1\delta^{(1)} + w$, where $P_1 \in [H; H]_+$, $\delta^{(1)}$ is as usual the first generalized derivative of the delta functional, w is the second generalized derivative of an $[H; H]$-valued function on R that is continuous with respect to the strong operator topology, and* $\operatorname{supp} w \subset [0, \infty)$.

(ii) *$y\,*$ is semipassive on $\mathscr{D}(H)$.*

8.15. A CONNECTION BETWEEN THE ADMITTANCE AND SCATTERING FORMULISMS

With $\mathfrak{N}$: $v \mapsto u$ and $\mathfrak{W}$: $q \mapsto r$ denoting respectively the admittance and scattering operators for a Hilbert port, the variables $v, u, q, r \in [\mathscr{D}; H]$ satisfy

$$v = q + r, \qquad u = q - r. \tag{1}$$

If a particular Hilbert port has an admittance operator, need it have a scattering operator as well? No. For, a substitution of (1) into $\mathfrak{N}v = u$ yields

$$\mathfrak{N}r + r = \mathfrak{N}q - q.$$

Upon setting $\mathfrak{N} = -\delta *$, we see that only $q = 0$ will satisfy this relation, whereas any r will do. This means that $\mathfrak{W}: q \mapsto r$ does not exist as an operator. A similar manipulation shows that $\mathfrak{N}$ will not exist as an operator when $\mathfrak{W} = -\delta *$.

However, when $\mathfrak{N}$ is a linear translation-invariant passive operator on $\mathscr{D}(H)$, $\mathfrak{W}$ exists and is a linear translation-invariant scatter-passive operator on $\mathscr{D}(H)$. The converse is not true in general, but it will be true if we assume in addition that $I + S(\zeta)$ possesses an inverse in $[H; H]$ for each $\zeta \in C_+$, where I is the identity operator on H and S is the scattering transform. To show these results, we first establish a connection between S and the admittance transform Y.

Lemma 8.15-1. *Let Y be positive*. Then, for every $\zeta \in C_+$, $I + Y(\zeta)$ possesses an inverse in $[H; H]$. Moreover,*

$$S(\cdot) \triangleq [I + Y(\cdot)]^{-1}[I - Y(\cdot)] \tag{2}$$

is bounded.*

PROOF. Fix $\zeta \in C_+$ and set $F \triangleq I + Y(\zeta)$. We wish to show that F^{-1} exists. Set $b = Fa$, where $a \in H$. Then,

$$2(a, a) \le 2(a, a) + (Ya, a) + (a, Ya) = (b, a) + (a, b) \le 2|(a, b)| \le 2\|a\|\,\|b\|,$$

so that

$$\|a\| \le \|Fa\|. \tag{3}$$

This implies that F is injective.

$F(H)$ is a closed linear subspace of H. Indeed, let b be a limit point of $F(H)$ and choose a sequence $\{b_n\} \subset F(H)$ that converges to b. Then, $b_n = Fa_n$, where $a_n \in H$. By (3),

$$\|a_n - a_m\| \le \|F(a_n - a_m)\| = \|b_n - b_m\| \to 0.$$

So, $\{a_n\}$ is a Cauchy sequence and converges therefore to an $a \in H$. Thus, $Fa_n \to Fa$ by the continuity of F, whereas $Fa_n = b_n \to b$. Hence, $b = Fa \in F(H)$.

If $F(H)$ were a proper subspace of H, there would be an $a \in H$ with $a \neq 0$ such that $(Fa, a) = 0$ (Berberian, 1961, p. 71). That is, $([I + Y(\zeta)]a, a) = 0$. But this cannot be since $\operatorname{Re}(Y(\zeta)a, a) \geq 0$. Hence, F is surjective. Thus, F^{-1} exists on H. Moreover, $F^{-1} \in [H; H]$ according to Appendix D14.

In the same way as in the scalar case, it can be shown that $[I + Y(\cdot)]^{-1}$ is analytic at every point ζ where $I + Y(\zeta)$ has an inverse and Y is analytic. (In this regard, see Problem 8.15-1.) Thus, $[I + Y(\cdot)]^{-1}$ and therefore S are $[H; H]$-valued analytic functions on C_+.

Upon premultiplying (2) by $I + Y$ and rearranging the result, we obtain $Y(I + S) = I - S$. Thus, for any $a \in H$,

$$\begin{aligned}\|a\|^2 - \|S(\zeta)a\|^2 &= \operatorname{Re}([I - S(\zeta)]a, [I + S(\zeta)]a) \\ &= \operatorname{Re}\big(Y(\zeta)[I + S(\zeta)]a, [I + S(\zeta)]a\big) \geq 0.\end{aligned}$$

This implies that $\|S(\zeta)\| \leq 1$ for all $\zeta \in C^+$. We have hereby shown that S is bounded*. ◇

Now, assume that $\mathfrak{N}$ is a linear translation-invariant passive operator on $\mathscr{D}(H)$. Therefore, $\mathfrak{N} = y\, *$, where $Y \triangleq \mathfrak{L}y$ is positive*. Moreover, for any $v \in \mathscr{D}(H)$, $u \triangleq y * v$ is a Laplace-transformable member of $\mathscr{D}_+(H)$ whose strip of definition contains C_+. Consequently, $q = \frac{1}{2}(v + u)$ and $r = \frac{1}{2}(v - u)$ have the same properties. Upon substituting (1) into $u = y * v$, taking Laplace transforms, and rearranging the result, we obtain

$$[I + Y(\zeta)]R(\zeta) = [I - Y(\zeta)]Q(\zeta), \qquad \zeta \in C_+. \tag{4}$$

In view of Lemma 8.15-1, this may be rewritten as

$$R(\zeta) = [I + Y(\zeta)]^{-1}[I - Y(\zeta)]Q(\zeta) = S(\zeta)Q(\zeta), \qquad \zeta \in C_+,$$

where S is bounded*. Applying the inverse Laplace transformation, we get

$$r = s * q, \tag{5}$$

where $s \triangleq \mathfrak{L}^{-1}S$ and r and q correspond to the given $v \in \mathscr{D}(H)$ as above. We use (5) to define the scattering operator $\mathfrak{W} \triangleq s\, *$ on other q and in particular on $\mathscr{D}(H)$. Since S is bounded*, the next theorem follows immediately from Theorem 7.5-1.

Theorem 8.15-1. *Let $\mathfrak{N}$ be a linear translation-invariant passive operator on $\mathscr{D}(H)$ and let Y be the corresponding admittance transform. Define S by* (2) *and set $s \triangleq \mathfrak{L}^{-1}S$. Then, $\mathfrak{W} \triangleq s\, *$ is a linear translation-invariant scatter-passive operator on $\mathscr{D}(H)$.*

Let us now go into the opposite direction, starting with a given scattering operator $\mathfrak{W}$ and deriving from it an admittance operator $\mathfrak{N}$. We first note that this may not always be possible even when $\mathfrak{W}$ is linear, translation-invariant, and scatter-passive on $\mathscr{D}(H)$. A counterexample is $\mathfrak{W} = -\delta *$.

Lemma 8.15-2. *Assume that S is bounded* and that, for every $\zeta \in C_+$, $I + S(\zeta)$ possesses an inverse in $[H; H]$. Then,*

$$Y(\cdot) \triangleq [I + S(\cdot)]^{-1}[I - S(\cdot)] \tag{6}$$

is positive and satisfies* (2).

PROOF. We again have that $[I + S(\cdot)]^{-1}$ is an $[H; H]$-valued analytic function on C_+. Therefore, so, too, is Y. Since $I + S$ commutes with $I - S$, (6) yields

$$Y(I + S) = I - S. \tag{7}$$

Now, let $b \in H$ be arbitrary and fix $\zeta \in C_+$. Since $[I + S(\zeta)]^{-1}$ exists, there is an $a \in H$ such that $[I + S(\zeta)]a = b$. Therefore,

$$\begin{aligned} \operatorname{Re}(Y(\zeta)b, b) &= \operatorname{Re}(Y[I + S(\zeta)]a, [I + S(\zeta)]a) \\ &= \operatorname{Re}([I - S(\zeta)]a, [I + S(\zeta)]a) \\ &= \|a\|^2 - \|S(\zeta)a\|^2 \geq 0. \end{aligned}$$

Thus, Y is positive*. Lemma 8.15-1 now shows that $[I + Y(\zeta)]^{-1}$ exists. This allows us to solve (7) to get (2). ◇

To obtain an admittance operator from a given scattering operator, we proceed in just about the same way as for Theorem 8.15-1 except that now we rely on Lemma 8.15-2.

Theorem 8.15-2. *Let $\mathfrak{W}$ be a linear translation-invariant scatter-passive operator on $\mathscr{D}(H)$, and let S be the corresponding scattering transform. Assume that $I + S(\zeta)$ possesses an inverse in $[H; H]$ for every $\zeta \in C_+$. Also, define Y by* (6) *and set $y \triangleq \mathfrak{L}^{-1}Y$. Then, $\mathfrak{N} \triangleq y *$ is a linear translation-invariant passive operator on $\mathscr{D}(H)$, and the scattering operator generated by $\mathfrak{N}$ in accordance with Theorem* 8.15-1 *coincides with $\mathfrak{W}$.*

Problem 8.15-1. Let G be an $[H; H]$-valued analytic function on C_+ such that $G(\zeta)^{-1}$ exists for all $\zeta \in C_+$. Choose any $z \in C_+$. Show that, for all $\zeta \in C_+$ such that

$$\|[G(z) - G(\zeta)]G(z)^{-1}\| < 1,$$

we have

$$G(\zeta)^{-1} - G(z)^{-1} = G(z)^{-1} \sum_{n=1}^{\infty} \{[G(z) - G(\zeta)]G(z)^{-1}\}^n.$$

Then, using this relation, show that $G(\cdot)^{-1}$ is a continuous function on C_+. Finally, show that $G(\cdot)^{-1}$ is analytic on C_+.

8.16. THE ADMITTANCE TRANSFORM OF A LOSSLESS HILBERT PORT

We saw in Section 7.7 how the losslessness of a scattering operator is characterized by the fact that the scattering transform has isometric boundary values almost everywhere on the imaginary axis. A similar connection exists between the losslessness of the scattering operator and the boundary values of the admittance operator. Now, however, the connection is not as complete, since the existence of the boundary values is taken as an assumption. The precise result is stated by the next theorem, (D'Amato, 1971).

Theorem 8.16-1. *Let H be a separable complex Hilbert space. Assume that S is a bounded* mapping of H into H and that $I + S(\zeta)$ has an inverse in $[H; H]$ for all $\zeta \in C_+$. Set $s \triangleq \mathfrak{L}^{-1}S$ and*

$$Y(\zeta) \triangleq [I + S(\zeta)]^{-1}[I - S(\zeta)], \qquad \zeta \in C_+. \tag{1}$$

Also, assume that, for almost all $\omega \in R$ and as $\sigma \to 0+$, $Y(\sigma + i\omega)$ converges in the strong operator topology to $Y(i\omega)$. If $Y(i\omega)$ is skew-adjoint for almost all ω, then $S(i\omega)$ is unitary for the same values of ω, and $s\,*$ is lossless on $\mathscr{D}(H)$. Conversely, if $s\,*$ is lossless on $\mathscr{D}(H)$, then $Y(i\omega)$ is skew-adjoint for almost all ω.

Proof. We may rewrite (1) as

$$S(\zeta)[I + Y(\zeta)] = I - Y(\zeta), \qquad \zeta \in C_+. \tag{2}$$

As was indicated in Section 7.7, as $\sigma \to 0+$, $S(\sigma + i\omega)$ converges in the strong operator topology to $S(i\omega)$ for almost all ω. It follows from our hypothesis on $Y(\zeta)$ and the principle of uniform boundedness that $S(\zeta)[I + Y(\zeta)]$ converges in the same way to $S(i\omega)[I + Y(i\omega)]$. Therefore, (2) holds almost everywhere when ζ is replaced by $i\omega$. By Lemma 8.15-2, Y is positive*. This implies that $\operatorname{Re}(Y(i\omega)a, a) \geq 0$ for all $a \in H$ and almost all ω. By the proof of Lemma 8.15-1, $[I + Y(i\omega)]^{-1}$ exists almost everywhere. We can now conclude from (2) that

$$S(i\omega) = [I - Y(i\omega)][I + Y(i\omega)]^{-1} \tag{3}$$

almost everywhere. After premultiplying and postmultiplying this by $I + Y(i\omega)$, we get

$$[I + Y(i\omega)]S(i\omega)[I + Y(i\omega)] = I - [Y(i\omega)]^2 = [I - Y(i\omega)][I + Y(i\omega)].$$

Hence,

$$S(i\omega) = [I + Y(i\omega)]^{-1}[I - Y(i\omega)] \tag{4}$$

almost everywhere.

We now employ the identity $(F')^{-1} = (F^{-1})'$ for any $F \in [H; H]$, where as always the prime denotes the adjoint operator. Setting $S'(i\omega) \triangleq [S(i\omega)]'$, we get from (4)

$$\begin{aligned} S'(i\omega) &= \{[I + Y(i\omega)]^{-1}[I - Y(i\omega)]\}' \\ &= [I - Y(i\omega)]'\{[I + Y(i\omega)]'\}^{-1} \\ &= [I + Y(i\omega)][I - Y(i\omega)]^{-1}. \end{aligned}$$

The last equality is due to the hypothesis that $Y(i\omega)$ is skew-adjoint. Combining this result with (3), we get

$$S'(i\omega)S(i\omega) = S(i\omega)S'(i\omega) = I$$

almost everywhere. This is precisely the condition that must be satisfied for $S(i\omega)$ to be unitary (Berberian, 1961, p. 145). Theorem 7.7-1 now shows that $s\,*$ is lossless on $\mathscr{D}(H)$.

Conversely, assume that $s\,*$ is lossless on $\mathscr{D}(H)$. By Theorem 7.7-2, $S(i\omega)$ is isometric for almost all ω, so that $S'(i\omega)S(i\omega) = I$. By (3) and the identity $(F')^{-1} = (F^{-1})'$,

$$[I + Y'(i\omega)]^{-1}[I - Y'(i\omega)][I - Y(i\omega)][I + Y(i\omega)]^{-1} = I.$$

This can be rearranged into

$$I - Y'(i\omega) - Y(i\omega) + Y'(i\omega)Y(i\omega) = I + Y'(i\omega) + Y(i\omega) + Y'(i\omega)Y(i\omega),$$

which is the same as $Y(i\omega) = -Y'(i\omega)$. In other words, $Y(i\omega)$ is skew-adjoint for almost all ω. ◇

Appendix A

Linear Spaces

Note. This and the following appendixes survey those standard definitions and results concerning topological linear spaces and the Bochner integral that are used in this book. No proofs are presented since all of them can be found in a variety of readily available books, such as those by Dunford and Schwartz (1966), Hille and Phillips (1957), Horváth (1966), Robertson and Robertson (1964), Rudin (1966), Schaefer (1966), Treves (1967), and Zaanen (1967). Much of the notation used here is explained in Section 1.2, and hence the reader may wish to look through that section before reading these appendixes.

The linear spaces occurring in this book are in almost all cases complex linear spaces. The only exceptions occur on the few occasions when we use real linear spaces. For this reason, we will employ the phrase "linear space" to mean a complex linear space, whose definition is as follows.

A1. A collection $\mathscr{V}$ of elements $\phi, \psi, \theta, \ldots$ is called a (*complex*) *linear space* if the following three axioms are satisfied.

1. There is an operation +, mapping $\mathscr{V} \times \mathscr{V}$ into $\mathscr{V}$ and called *addition*, by which any pair $\{\phi, \psi\} \in \mathscr{V} \times \mathscr{V}$ can be combined to yield an element $\phi + \psi \in \mathscr{V}$. In addition, the following rules hold:

1a. $\phi + \psi = \psi + \phi$ (commutativity).
1b. $(\phi + \psi) + \theta = \phi + (\psi + \theta)$ (associativity).
1c. There is a unique element $\varnothing \in \mathscr{V}$ such that $\phi + \varnothing = \phi$ for every $\phi \in \mathscr{V}$.
1d. For each $\phi \in \mathscr{V}$, there exists a unique element $-\phi \in \mathscr{V}$ such that $\phi + (-\phi) = \varnothing$.

2. There is an operation mapping $C \times \mathscr{V}$ into $\mathscr{V}$ and called *multiplication by a complex number*, by which any $\alpha \in C$ and $\phi \in \mathscr{V}$ can be combined to yield an element $\alpha\phi \in \mathscr{V}$. Moreover, the following rules hold for all $\alpha, \beta \in C$:

2a. $\alpha(\beta\phi) = (\alpha\beta)\phi$.
2b. $1\phi = \phi$ (1 denotes the number one).

3. The following distributive laws hold:

3a. $\alpha(\phi + \psi) = \alpha\phi + \alpha\psi$.
3b. $(\alpha + \beta)\phi = \alpha\phi + \beta\phi$.

A2. The element $-\psi$ is called the *negative of* ψ. The *subtraction of* ψ *from* ϕ is defined as $\phi - \psi \triangleq \phi + (-\psi)$. Also, $\varnothing$ is called the *zero element* or the *origin of* $\mathscr{V}$; we usually denote $\varnothing$ by 0. The following rules are consequences of the definition A1:

(i) $\phi + \psi = \phi + \theta$ implies that $\psi = \theta$.
(ii) $\alpha\varnothing = \varnothing$.
(iii) $0\phi = \varnothing$ (here, 0 denotes the number zero).
(iv) $(-1)\phi = -\phi$.
(v) If $\alpha\phi = \beta\phi$ and $\phi \neq \varnothing$, then $\alpha = \beta$.
(vi) If $\alpha\phi = \alpha\psi$ and $\alpha \neq 0$, then $\phi = \psi$.

A3. A subset $\mathscr{U}$ of a linear space $\mathscr{V}$ is said to be a *linear subspace* (or simply a *subspace*) *of* $\mathscr{V}$ if, for every $\phi, \psi \in \mathscr{V}$ and $\alpha \in C$, we have that $\phi + \psi \in \mathscr{U}$ and $\alpha\phi \in \mathscr{U}$. In this case, it follows that $\mathscr{U}$ is also a linear space under $\mathscr{V}$'s rules for addition and multiplication by a complex number. Also, the intersection of any collection of linear subspaces of $\mathscr{V}$ is a linear subspace of $\mathscr{V}$.

A *linear combination* of elements in $\mathscr{V}$ is a sum $\sum \alpha_k \phi_k$, where $\alpha_k \in C$, $\phi_k \in \mathscr{V}$, and the summation is over a finite number of terms. The *span of any given subset* $\Omega \subset \mathscr{V}$ is the set of all linear combinations of elements in Ω. Any such span is a linear subspace of $\mathscr{V}$. A subspace $\mathscr{U}$ of $\mathscr{V}$ is said to be *finite-dimensional* if $\mathscr{U}$ is the span of a finite number of members of $\mathscr{U}$.

A finite set $\{e_j\}_{j=1}^n \subset \mathscr{V}$ is called *linearly independent* if the equation $\sum \alpha_j e_j = \sum \beta_j e_j$, where $\alpha_j, \beta_j \in C$, implies that $\alpha_j = \beta_j$ for every j. Every finite-dimensional subspace $\mathscr{U}$ contains a linearly independent finite set $\{e_j\}$ whose span is $\mathscr{U}$.

A4. A set Ω in a complex linear space $\mathscr{V}$ is called *convex* if $\lambda\phi + (1-\lambda)\psi \in \Omega$ whenever $\phi, \psi \in \Omega$ and $\lambda \in R$ is such that $0 \le \lambda \le 1$. Ω is called *balanced* if $\alpha\phi \in \Omega$ whenever $\phi \in \Omega$ and $\alpha \in C$ is such that $|\alpha| \le 1$. If Ω is both balanced and convex, it is also called *absolutely convex*; this occurs if and only if $\alpha\phi + \beta\psi \in \Omega$ whenver $\phi, \psi \in \Omega$ and $\alpha, \beta \in C$ are such that $|\alpha| + |\beta| \le 1$.

Let Ω be any nonvoid set in the complex linear space $\mathscr{V}$. The *convex hull of* Ω is the set of all sums $\sum \lambda_k \phi_k$, where $\phi_k \in \Omega$, the $\lambda_k \in R$ are such that $\lambda_k > 0$ and $\sum \lambda_k = 1$, and the summation is over a finite number of terms. A convex set coincides with its convex hull. The *balanced convex hull of* Ω is the set of all sums $\sum \alpha_k \phi_k$, where $\phi_k \in \Omega$, the $\alpha_k \in C$ are such that $\sum |\alpha_k| \le 1$, and the summation is over a finite number of terms.

The set Ω is said to be *absorbent* if, for any given $\phi \in \mathscr{V}$, there exists a $\lambda \in R$ with $\lambda > 0$ such that $\phi \in \alpha\Omega$ for all $\alpha \in C$ with $|\alpha| \ge \lambda$.

The intersection of any finite collection of balanced convex absorbent sets is also balanced, convex, and absorbent.

A5. Let $\mathscr{V}$ and $\mathscr{W}$ be linear spaces. A mapping f of $\mathscr{V}$ into $\mathscr{W}$ is called *linear* if, for every $\phi, \psi \in \mathscr{V}$ and $\alpha, \beta \in C$, we always have that

$$f(\alpha\phi + \beta\psi) = \alpha f(\phi) + \beta f(\psi).$$

The set of all such mappings is denoted by $L(\mathscr{V}; \mathscr{W})$. For any $f, g \in L(\mathscr{V}; \mathscr{W})$ and any $\alpha \in C$, we define $f + g$ and αf as follows. For any $\phi \in \mathscr{V}$, $(f+g)(\phi) \triangleq f(\phi) + g(\phi)$ and $(\alpha f)(\phi) \triangleq \alpha f(\phi)$. As a consequence, $L(\mathscr{V}; \mathscr{W})$ is a linear space.

When $\mathscr{W}$ is the complex plane C, $L(\mathscr{V}; C)$ is called the *algebraic dual of* $\mathscr{V}$ and is also denoted by $\mathscr{V}^*$. Each member of $L(\mathscr{V}; C)$ is called a *linear form on* $\mathscr{V}$.

If f is a linear bijection of $\mathscr{V}$ onto $\mathscr{W}$, its inverse f^{-1}, which by definition exists, is also linear.

A6. Let $\mathscr{U}$, $\mathscr{V}$, and $\mathscr{W}$ be complex linear spaces. A mapping f of $\mathscr{U} \times \mathscr{V}$ into $\mathscr{W}$ is called *bilinear* if $\phi \mapsto f(\phi, \psi)$ is linear on $\mathscr{U}$ for each fixed $\psi \in \mathscr{V}$ and $\psi \mapsto f(\phi, \psi)$ is linear on $\mathscr{V}$ for each fixed $\phi \in \mathscr{U}$. $B(\mathscr{U}, \mathscr{V}; \mathscr{W})$ denotes the set of all bilinear mappings of $\mathscr{U} \times \mathscr{V}$ into $\mathscr{W}$. It becomes a linear space when, for any $f, g \in B(\mathscr{U}, \mathscr{V}; \mathscr{W})$ and any $\alpha \in C$, we define $f + g$ and αf by $(f+g)(\phi, \psi) = f(\phi, \psi) + g(\phi, \psi)$ and $(\alpha f)(\phi, \psi) = \alpha f(\phi, \psi)$ for all $\phi \in \mathscr{U}$ and $\psi \in \mathscr{V}$. We set $B(\mathscr{U}, \mathscr{V}) \triangleq B(\mathscr{U}, \mathscr{V}; C)$. The members of $B(\mathscr{U}, \mathscr{V})$ are called *bilinear forms on* $\mathscr{U} \times \mathscr{V}$.

A mapping h of $\mathscr{V}$ into $\mathscr{W}$ is called *antilinear* if, for any $\psi, \theta \in \mathscr{V}$ and $\alpha, \beta \in C$, we have that

$$h(\alpha\psi + \beta\theta) = \bar{\alpha}h(\psi) + \bar{\beta}h(\theta).$$

A mapping f of $\mathscr{U} \times \mathscr{V}$ into $\mathscr{W}$ is called *sesquilinear* if $\phi \mapsto f(\phi, \psi)$ is linear on $\mathscr{U}$ for each fixed $\psi \in \mathscr{V}$ and $\psi \mapsto f(\phi, \psi)$ is antilinear on $\mathscr{V}$ for each fixed $\phi \in \mathscr{U}$. f is called a *sesquilinear form on* $\mathscr{U} \times \mathscr{V}$ if $\mathscr{W} = C$. f is called a *positive sesquilinear form on* $\mathscr{V} \times \mathscr{V}$ if $f(\phi, \phi) \geq 0$ for all $\phi \in \mathscr{V}$; in this case, we have the *Schwarz inequality*:

$$|f(\phi, \psi)|^2 \leq f(\phi, \phi)f(\psi, \psi)$$

for all $\phi, \psi \in \mathscr{V}$.

A7. For any sesquilinear mapping f of $\mathscr{V} \times \mathscr{V}$ into $\mathscr{W}$, where $\mathscr{V}$ and $\mathscr{W}$ are complex linear spaces, we have the *polarization identity*:

$$\begin{aligned} f(\phi, \psi) = \tfrac{1}{4}[&f(\phi + \psi, \phi + \psi) - f(\phi - \psi, \phi - \psi) \\ &+ if(\phi + i\psi, \phi + i\psi) - if(\phi - i\psi, \phi - i\psi)]. \end{aligned}$$

A8. When the complex plane C is replaced by the real line R in axioms 2 and 3 of Appendix A1, the definition of a *real linear space* is obtained. Except for the definitions of antilinear and sesquilinear mappings and the polarization identity, the preceding discussion carries over to real linear spaces.

Appendix B

Topological Spaces

B1. A *topological space* is a set $\mathscr{V}$ for which a collection $\mathscr{O}$ of subsets of $\mathscr{V}$ is specified and has the following properties.

(i) $\mathscr{V}$ and the empty set are members of $\mathscr{O}$.
(ii) Every union of members of $\mathscr{O}$ is a member of $\mathscr{O}$.
(iii) The intersection of any finite number of members of $\mathscr{O}$ is a member of $\mathscr{O}$.

The members of $\mathscr{O}$ are called *open sets*, and $\mathscr{O}$ is said to be a *topology on* $\mathscr{V}$. The complement of any open set in $\mathscr{V}$ is called a *closed set*. It follows that any intersection of closed sets is closed, and so, too, is the union of any finite number of closed sets. Also, $\mathscr{V}$ and the empty set are closed. Given any set Ω in $\mathscr{V}$, the largest open set contained in Ω is the *interior* $\mathring{\Omega}$ *of* Ω, and the points of $\mathring{\Omega}$ are called the *interior points of* Ω. The smallest closed set containing Ω is the *closure* $\overline{\Omega}$ *of* Ω. Let $C\Omega$ denote the complement of Ω. Then, $\overline{\Omega} \cap \overline{C\Omega}$ is called the *boundary of* Ω. If Λ is another set in $\mathscr{V}$ and if $\overline{\Lambda} \supset \Omega$, then Λ is said to be *dense in* Ω. $\mathscr{V}$ is called *separable* or *of countable type* if it contains a countable dense subset. On the other hand, $\mathscr{V}$ is called *separated* or *Hausdorff* and $\mathscr{O}$ is said to *separate* $\mathscr{V}$ if, for every pair of points $\phi, \psi \in \mathscr{V}$, $\phi \neq \psi$, there exist open sets Φ and Ψ such that $\phi \in \Phi$, $\psi \in \Psi$, and $\Phi \cap \Psi$ is empty.

B2. A subset Ω of $\mathscr{V}$ is called a *neighborhood of a point* $\phi \in \mathscr{V}$ (or a *neighborhood of a subset* $\Psi \subset \mathscr{V}$) if there exists an open set Λ such that $\phi \in \Lambda \subset \Omega$ (respectively $\Psi \subset \Lambda \subset \Omega$). Let $\mathscr{N}_\phi$ denote the collection of all neighborhoods of a fixed point $\phi \in \mathscr{V}$. Then, $\mathscr{N}_\phi$ has the following properties.

(i) $\phi \in \Omega$ for all $\Omega \in \mathscr{N}_\phi$.
(ii) If $\mathscr{V} \supset \Lambda \supset \Omega \in \mathscr{N}_\phi$, then $\Lambda \in \mathscr{N}_\phi$.
(iii) If $\Omega, \Lambda \in \mathscr{N}_\phi$, then $\Omega \cap \Lambda \in \mathscr{N}_\phi$.
(iv) If $\Omega \in \mathscr{N}_\phi$, then there exists a $\Lambda \in \mathscr{N}_\phi$ such that $\Omega \in \mathscr{N}_\psi$ for all $\psi \in \Lambda$.

The following is a fact: Given any space $\mathscr{V}$ and, for each $\phi \in \mathscr{V}$, given a nonempty collection $\mathscr{N}_\phi$ of subsets of $\mathscr{V}$, if conditions (i)–(iv) are satisfied by every $\mathscr{N}_\phi$, then there exists a unique topology in $\mathscr{V}$ that makes $\mathscr{N}_\phi$ the collection of neighborhoods of ϕ for every $\phi \in \mathscr{V}$. Because of this, the collection of all neighborhoods of all points of $\mathscr{V}$ can be used as the definition of the topology on $\mathscr{V}$.

A subset $\mathscr{B}_\phi$ of the collection $\mathscr{N}_\phi$ of neighborhoods of $\phi \in \mathscr{V}$ said to be a *base of neighborhoods of* ϕ if, given any $\Omega \in \mathscr{N}_\phi$, there exists a $\Lambda \in \mathscr{B}_\phi$ such that $\Lambda \subset \Omega$. A specification of $\mathscr{B}_\phi$ for every $\phi \in \mathscr{V}$ uniquely determines the topology on $\mathscr{V}$.

B3. Given two topologies $\mathscr{O}_1$ and $\mathscr{O}_2$ on $\mathscr{V}$, $\mathscr{O}_1$ is said to be *stronger* or *finer than* $\mathscr{O}_2$ if $\mathscr{O}_1 \supset \mathscr{O}_2$. In this case, $\mathscr{O}_2$ is said to be *weaker* or *coarser than* $\mathscr{O}_1$. Also, the special case $\mathscr{O}_1 = \mathscr{O}_2$ is allowed here. For each $\phi \in \mathscr{V}$, let $\mathscr{B}_\phi{}^1$ and $\mathscr{B}_\phi{}^2$ be bases of neighborhoods of ϕ for $\mathscr{O}_1$ and $\mathscr{O}_2$, respectively. Then, $\mathscr{O}_1$ is stronger than $\mathscr{O}_2$ if and only if, given any $\Lambda \in \mathscr{B}_\phi{}^2$, there exists an $\Omega \in \mathscr{B}_\phi{}^1$ such that $\Omega \subset \Lambda$.

B4. Let $\mathscr{V}$ and $\mathscr{W}$ be two topological spaces and let f be a mapping of $\mathscr{V}$ into $\mathscr{W}$. f is said to be *continuous at* $\phi \in \mathscr{V}$ if, for any neighborhood Λ of $f(\phi) \in \mathscr{W}$, there exists a neighborhood Ω of $\phi \in \mathscr{V}$ such that $f(\Omega) \subset \Lambda$. f is called *continuous* if it is continuous at every point of $\mathscr{V}$. The following three conditions are equivalent.

(i) f is continuous.
(ii) $f^{-1}(\Lambda)$ is open in $\mathscr{V}$ for every open set Λ in $\mathscr{W}$.
(iii) $f^{-1}(\Lambda)$ is closed in $\mathscr{V}$ for every closed set Λ in $\mathscr{W}$.

B5. A sequence $\{\phi_k\}_{k=1}^\infty$ in a topological space $\mathscr{V}$ is said to *converge in* $\mathscr{V}$ *to a limit* $\phi \in \mathscr{V}$ if, given any neighborhood Ω of ϕ, there exists an integer

K such that $\phi_k \in \Omega$ for all $k > K$. In this case, we write $\phi_k \to \phi$ or $\lim \phi_k = \phi$. A mapping f from $\mathscr{V}$ into another topological space $\mathscr{W}$ is called *sequentially continuous* if, for every convergent sequence $\{\phi_k\}$ with limit $\phi \in \mathscr{V}$, we have that $f(\phi_k) \to f(\phi)$ in $\mathscr{W}$. The continuity of f implies its sequential continuity, but the converse is not true in general.

If $\mathscr{U}$ is a subset of $\mathscr{V}$, $\mathscr{U}$ is said to be *sequentially dense in* $\mathscr{V}$ if, for every $\phi \in \mathscr{V}$, there exists a sequence $\{\phi_k\} \subset \mathscr{U}$ such that $\phi_k \to \phi$ in $\mathscr{V}$. The sequential density of $\mathscr{U}$ in $\mathscr{V}$ implies the density of $\mathscr{U}$ in $\mathscr{V}$, but again the converse is not true in general.

B6. Let $\mathscr{W}$ be a topological space and $\mathscr{V}$ a subset of $\mathscr{W}$. The *induced topology* $\mathscr{O}_i$ *on* $\mathscr{V}$ is the collection of all sets of the form $\mathscr{V} \cap \Lambda$, where Λ is any open set in $\mathscr{W}$. $\mathscr{V}$ with the induced topology is separated whenever $\mathscr{W}$ is. The *canonical injection of* $\mathscr{V}$ *into* $\mathscr{W}$ is the mapping that assigns to each $\phi \in \mathscr{V}$ the element $\phi \in \mathscr{W}$. If $\mathscr{V}$ has its own topology, say $\mathscr{O}$, then $\mathscr{O}$ is stronger than $\mathscr{O}_i$ if and only if the canonical injection of $\mathscr{V}$ into $\mathscr{W}$ is continuous.

B7. Let $\mathscr{V}_1, \ldots, \mathscr{V}_n$ be a finite collection of topological spaces. *The Cartesian product* $\mathscr{V} \triangleq \mathscr{V} \times \cdots \times \mathscr{V}_n$ *of these spaces* is the set of all ordered n-tuples $\phi \triangleq \{\phi_1, \ldots, \phi_n\}$, where $\phi_k \in \mathscr{V}_k$ for each $k = 1, \ldots, n$. Now, let P_k denote the mapping $\phi \mapsto \phi_k$. Also, for any given $\phi \in \mathscr{V}$, let $\mathscr{B}_\phi$ be the collection of all subsets $\Omega \subset \mathscr{V}$ for which $P_k(\Omega)$ is a neighborhood of ϕ_k for every k. $\mathscr{B}_\phi$ is a base of neighborhoods for a unique topology on $\mathscr{V}$, called the *product topology*. When $\mathscr{V}$ is assigned this topology, it is called the *topological product of the* $\mathscr{V}_k$. In this case, $\mathscr{V}$ is separated whenever each $\mathscr{V}_k$ is.

B8. A *metric* ρ on an arbitrary set $\mathscr{V}$ is a mapping from $\mathscr{V} \times \mathscr{V}$ into the real line R such that the following three rules are satisfied whatever be the elements $\phi, \psi, \theta, \in \mathscr{V}$:

(i) $\rho(\phi, \psi) \geq 0$; also, $\rho(\phi, \psi) = 0$ if and only if $\phi = \psi$.
(ii) $\rho(\phi, \psi) = \rho(\psi, \phi)$.
(iii) $\rho(\phi, \psi) \leq \rho(\phi, \theta) + \rho(\theta, \psi)$.

A *metric space* $\mathscr{V}$ is a set with a metric defined upon it. Given any $\rho \in \mathscr{V}$ and a real number $r > 0$, the *open sphere* $O(\phi, r)$ *centered at* ϕ *and of radius* r is the set $\{\psi \in \mathscr{V} : \rho(\phi, \psi) < r\}$. We specify a topology on $\mathscr{V}$ by defining each neighborhood of any $\phi \in \mathscr{V}$ as a set containing an open sphere centered at ϕ. A topological space is said to be *metrizable* if its topology can be obtained in this way from a metric. Every metrizable topological space is separated and has a countable base of neighborhoods of each of its points ϕ, namely $\{O(\phi, 1/n)\}_{n=1}^{\infty}$.

A mapping f from a metric space $\mathscr{V}$ into a topological space $\mathscr{W}$ is continuous if and only if it is sequentially continuous. Also, a subset $\mathscr{U}$ of a metric space $\mathscr{V}$ is dense in $\mathscr{V}$ if and only if it is sequentially dense in $\mathscr{V}$.

Appendix C

Topological Linear Spaces

Note. In this and the subsequent appendices, we will continue to fix our attention on complex linear spaces. Nevertheless, all the results listed here become valid for real linear spaces upon making the obvious alterations. See Appendix A7.

C1. By a *topological linear space*, we mean a linear space $\mathscr{V}$ having a topology such that the algebraic operations of addition and multiplication by a complex number are continuous. That is, the mapping $\{\phi, \psi\} \mapsto \phi + \psi$ is continuous from $\mathscr{V} \times \mathscr{V}$ into $\mathscr{V}$, and the mapping $\{\alpha, \phi\} \mapsto \alpha\phi$ is continuous from $C \times \mathscr{V}$ into $\mathscr{V}$ when $\mathscr{V} \times \mathscr{V}$ and $C \times \mathscr{V}$ carry the product topologies.

If $\mathscr{B}$ is a collection of subsets of $\mathscr{V}$ and $\phi \in \mathscr{V}$, $\mathscr{B} + \phi$ denotes the collection of sets obtained by adding ϕ to every element in each set of $\mathscr{B}$. In a topological linear space, $\mathscr{B}$ is a base of neighborhoods of 0 (0 denotes the origin) if and only if $\mathscr{B} + \phi$ is a base of neighborhoods of ϕ. Thus, to determine the topology of $\mathscr{V}$, we need merely specify a base of neighborhoods of 0.

C2. Let $\mathscr{B}$ be a base of neighborhoods of 0 in a topological linear space $\mathscr{V}$. The following three assertions are equivalent.

(i) $\mathscr{V}$ is separated.
(ii) $\bigcap_{\Omega \in \mathscr{B}} \Omega = \{0\}$.
(iii) $\{0\}$ is a closed set in $\mathscr{V}$.

C3. A *locally convex space* is a topological linear space having a base of convex neighborhoods of 0. Its topology is also called *locally convex*. Such spaces can be characterized as follows.

If $\mathscr{V}$ is a locally convex space, it has a base $\mathscr{B}$ of neighborhoods of 0 with the following properties.

(i) If $\Omega, \Lambda \in \mathscr{B}$, then there exists a $\Xi \in \mathscr{B}$ such that $\Xi \subset \Omega \cap \Lambda$.
(ii) If $\Omega \in \mathscr{B}$ and $\alpha \in C$, where $\alpha \neq 0$, then $\alpha\Omega \in \mathscr{B}$.
(iii) Every $\Omega \in \mathscr{B}$ is balanced, convex, and absorbent.

Conversely, if $\mathscr{V}$ is a linear space and $\mathscr{B}$ is a nonempty collection of subsets of $\mathscr{V}$ having these three properties, then there exists a unique topology that renders $\mathscr{V}$ into a locally convex space with $\mathscr{B}$ as a base of neighborhoods of 0.

C4. Let $\mathscr{V}$ be a linear space and $\mathscr{A}$ any set of balanced convex absorbent subsets of $\mathscr{V}$. Let $\mathscr{V}$ have the topology $\mathscr{O}$ generated by the base of neighborhoods of 0 consisting of all sets of the form $\varepsilon \bigcap_k \Lambda_k$, where $\varepsilon \in R$, $\varepsilon > 0$, and $\bigcap_k \Lambda_k$ is the intersection of a finite collection of $\Lambda_k \in \mathscr{A}$. Then, $\mathscr{V}$ is a locally convex space. Moreover, $\mathscr{O}$ is the weakest topology under which the algebraic operations of addition and multiplication by a complex number are continuous and every member of $\mathscr{A}$ is a neighborhood of 0. Conversely, the topology $\mathscr{O}$ of any locally convex space $\mathscr{V}$ can be produced in this way from a collection $\mathscr{A}$ of balanced convex absorbent subsets of $\mathscr{V}$.

C5. Let $\mathscr{V}$ be a complex linear space. A *seminorm* γ *on* $\mathscr{V}$ is a mapping of $\mathscr{V}$ into R such that, for every $\phi, \psi \in \mathscr{V}$ and every $\alpha \in C$, we have that $\gamma(\alpha\phi) = |\alpha|\gamma(\phi)$ and $\gamma(\phi + \psi) \leq \gamma(\phi) + \gamma(\psi)$. It follows that $\gamma(0) = 0$, $\gamma(\phi) \geq 0$, and $|\gamma(\phi) - \gamma(\psi)| \leq \gamma(\phi - \psi)$. If, in addition, $\gamma(\phi) = 0$ implies that $\phi = 0$, γ is called a *norm*. For two seminorms γ and ρ on $\mathscr{V}$, we write $\gamma \leq \rho$ to mean $\gamma(\phi) \leq \rho(\phi)$ for all $\phi \in \mathscr{V}$.

If $\mathscr{V}$ is a locally convex space, a seminorm on $\mathscr{V}$ is a continuous mapping if and only if it is continuous at the origin.

For any finite collection $\{\gamma_k\}$ of seminorms on $\mathscr{V}$, $\max_k \gamma_k$ is defined by $(\max_k \gamma_k)(\phi) \triangleq \max_k \gamma_k(\phi)$ and is also a seminorm on $\mathscr{V}$; moreover, it is continuous whenever each γ_k is continuous. The same is true for sums $\sum_k \gamma_k$ of seminorms.

C6. Let $\mathscr{V}$ be a linear space and Γ any collection of seminorms on $\mathscr{V}$. Let $\mathscr{V}$ have a topology $\mathscr{O}$ generated by the base of neighborhoods of 0 consisting of all sets of the form

$$\left\{\phi \in \mathscr{V} : \max_k \gamma_k(\phi) \leq \varepsilon\right\},$$

where $\varepsilon \in R$, $\varepsilon > 0$, and the γ_k comprise a finite set of seminorms in Γ. Then, $\mathscr{V}$ is a locally convex space. Moreover, $\mathscr{O}$ is the weakest topology under which the algebraic operations of addition and multiplication by a complex number are continuous and every seminorm in Γ is a continuous mapping. Conversely, the topology $\mathscr{O}$ of any locally convex space $\mathscr{V}$ can be produced in this way from a collection Γ of seminorms on $\mathscr{V}$.

We say that $\mathscr{O}$ *is generated by* Γ and call Γ a *generating family of seminorms for the topology of* $\mathscr{V}$ or simply a *generating family of seminorms.* If $\mathscr{V}$ is separated, Γ is called a *multinorm for* $\mathscr{V}$. Actually, $\mathscr{V}$ will be separated when and only when, for every $\phi \in \mathscr{V}$ with $\phi \neq 0$, there exists some $\gamma \in \Gamma$ such that $\gamma(\phi) > 0$.

A sequence $\{\phi_k\}_{k=1}^{\infty}$ converges in the locally convex space $\mathscr{V}$ to a limit ϕ if and only if $\gamma(\phi_k - \phi) \to 0$ as $k \to \infty$ for every γ in any given generating family of seminorms.

A *base of continuous seminorms for* $\mathscr{V}$ is any collection P of seminorms ρ on $\mathscr{V}$ obtained from any generating family Γ of seminorms for $\mathscr{O}$ by setting $\rho = \max_k \gamma_k$, where $\{\gamma_k\}$ traverses all finite subsets of Γ. P is also a generating family of seminorms for 0. Moreover, P has the property that, given any continuous seminorm η on $\mathscr{V}$, there exist a constant $M > 0$ and a $\rho \in$ P such that $\eta \leq M\rho$.

C7. Let $\mathscr{V}$ be a locally convex space. The following three assertions are equivalent.

(i) $\mathscr{V}$ is metrizable (see Appendix B8).
(ii) $\mathscr{V}$ is separated and has a countable base of neighborhoods of 0.
(iii) The topology of $\mathscr{V}$ is generated by a countable multinorm (i.e., $\mathscr{V}$ is separated, and its topology is produced from a countable generating family of seminorms).

C8. A subset Ω of a locally convex space $\mathscr{V}$ is called *bounded* if $\sup_{\phi \in \Omega} \gamma(\phi) < \infty$ for every seminorm γ in a generating family of seminorms. Ω is a bounded set in $\mathscr{V}$ if and only if, given any balanced convex neighborhood Λ of 0 in $\mathscr{V}$, there exists a $p \in R$ such that $\Omega \subset p\Lambda$. The union of any convergent sequence and its limit is a bounded set.

C9. A subset of a locally convex space $\mathscr{V}$ is said to be *total* if its span is a dense subspace of $\mathscr{V}$.

C10. Let $\mathscr{V}$ be a nonempty set. A *filter* $\mathscr{F}$ *on* $\mathscr{V}$ is a nonempty collection of subsets of $\mathscr{V}$ having the following three properties.

(i) The empty set is not a member of $\mathscr{F}$.
(ii) If $\Omega, \Lambda \in \mathscr{F}$, then $\Omega \cap \Lambda \in \mathscr{F}$.
(iii) If $\Omega \in \mathscr{F}$ and $\Lambda \supset \Omega$, then $\Lambda \in \mathscr{F}$.

Now, let $\mathscr{V}$ be a topological space. The collection of all neighborhoods of a nonempty subset of $\mathscr{V}$ is an example of a filter. A filter is said to *converge to a limit* $\phi \in \mathscr{V}$ if every neighborhood of ϕ is a member of $\mathscr{F}$.

C11. Assume that $\mathscr{V}$ is a topological linear space. A filter $\mathscr{F}$ on $\mathscr{V}$ is called a *Cauchy filter* if, for every neighborhood Ω of 0 in $\mathscr{V}$, there is a $\Lambda \in \mathscr{F}$ such that $\Lambda - \Lambda \subset \Omega$. Every convergent filter is a Cauchy filter. $\mathscr{V}$ is called *complete* if every Cauchy filter on $\mathscr{V}$ converges to a limit $\phi \in \mathscr{V}$. On the other hand, a sequence $\{\phi_k\}$ in $\mathscr{V}$ is called a *Cauchy sequence* if, given any neighborhood Ω of 0 in $\mathscr{V}$, there exists an integer K such that $\phi_k - \phi_m \in \Omega$ for all $k, m > K$. This condition is equivalent to the requirement that, as k and m tend to infinity independently, $\gamma(\phi_k - \phi_m) \to 0$ for every γ in any given generating family of seminorms for the topology of $\mathscr{V}$. $\mathscr{V}$ is called *sequentially complete* if every Cauchy sequence in $\mathscr{V}$ converges to a limit $\phi \in \mathscr{V}$. The completeness of $\mathscr{V}$ implies its sequential completeness, but the converse is not true in general. The converse is true for metrizable spaces. A complete metrizable locally convex space is called a *Fréchet space*.

If $\mathscr{V}$ is complete and $\mathscr{U}$ is a closed linear subspace of $\mathscr{V}$ supplied with the induced topology, then $\mathscr{U}$ is also complete.

C12. Let $\mathscr{V}$ be a locally convex space, and Γ a generating family of seminorms for $\mathscr{V}$. A series $\sum_{j=1}^{\infty} \phi_j$, where $\phi_j \in \mathscr{V}$, is said to *converge in* $\mathscr{V}$ *to a limit* $\phi \in \mathscr{V}$ if the sequence $\{\sum_{j=1}^{k} \phi_j\}_{k=1}^{\infty}$ converges in $\mathscr{V}$ to ϕ. On the other hand, $\sum_{j=1}^{\infty} \phi_j$ is said to *converge absolutely* if $\sum_j \gamma(\phi_j)$ converges for every $\gamma \in \Gamma$. An absolutely convergent series converges to some limit ϕ, and every rearrangement of that series converges to the same limit ϕ.

C13. A topological linear space A whose topology is generated by a single norm is called a *normed linear space*. We usually denote that norm by $\|\cdot\|$ or $\|\cdot\|_A$, but other symbols are also used. If, in addition, A is complete (or, equivalently, sequentially complete), A is called a (*complex*) *Banach space*.

A set $F \subset A$ is called *nowhere dense* if $\bar{F}$ has no interior points. No Banach space is equal to the union of a countable collection of nowhere dense sets. This is *Baire's category theorem* for Banach spaces.

C14. Let H be a complex linear space and let there exist a mapping $(\cdot, \cdot): \{\phi, \psi\} \mapsto (\phi, \psi)$ of $H \times H$ into C such that the following conditions are satisfied.

(i) $(\phi, \psi) = \overline{(\psi, \phi)}$, where the bar denotes the complex conjugate.
(ii) For each fixed $\psi \in H$, $\phi \mapsto (\phi, \psi)$ is a linear mapping on H.
(iii) $(\phi, \phi) > 0$ if $\phi \neq 0$.

Then, $(\cdot, \cdot)$ is called an *inner product on H*. It follows that $(\cdot, \cdot)$ is a sesquilinear form. Moreover, with $\|\phi\| \triangleq [(\phi, \phi)]^{1/2}$, $\|\cdot\|$ is a norm on H. When H is assigned the topology generated by this norm, it is called a *(complex) inner-product space*. If in this case H is complete, it is called a *(complex) Hilbert space*. [A *real Hilbert space* is defined in the same way except that H is a real linear space and the range of $(\cdot, \cdot)$ is contained in R.]

C15. The *Schwarz inequality* for the inner product $(\cdot, \cdot)$ on the inner product space H states that $|(\phi, \psi)| \leq \|\phi\| \, \|\psi\|$. This implies that the inner product is a continuous mapping when $H \times H$ is equipped with the product topology.

C16. Let H be a separable Hilbert space. Then, there exists at least one sequence $\{e_k\}_{k=1}^{\infty} \subset H$ satisfying the following conditions.

(i) $(e_k, e_k) = 1$, and $(e_k, e_j) = 0$ if $k \neq j$.
(ii) Every $f \in H$ can be expanded into the series

$$f = \sum_{k=1}^{\infty} (f, e_k) e_k, \tag{1}$$

which converges in H.

When $\{e_k\}$ satisfies (i), it is called *orthonormal* and when it satisfies both (i) and (ii), it is called a *complete orthonormal set*. (Thus, the meaning of the adjective "complete" in this case is different from that of Appendix C11.) The expansion (1) is unique in the sense that the alteration of any of the coefficients of the e_k will alter the sum of the series.

C17. Let $\{e_k\}$ be a complete orthonormal sequence in the separable Hilbert space H. Given any $f \in \{e_k\}$, we have *Parseval's equation*:

$$\sum_{k=1}^{\infty} |(f, e_k)|^2 = \|f\|^2.$$

Conversely, the *Riesz–Fischer theorem* states that, for any sequence $\{\alpha_k\}_{k=1}^{\infty} \subset C$ such that $\sum_{k=1}^{\infty} |\alpha_k|^2 < \infty$, there exists a unique $f \in H$ such that $\alpha_k = (f, e_k)$ for all k.

Appendix D

Continuous Linear Mappings

Throughout Appendix D, $\mathscr{V}$ and $\mathscr{W}$ will denote locally convex spaces, and Γ and P will be generating families of seminorms for the topologies of $\mathscr{V}$ and $\mathscr{W}$, respectively.

D1. Continuous mappings and sequentially continuous mappings from one topological space into another topological space have been defined in Appendix B, Sections B4 and B5. If $\mathscr{V}$ is metrizable, then the sequential continuity of a mapping of $\mathscr{V}$ into $\mathscr{W}$ is equivalent to its continuity.

D2. Let f be a linear mapping of $\mathscr{V}$ into $\mathscr{W}$. The following four assertions are equivalent.

(i) f is continuous.
(ii) f is continuous at the origin.
(iii) For every continuous seminorm ρ on $\mathscr{W}$, there exists a continuous seminorm γ on $\mathscr{V}$ such that $\rho(f(\phi)) \leq \gamma(\phi)$ for all $\phi \in V$ [or, equivalently, $\gamma(\phi) < 1$ implies that $\rho(f(\phi)) \leq 1$].

(iv) For every $\rho \in \mathrm{P}$, there exist a constant $M > 0$ and a finite collection $\{\gamma_1, \ldots, \gamma_m\} \subset \Gamma$ such that

$$\rho(f(\phi)) \leq M \max_{1 \leq k \leq m} \gamma_k(\phi)$$

for all $\phi \in \mathscr{V}$.

D3. The spaces $\mathscr{V}$ and $\mathscr{W}$ are said to be *isomorphic* if there exists a bijection ι of $\mathscr{V}$ onto $\mathscr{W}$ such that ι is continuous and linear and its inverse ι^{-1} is also continuous and linear. In this case, ι is called an *isomorphism of* $\mathscr{V}$ *onto* $\mathscr{W}$. If $\mathscr{V}$ and $\mathscr{W}$ are the same space, then ι is called an *automorphism on* $\mathscr{V}$.

D4. If $\mathscr{V}$ is separated, there exists a complete locally convex space $\hat{\mathscr{V}}$ and a mapping ι of $\mathscr{V}$ into $\hat{\mathscr{V}}$ such that the following three conditions are satisfied.

(i) ι is an isomorphism of $\mathscr{V}$ onto the image of $\mathscr{V}$ in $\hat{\mathscr{V}}$ supplied with the topology induced by $\hat{\mathscr{V}}$.
(ii) The image of $\mathscr{V}$ is dense in $\hat{\mathscr{V}}$.
(iii) Given any complete separated topological linear space $\mathscr{W}$ and any continuous linear mapping f of $\mathscr{V}$ into $\mathscr{W}$, there exists a unique continuous linear mapping $\hat{f}$ of $\hat{\mathscr{V}}$ into $\mathscr{W}$ such that f is equal to the composite mapping $\hat{f}\iota$ obtained by first applying ι and then applying $\hat{f}$.

Moreover, if $\tilde{\mathscr{V}}$ is any other complete locally convex space for which there exists a mapping $\tilde{\iota}$ satisfying conditions (i) and (ii), then $\tilde{\mathscr{V}}$ and $\hat{\mathscr{V}}$ are isomorphic.

The space $\hat{\mathscr{V}}$ is called the *completion of* $\mathscr{V}$.

D5. Let $\mathscr{V}$ be a Fréchet space, $\mathscr{U}$ a dense linear subspace of $\mathscr{V}$ supplied with the induced topology, $\mathscr{W}$ a sequentially complete separated space, and f a continuous linear mapping of $\mathscr{U}$ into $\mathscr{W}$. Then, there exists a unique continuous linear mapping g of $\mathscr{V}$ into $\mathscr{W}$ such that $g(\phi) = f(\phi)$ for all $\phi \in \mathscr{U}$.

A similar result for sesquilinear forms is the following. Let $\mathscr{W}$ be as before, let $\mathscr{V}_1$ and $\mathscr{V}_2$ be Fréchet spaces, and let $\mathscr{U}_1$ and $\mathscr{U}_2$ be dense linear subspaces of $\mathscr{V}_1$ and $\mathscr{V}_2$ respectively. Supply $\mathscr{V}_1 \times \mathscr{V}_2$ with the product topology and $\mathscr{U}_1 \times \mathscr{U}_2$ with the induced topology. Assume that f is a continuous sesquilinear mapping of $\mathscr{U}_1 \times \mathscr{U}_2$ into $\mathscr{W}$. The continuity property is equivalent to the condition that, given any $\rho \in \mathrm{P}$, there is a constant $M > 0$ and two continuous seminorms γ_1 and γ_2 on $\mathscr{V}_1$ and $\mathscr{V}_2$, respectively, for which

$$\rho[f(\phi, \psi)] \leq M\gamma_1(\phi)\gamma_2(\psi), \qquad \phi \in \mathscr{U}_1, \quad \psi \in \mathscr{U}_2. \tag{1}$$

We can conclude that there exists a unique continuous sesquilinear mapping g of $\mathscr{V}_1 \times \mathscr{V}_2$ into $\mathscr{W}$ such that $g(\phi, \psi) = f(\phi, \psi)$ for all $\phi \in \mathscr{U}_1$ and $\psi \in \mathscr{U}_2$. Moreover, (1) holds again for f replaced by g and for all $\phi \in \mathscr{V}_1$ and $\psi \in \mathscr{V}_2$.

D6. Let f be a linear mapping of $\mathscr{V}$ into $\mathscr{W}$. f is called a *bounded mapping* if it takes bounded sets in $\mathscr{V}$ into bounded sets in $\mathscr{W}$. If f is continuous, it is bounded, but the converse is not true in general. However, the converse is true when $\mathscr{V}$ is a metrizable space.

D7. A *functional on* $\mathscr{V}$ or a *form on* $\mathscr{V}$ is a mapping of $\mathscr{V}$ into C. The space of all continuous linear functionals on $\mathscr{V}$ is denoted by $\mathscr{V}'$ or by $[\mathscr{V}; C]$ and is called the *dual of* $\mathscr{V}$. Given any $f \in \mathscr{V}'$, the mapping $\gamma_f\colon \phi \mapsto |f(\phi)|$ is a seminorm on $\mathscr{V}$. The topology generated on $\mathscr{V}$ by the collection $\{\gamma_f\}_{f \in \mathscr{V}'}$ of all such seminorms is called the *weak topology of* $\mathscr{V}$. If $\mathscr{V}$ is separated and ϕ is any member of $\mathscr{V}$ such that $\phi \neq 0$, there exists an $f \in \mathscr{V}'$ such that $f(\phi) \neq 0$. Thus, under its weak topology, $\mathscr{V}$ is again a separated locally convex space.

Moreover, if A is a Banach space and $\{a_k\}$ is a sequence that converges under the weak topology of A, then $\sup_k \|a_k\| < \infty$.

D8. The symbol $[\mathscr{V}; \mathscr{W}]$ denotes the set of all continuous linear mappings of $\mathscr{V}$ into $\mathscr{W}$. The addition and multiplication by a complex number of members of $[\mathscr{V}; \mathscr{W}]$ is defined as in Appendix A5. As a consequence, $[\mathscr{V}; \mathscr{W}]$ is a linear space. Given any bounded set $\Omega \subset \mathscr{V}$ and any $\rho \in \mathrm{P}$, the mapping

$$\eta_{\Omega, \rho}\colon f \mapsto \sup_{\phi \in \Omega} \rho(f(\phi)), \qquad f \in [\mathscr{V}; \mathscr{W}]$$

defines a seminorm on $[\mathscr{V}; \mathscr{W}]$. The collection of all such seminorms $\{\eta_{\Omega, \rho}\}$, where Ω traverses the bounded sets in $\mathscr{V}$ and ρ traverses P, defines a locally convex topology on $\mathscr{V}$, which is called the *topology of uniform convergence on the bounded sets of* $\mathscr{V}$ or simply the *bounded topology*. On the other hand, when Ω traverses only the finite sets in $\mathscr{V}$ and ρ traverses P, the topology generated by $\{\eta_{\Omega, \rho}\}$ is called the *topology of pointwise convergence on* $\mathscr{V}$ or simply the *pointwise topology*. The pointwise topology is weaker than the bounded topology. Unless the opposite is explicitly stated, $[\mathscr{V}; \mathscr{W}]$ is understood to have the bounded topology. When it has the pointwise topology, it is denoted by $[\mathscr{V}; \mathscr{W}]^\sigma$.

D9. A subset Ξ of $[\mathscr{V}; \mathscr{W}]$ is said to be *equicontinuous* if, given any $\rho \in \mathrm{P}$, there exists a constant $M > 0$ and a finite collection $\{\gamma_1, \ldots, \gamma_m\} \subset \Gamma$ such that

$$\sup_{f \in \Xi} \rho(f(\phi)) \leq M \max_{1 \leq k \leq m} \gamma_k(\phi)$$

for all $\phi \in \mathscr{V}$. If $\mathscr{V}$ is a Fréchet space, the following three assertions are equivalent:

(i) Ξ is bounded in $[\mathscr{V}; \mathscr{W}]$.
(ii) Ξ is bounded in $[\mathscr{V}; \mathscr{W}]^{\sigma}$.
(iii) Ξ is equicontinuous.

D10. The following is one version of the *closed graph theorem*: Let $\mathscr{V}$ and $\mathscr{W}$ be Fréchet spaces and f a linear mapping of $\mathscr{V}$ into $\mathscr{W}$. Assume that, for every sequence $\{\phi_k\}$ which converges to zero in $\mathscr{V}$ and for which $f(\phi_k)$ converges to ψ in $\mathscr{W}$, we always have that $\psi = 0$. Then, f is continuous.

D11. Let A and B be normed linear spaces with norms $\|\cdot\|_A$ and $\|\cdot\|_B$, respectively. The bounded topology of $[A; B]$ is equal to the topology generated by the single norm $\|\cdot\|_{[A;B]}$, where

$$\|f\|_{[A;B]} \triangleq \sup_{\|a\|=1} \|f(a)\|_B, \qquad f \in [A; B].$$

(The supremum notation means that the supremum is taken over all $a \in A$ for which $\|a\|_A = 1$.) This topology is called the *uniform operator topology*. When B is a Banach space, $[A; B]$ under the uniform operator topology is a Banach space, too.

For any fixed $a \in A$, the mapping $\gamma_a: f \mapsto \|f(a)\|_B$ is a seminorm on $[A; B]$. The topology generated by the collection $\{\gamma_a: a \in A\}$ of seminorms is called the *strong operator topology of* $[A; B]$. It is the same as the pointwise topology of $[A; B]$, defined in Appendix D8. In this case, if A is a Banach space, $[A; B]$ under the strong operator topology is a sequentially complete separated space; if, in addition, $\{f_k\}_{k=1}^{\infty}$ converges in this topology to f, then $\|f\| \leq \underline{\lim}_{k\to\infty}\|f_k\|$.

Finally, for any given $a \in A$ and $b' \in B'$ (B' is the dual of B), $\gamma_{a,b'}: f \mapsto |b'(f(a))|$ is also a seminorm on $[A; B]$. The topology generated by the collection $\{\gamma_{a,b'}: a \in A, b' \in B'\}$ of seminorms is called the *weak operator topology of* $[A; B]$.

Each of these topologies separates $[A; B]$.

D12. Upon applying Appendix D9 in the context of Appendix D11, we obtain the *principle of uniform boundedness*: If Ξ is a subset of $[A; B]$ and if $\sup_{f\in\Xi}\|f(a)\|_B < \infty$ for each $a \in A$, then $\sup_{f\in\Xi}\|f\|_{[A;B]} < \infty$.

Upon combining this principle with the next paragraph we get the following result: If Ω is a subset of A and if $\sup_{a\in\Omega}|a'(a)| < \infty$ for each $a' \in A'$, then $\sup_{a\in\Omega}\|a\|_A < \infty$.

D13. The *second dual* A'' of a Banach space A is the dual of the dual A' of A. Every $a \in A$ generates a functional a'' on A' through the definition $\langle a'', b\rangle \triangleq \langle b, a\rangle$, where $b \in A'$. It follows that $a'' \in A''$ and that $\|a''\| = \|a\|$.

D14. Let A and B be Banach spaces. If $f \in [A; B]$ is bijective (so that its inverse f^{-1} exists and B is the domain of f^{-1}), then $f^{-1} \in [B; A]$.

D15. Let H be a Hilbert space. Corresponding to each $g \in H'$, where H' is the dual of H, there exists a unique $b \in H$ such that $g(a) = (a, b)$ for every $a \in H$; the converse is also true. Moreover, $\|g\|_{[H;C]} = \|b\|_H$.

Next, let f be a sequilinear form on $H \times H$, which is continuous when $H \times H$ is supplied the product topology [i.e., there exists a constant M for which $|f(a, b)| \leq M\|a\| \|b\|$ for all $a, b \in H$]. Then, there exists a unique $P \in [H; H]$ such that $(Pa, b) = f(a, b)$ for all $a, b \in H$.

An operator P mapping H into H is called *positive* if $(Pa, a) \geq 0$ for all $a \in H$. The set of positive continuous linear mappings of H into H is denoted by $[H; H]_+$. Similarly, a sesquilinear form f on $H \times H$ is called *positive* if $f(a, a) \geq 0$ for all $a \in H$. If f is a positive sesquilinear form on $H \times H$ such that $f(a, a) \leq M\|a\|^2$, where M is a constant not depending on a, then there exists a unique $P \in [H; H]_+$ such that $(Pa, b) = f(a, b)$ for all $a, b \in H$.

Furthermore, if $Q \in [H; J]$, where J is another Hilbert space, its *adjoint* Q' is defined by $(Q'a, b) \triangleq (a, Qb)$, where $a \in J$ and $b \in H$. We have that $Q' \in [J; H]$ and $\|Q'\| = \|Q\|$. A $Q \in [H; H]$ is called *self-adjoint* if $Q = Q'$; this is the case if and only if (Qa, a) is real for all $a \in H$. For such an operator, we have

$$\|Q\|_{[H;H]} = \sup_{\|a\|=1} |(Qa, a)|.$$

$Q \in [H; H]$ is called *skew-adjoint* if $Q = -Q'$; this is so if and only if iQ is self-adjoint.

Appendix E

Inductive-Limit Spaces

E1. Let N be any index set. Also, let $\mathscr{V}$ be a linear space such that $\mathscr{V} = \bigcup_{n \in N} \mathscr{V}_n$, where each $\mathscr{V}_n$ is a locally convex space. We define an *inductive-limit topology on* $\mathscr{V}$ by assigning the following base $\mathscr{B}$ of neighborhoods of 0 to $\mathscr{V}$: $\Lambda \in \mathscr{B}$ if and only if Λ is a balanced convex absorbent set in $\mathscr{V}$ and, for every $n \in N$, $\Lambda \cap \mathscr{V}_n$ is a neighborhood of 0 in $\mathscr{V}_n$. In this case, $\mathscr{V}$ is a locally convex space. It is called an *inductive-limit space* or the *inductive limit of* $\{\mathscr{V}_n\}$. Moreover, the collection $\mathscr{A}$ of all convex subsets of $\mathscr{V}$ such that each $\Lambda \in \mathscr{A}$ intersects every $\mathscr{V}_n$ in a neighborhood of 0 in $\mathscr{V}_n$ is also a base of neighborhoods of 0 in $\mathscr{V}$.

It is convenient to allow the following trivial situation as a special case of an inductive-limit space. Every locally convex space $\mathscr{V}$ can be considered to be the inductive limit of itself by letting N be any index set and setting $\mathscr{V}_n = \mathscr{V}$ for every $n \in N$.

E2. Let $\mathscr{V} = \bigcup_{n \in N} \mathscr{V}_n$ be an inductive-limit space, let $\mathscr{W}$ be a locally convex space, and let f be a linear mapping of $\mathscr{V}$ into $\mathscr{W}$. Then, f is continuous if and only if, for every $n \in N$, the restriction of f to $\mathscr{V}_n$ is continuous.

In particular, if each $\mathscr{V}_n$ is a metrizable locally convex space and if the restriction of f to each $\mathscr{V}_n$ is a bounded mapping, then f is continuous on $\mathscr{V}$.

E3. Let $\mathscr{V}$ be a linear space such that $\mathscr{V} = \bigcup_{n=1}^{\infty} \mathscr{V}_n$, where, for each n, the following three conditions hold:

(i) Each $\mathscr{V}_n$ is a locally convex space.
(ii) $\mathscr{V}_n \subset \mathscr{V}_{n+1}$.
(iii) The topology of $\mathscr{V}_n$ is equal to the topology induced on $\mathscr{V}_n$ by $\mathscr{V}_{n+1}$.

Finally, let $\mathscr{V}$ have the inductive-limit topology. Under these conditions, $\mathscr{V}$ is called a *strict inductive-limit space* or the *strict inductive limit of* $\{\mathscr{V}_n\}_{n=1}^{\infty}$.

As a consequence of this definition, the topology of every $\mathscr{V}_n$ is equal to the topology induced on $\mathscr{V}_n$ by $\mathscr{V}$. Moreover, if each $\mathscr{V}_n$ is complete and separated, then so, too, is $\mathscr{V}$.

E4. Let $\mathscr{V}$ be the strict inductive limit of $\{\mathscr{V}_n\}_{n=1}^{\infty}$. If, for each n, $\mathscr{V}_n$ is a closed subset of $\mathscr{V}_{n+1}$, $\mathscr{V}$ is said to *possess the closure property*. Under this circumstance, a subset Ω of $\mathscr{V}$ is bounded in $\mathscr{V}$ if and only if, for some n, $\Omega \subset \mathscr{V}_n$ and Ω is bounded in $\mathscr{V}_n$. Also, a sequence converges in $\mathscr{V}$ if and only if it is contained and converges in $\mathscr{V}_n$ for some n.

It is a fact that $\mathscr{V}$ possesses the closure property if every $\mathscr{V}_n$ is a complete separated space.

Appendix F

Bilinear Mappings and Tensor Products

F1. Let $\mathscr{U}$, $\mathscr{V}$, and $\mathscr{W}$ be topological linear spaces. A bilinear or sesquilinear mapping f on $\mathscr{U} \times \mathscr{V}$ into $\mathscr{W}$ is called *separately continuous* if $\phi \mapsto f(\phi, \psi)$ is continuous on $\mathscr{U}$ for each fixed $\psi \in \mathscr{V}$ and $\psi \mapsto f(\phi, \psi)$ is continuous on $\mathscr{V}$ for each fixed $\phi \in \mathscr{U}$. If f is continuous from $\mathscr{U} \times \mathscr{V}$ into $\mathscr{W}$ when $\mathscr{U} \times \mathscr{V}$ is supplied with the product topology, then f is separately continuous.

F2. Let $\mathscr{U}$ be a metrizable locally convex space or the inductive limit of such spaces, and let $\mathscr{V}$ be a Fréchet space or the inductive limit of Fréchet spaces. Assume that f is a separately continuous bilinear or sesquilinear mapping of $\mathscr{U} \times \mathscr{V}$ into a locally convex space $\mathscr{W}$. Then, f is continuous on $\mathscr{U} \times \mathscr{V}$ when $\mathscr{U} \times \mathscr{V}$ is equipped with its product topology.

Let $\mathscr{U}$ and $\mathscr{V}$ be Fréchet spaces. The topologies of $\mathscr{U}$ and $\mathscr{V}$ can always be obtained from two multinorms $\{\gamma_n\}_{n=1}^{\infty}$ and $\{\zeta_n\}_{n=1}^{\infty}$ respectively having the monotonic properties: $\gamma_1 \le \gamma_2 \le \gamma_3 \le \cdots$ and $\zeta_1 \le \zeta_2 \le \zeta_3 \le \cdots$. Assume that $\mathscr{W}$ is a complex Banach space B. Then, there exist a nonnegative integer m and a constant $M > 0$ such that

$$\|f(\phi, \psi)\|_B \leq M\gamma_m(\phi)\zeta_m(\psi)$$

for all $\phi \in \mathscr{U}$ and $\psi \in \mathscr{V}$.

F3. In this paragraph, $\mathscr{U}$ and $\mathscr{V}$ are complex linear spaces without any topologies. f denotes an arbitrary bilinear form on $\mathscr{U} \times \mathscr{V}$; i.e., $f \in B(\mathscr{U}, \mathscr{V})$. For any given $\phi \in \mathscr{U}$ and $\psi \in \mathscr{V}$, the mapping $f \mapsto f(\phi, \psi)$ is a linear form on $B(\mathscr{U}, \mathscr{V})$. We denote this mapping by $\phi \otimes \psi$ and therefore have $\phi \otimes \psi \in B(\mathscr{U}, \mathscr{V})^*$. It follows easily that the operator $\mathfrak{X}$: $\{\phi, \psi\} \mapsto \phi \otimes \psi$ is a bilinear mapping of $\mathscr{U} \times \mathscr{V}$ into $B(\mathscr{U}, \mathscr{V})^*$. The span of $\mathfrak{X}(\mathscr{U} \times \mathscr{V})$ is denoted by $\mathscr{U} \otimes \mathscr{V}$ and is called the *tensor product of* $\mathscr{U}$ *and* $\mathscr{V}$. Also, $\mathfrak{X}$ is called the *canonical bilinear mapping of* $\mathscr{U} \times \mathscr{V}$ *into* $\mathscr{U} \otimes \mathscr{V}$. Every element $\theta \in \mathscr{U} \otimes \mathscr{V}$ has a nonunique representation of the form $\theta = \sum_{k=1}^{n} \phi_k \otimes \psi_k$, where $\phi_k \in \mathscr{U}$ and $\psi_k \in \mathscr{V}$.

F4. Now, let $\mathscr{U}$, $\mathscr{V}$, and $\mathscr{W}$ be three complex linear spaces without topologies. $L(\mathscr{U} \otimes \mathscr{V}, \mathscr{W})$ denotes the linear space of all linear mappings of $\mathscr{U} \otimes \mathscr{V}$ into $\mathscr{W}$, and $B(\mathscr{U}, \mathscr{V}; \mathscr{W})$ denotes the linear space of all bilinear mappings of $\mathscr{U} \times \mathscr{V}$ into $\mathscr{W}$. Given any $h \in L(\mathscr{U} \otimes \mathscr{V}, \mathscr{W})$, $h\mathfrak{X}$ denotes the composite mapping on $\mathscr{U} \times \mathscr{V}$ into $\mathscr{V}$ obtained by first applying $\mathfrak{X}$ and then applying h. We can now state a theorem: The mapping $h \mapsto h\mathfrak{X}$ is a linear bijection of $L(\mathscr{U} \otimes \mathscr{V}, \mathscr{W})$ onto $B(\mathscr{U}, \mathscr{V}; \mathscr{W})$.

The significance of this theorem is the following. Tensor products allow us to replace linear spaces of bilinear mappings by linear spaces of linear mappings. This is one motivation for introducing tensor products.

F5. Let $\mathscr{U}$ and $\mathscr{V}$ be locally convex spaces and let Γ and H be bases of continuous seminorms for $\mathscr{U}$ and $\mathscr{V}$, respectively. Given any $\gamma \in \Gamma$ and $\eta \in \mathrm{H}$, define the function ρ on any $\theta \in \mathscr{U} \otimes \mathscr{V}$ by

$$\rho(\theta) \triangleq \inf\left\{\sum_k \gamma(\phi_k)\eta(\psi_k): \quad \theta = \sum_k \phi_k \otimes \psi_k\right\}. \tag{1}$$

That is, the infimum is taken over all representations of θ of the form $\theta = \sum_k \phi_k \otimes \psi_k$. Then, ρ is a seminorm on $\mathscr{U} \otimes \mathscr{V}$, and $\rho(\phi \otimes \psi) = \gamma(\phi)\eta(\psi)$ for all $\phi \in \mathscr{U}$ and $\psi \in \mathscr{V}$. Also, ρ is a norm if and only if both γ and η are norms. The collection P of all such ρ is taken to be a generating family of seminorms for a topology $\mathcal{O}_\pi$ on $\mathscr{U} \otimes \mathscr{V}$, called the *projective tensor-product topology* or the *π-topology*. In fact, P is a base of continuous seminorms for $\mathcal{O}_\pi$. $\mathcal{O}_\pi$ is the strongest locally convex topology on $\mathscr{U} \otimes \mathscr{V}$ under which the canonical bilinear mapping $\mathfrak{X}$ defined in Appendix F3 is continuous.

Also, $\mathscr{U} \otimes \mathscr{V}$ is separated if and only if both $\mathscr{U}$ and $\mathscr{V}$ are separated. Henceforth, it is understood that $\mathscr{U} \otimes \mathscr{V}$ is supplied with the topology $\mathscr{O}_\pi$.

F6. It is a fact that a linear mapping h of $\mathscr{U} \otimes \mathscr{V}$ into $\mathscr{W}$ is continuous if and only if the bilinear mapping $h\mathfrak{X}$ of $\mathscr{U} \times \mathscr{V}$ into $\mathscr{W}$ is continuous. (Here, $\mathscr{U} \times \mathscr{V}$ is supplied with the product topology.)

F7. Let $\mathscr{U}$ and $\mathscr{V}$ be normed linear spaces with norms γ and η, respectively. Then, $\mathscr{U} \otimes \mathscr{V}$ is a normed linear space whose norm ρ is defined on any $\theta \in \mathscr{U} \otimes \mathscr{V}$ by (1). The completion of $\mathscr{U} \otimes \mathscr{V}$ (see Appendix D4) is denoted by $\mathscr{U} \hat{\otimes} \mathscr{V}$ and is the collection of all equivalence classes (equivalence taken in the sense of Cauchy sequences under the norm ρ) of series of the form $\sum_{k=1}^{\infty} \phi_k \otimes \psi_k$, where $\phi_k \in \mathscr{U}$, $\psi_k \in \mathscr{V}$, and

$$\sum_{k=1}^{\infty} \gamma(\phi_k)\eta(\psi_k) < \infty.$$

Thus, $\mathscr{U} \hat{\otimes} \mathscr{V}$ is a Banach space. Its norm ρ is given on any $\theta \in \mathscr{U} \hat{\otimes} \mathscr{V}$ by

$$\rho(\theta) = \inf\left\{\sum_{k=1}^{\infty} \gamma(\phi_k)\eta(\psi_k): \quad \theta = \sum_{k=1}^{\infty} \phi_k \otimes \psi_k\right\}.$$

Appendix G

The Bochner Integral

G1. Let T be a nonvoid set. A nonvoid collection $\mathfrak{C}$ of subsets of T is called a *σ-algebra of subsets of T* or simply a *σ-algebra in T* if the following two conditions hold:

(i) If $E \in \mathfrak{C}$, then $T \backslash E \in \mathfrak{C}$.
(ii) $\bigcup_{k=1}^{\infty} E_k \in \mathfrak{C}$ if every $E_k \in \mathfrak{C}$.

It follows that T and the empty set are both members of $\mathfrak{C}$, and that $\bigcap_{k=1}^{\infty} E_k \in \mathfrak{C}$ if every $E_k \in \mathfrak{C}$. Also, $\Omega \backslash \Lambda \in \mathfrak{C}$ if $\Omega, \Lambda \in \mathfrak{C}$. The members of $\mathfrak{C}$ are called *measurable sets.*

G2. Given any collection $\mathfrak{S}$ of subsets of T, there exists a smallest σ-algebra $\mathfrak{C}$ of subsets of T such that $\mathfrak{S} \subset \mathfrak{C}$. $\mathfrak{C}$ is said to be the *σ-algebra generated by* $\mathfrak{S}$. Now, assume that T is a topological space. Let $\mathfrak{C}_B$ be the σ-algebra generated by the collection of all open sets in T. The members of $\mathfrak{C}_B$ are called the *Borel sets in T.*

G3. Let T and X be two nonvoid sets and let $\mathfrak{C}$ and $\mathfrak{C}'$ be σ-algebras of subsets of T and X, respectively. The *product σ-algebra in* $T \times X$ is the smallest σ-algebra of subsets of the Cartesian product $T \times X$ that contains every set of the form $E \times E'$, where $E \in \mathfrak{C}$ and $E' \in \mathfrak{C}'$. This σ-algebra is denoted by $\mathfrak{C} \times \mathfrak{C}'$. (Thus, in this case, $\mathfrak{C} \times \mathfrak{C}'$ does not represent the Cartesian product of $\mathfrak{C}$ and $\mathfrak{C}'$.)

G4. A *positive measure* μ (or, on the other hand, a *complex measure*) is a mapping of a σ-algebra $\mathfrak{C}$ in T whose range is in $[0, \infty] \subset R_e^1$ (or, respectively, in C) and which is *σ-additive*. The last phrase is defined as follows. μ is called *σ-additive on* $\mathfrak{C}$ if, for every sequence $\{E_k\}_{k=1}^{\infty}$ of pairwise disjoint sets $E_k \in \mathfrak{C}$, we have that $\mu(\bigcup_k E_k) = \sum_k \mu(E_k)$. However, if this condition is required to hold only for finite collections $\{E_k\}_{k=1}^{n}$, then μ is called *additive on* $\mathfrak{C}$. μ is σ-additive on $\mathfrak{C}$ if and only if it is additive on $\mathfrak{C}$ and $\mu(\bigcup_{k=1}^{\infty} E_k) = \lim_{k \to \infty} \mu(E_k)$ for every increasing sequence $\{E_k\}_{k=1}^{\infty}$ in $\mathfrak{C}$ (i.e., $E_k \in \mathfrak{C}$ and $E_k \subset E_{k+1}$ for every k).

A positive measure μ is called *finite* if $\mu(T) < \infty$.

Throughout Appendix G, μ will be allowed to be either a positive or complex measure unless it is explicitly restricted to being just one of these types of measures, and we shall refer to μ as a *measure*.

G5. An important special case is *Lebesgue measure on the Borel subsets of* R^n. This is the measure that assigns to each interval $[x, y] \triangleq \{t \in R^n : x \leq t \leq y\}$ the value $\mathrm{vol}[x, y] \triangleq \prod_{k=1}^{n} (y_k - x_k)$, where $x = \{x_k\}_{k=1}^{n} \in R^n$ and $y = \{y_k\}_{k=1}^{n} \in R^n$. Lebesgue measure is a positive measure and is unique in the following sense. No other positive measure on the Borel subsets of R^n assigns the value $\mathrm{vol}[x, y]$ to every interval $[x, y] \subset R^n$.

G6. A set $N \in \mathfrak{C}$ such that $\mu(N) = 0$ is called a *μ-null set* and is said to be of *measure zero*. Let P denote a property that each point $t \in E \in \mathfrak{C}$ either has or does not have depending on the choice of t. We say that P holds *almost everywhere on* E or *for almost all* $t \in E$ when there exists a μ-null set N such that P holds for every point of $E \backslash N$.

G7. A *partition* π *of* $E \in \mathfrak{C}$ is a finite collection $\{E_k\}_{k=1}^{r}$ of pairwise disjoint sets $E_k \in \mathfrak{C}$ such that $E = \bigcup_{k=1}^{r} E_k$. Let μ be a complex measure on $\mathfrak{C}$. We define a function $|\mu|$ on $\mathfrak{C}$ into $[0, \infty]$ by

$$|\mu|(E) \triangleq \sup_{\pi} \sum_{k=1}^{r} |\mu(E_k)|, \qquad E \in \mathfrak{C},$$

where the supremum is taken over all partitions π of E. It is a fact that $|\mu|$ is a finite positive measure on $\mathfrak{C}$. Moreover, $|\mu(E)| \leq |\mu|(E) \leq |\mu|(T) < \infty$ for every $E \in \mathfrak{C}$. $|\mu|$ is called the *total-variation measure for* μ, and $|\mu|(T) \triangleq \mathrm{Var}\, \mu$ is called the *total variation of* μ *on* T. When the range of μ happens to be real and contained in $[0, \infty)$, we have that $|\mu| = \mu$.

Now, let μ be any positive measure on $\mathfrak{C}$. Thus, μ is now allowed to assign ∞ to some of the members of $\mathfrak{C}$. In this case, we set $|\mu| \triangleq \mu$.

G8. Let S be any subset of T. The *characteristic function* χ_S *of* S is the function on T defined by $\chi_S(t) = 1$ if $t \in S$ and $\chi_S(t) = 0$ if $t \notin S$. Let A be a complex Banach space. An *A-valued simple function f on T* is any function of the form $f = \sum_{k=1}^{r} a_k \chi_{E_k}$, where $a_k \in A$ and $\{E_k\}_{k=1}^{r}$ is a partition of T. The simple function f is said to be *integrable* or *μ-integrable* if $a_k = 0$ whenever $|\mu|(E_k) = \infty$. The *Bochner integral of f on T* is denoted by $\mathfrak{I}f \triangleq \int_T f\,d\mu \triangleq \int_T d\mu f$ and defined to be $\sum_{k=1}^{r} a_k \mu(E_k)$, where now we use the convention that, if a_k is the zero element $0 \in A$ and $\mu(E_k) = \infty$, then the product 0∞ is the zero element $0 \in A$.

G9. A function f on T into A is called *measurable* or *μ-measurable* if, for any $E \in \mathfrak{C}$ such that $|\mu|(E) \neq \infty$, there exists a sequence of A-valued simple functions on T that converges pointwise to $f(t)$ for almost all $t \in E$. In this case, $\|f(\cdot)\|_A$ is also measurable as a function on T into R.

In the special case where $T = R^n$, we have the following result. Every weakly continuous function on R^n [i.e., every function f from R^n into A such that $a'f(\cdot)$ is continuous for all $a' \in A'$, where A' is the dual of A] is measurable.

G10. Let $\{f_k\}_{k=1}^{\infty}$ be a sequence of integrable A-valued simple functions on T such that

$$\int_T \|f_k(\cdot) - f_j(\cdot)\|_A \, d|\mu| \to 0 \tag{1}$$

as k and j tend to ∞ independently. It is a fact that there exists an A-valued function f on T and a subsequence $\{h_k\}$ of $\{f_k\}$ such that $h_k(t) \to f(t)$ for almost all $t \in T$. There will be other such functions and subsequences. However, it is also a fact that g is another such function if and only if $f(t) = g(t)$ for almost all $t \in T$. Thus, the sequence $\{f_k\}$ determines a class F consisting of all functions each of which differs from f on no more than a μ-null set.

Still more is true. The sequence $\{\mathfrak{I}f_k\}_{k=1}^{\infty}$ of Bochner integrals of the f_k converges in A. Its limit is called the *Bochner integral $\mathfrak{I}f$ of f on T with respect to μ*, where f is any member of F. We also use the notation $\mathfrak{I}f \triangleq \int_T f\,d\mu \triangleq \int_T d\mu f$ and say that f is *integrable* or *μ-integrable on T*. The subscript T may be dropped when this leads to no confusion. The notation

$$\mathfrak{I}f \triangleq \int f(t)\,d\mu_t \triangleq \int d\mu_t\, f(t)$$

is used when it is useful to display the independent variable of the function f. When T is a Borel subset of R^n and μ is Lebesgue measure on the Borel subsets of T, we write $\mathfrak{I}f \triangleq \int_T f(t)\,dt$.

If f and g are both members of the same class F (i.e., if they differ on no more than a μ-null set, then $\mathfrak{I}f = \mathfrak{I}g$.

Furthermore, if $\{g_k\}$ is any other sequence of integrable A-valued simple functions on T satisfying

$$\int_T \|g_k(\cdot) - g_j(\cdot)\|_A \, d|\mu| \to 0$$

as $k, j \to \infty$ independently and determining f as above, then $\{\mathfrak{I}g_k\}$ also converges to $\mathfrak{I}f$. Thus, $\mathfrak{I}f$ is independent of the choice of the sequence $\{f_k\}$.

Finally, we mention that a function is μ-integrable if and only if it is $|\mu|$-integrable.

G11. If f is a μ-integrable A-valued function on T, then $\|f(\cdot)\|_A$ is a $|\mu|$-integrable nonnegative function on T. Moreover, we have the useful estimate

$$\|\mathfrak{I}f\|_A \leq \int_T \|f(\cdot)\|_A \, d|\mu|.$$

The right-hand side remains the same for every member f of a given class F.

G12. Let $\mathscr{D}$ denote the space of all complex-valued smooth functions on R^n of compact support. Also, let v be a function that assigns a complex number to each bounded Borel set in R^n in such a way that, for each sphere $S_m \triangleq \{t \in R^n \colon |t| \leq m\}$, where $m = 1, 2, \ldots,$ we have that v is a complex measure on the σ-algebra of all Borel subsets of S_m. Such a function v is called a *σ-finite complex measure*. Given any $\phi \in \mathscr{D}$, choose m such that supp $\phi \subset S_m$ and set $\int_{R^n} \phi \, dv \triangleq \int_{S_m} \phi \, dv$. Then, $\int_{R^n} \phi \, dv$ exists for each $\phi \in \mathscr{D}$ and is independent of the choice of S_m. Moreover, a knowledge of the values of $\int_{R^n} \phi \, dv$ for every $\phi \in \mathscr{D}$ uniquely determines the values that v assigns to all the bounded Borel sets in R^n. (See Schwartz, 1966, p. 25.)

Similarly, if both g and h are v-integrable A-valued functions on S_m for every m, then so, too, are $g\phi$ and $h\phi$ whenever $\phi \in \mathscr{D}$. We again set $\int_{R^n} g\phi \, dv \triangleq \int_{S_m} g\phi \, dv$, where S_m is chosen for the given ϕ as before. If $\int_{R^n} g\phi \, dv = \int_{R^n} h\phi \, dv$ for all $\phi \in \mathscr{D}$, then $g(t) = h(t)$ for almost all $t \in R^n$.

G13. Given T, $\mathfrak{C}$, μ, and A, the set of all integrable A-valued functions f on T can be partitioned into equivalence classes F as indicated in Appendix G10. If f and g are members of the same F, then $\int f \, d\mu = \int g \, d\mu$ and, in addition, $\int \|f(\cdot)\| \, d|\mu| = \int \|g(\cdot)\| \, d|\mu|$. We define $\int F \, d\mu \triangleq \int f \, d\mu$, where f is any member of F. The notation $L_1(T, \mathfrak{C}; \mu; A) = L_1(\mu; A)$ denotes the linear space of all equivalence classes F, and this space is a Banach space under the norm $\|\cdot\|_{L_1}$, where

$$\|F\|_{L_1} \triangleq \int_T \|f(\cdot)\|_A \, d|\mu|, \qquad f \in F.$$

When T is R^n, $\mathfrak{C}$ the collection of all Borel subsets of R^n, and μ Lebesgue measure, we denote $L_1(\mu, A)$ simply by $L_1(A)$, and $L_1(C)$ by L_1.

It is customary to represent F by any member $f \in F$ and to replace F by f in all manipulations on the members F of $L_1(\mu; A)$. In fact, we shall say that the members of $L_1(\mu; A)$ *are* the functions f and will maintain the tacit understanding that f should really be replaced by the equivalence class F to which f belongs.

G14. Let $\mathscr{G}(T, \mathfrak{C}; C)$ be the space of all complex-valued functions g on T that are the limits under the norm $\|\cdot\|_G$, where $\|g\|_G \triangleq \sup_{t \in T}|g(t)|$, of sequences of simple functions on T. Thus, every member of $\mathscr{G}(T, \mathfrak{C}; C)$ is a bounded function. If $f \in L_1(T, \mathfrak{C}; \mu; A)$ and $g \in \mathscr{G}(T, \mathfrak{C}; C)$, then $fg \in L_1(T, \mathfrak{C}; \mu; A)$.

Furthermore, if T, X, $\mathfrak{C}$, and $\mathfrak{C}'$ are as in Appendix G3 and if $l \in G(T \times X, \mathfrak{C} \times \mathfrak{C}'; C)$, then, for each fixed $x \in X$, we have $l(\cdot, x) \in G(T, \mathfrak{C}; C)$.

G15. If f is a μ-measurable function on T into A and if $\|f(t)\|_A \leq g(t)$, where $g \in L_1(T, \mathfrak{C}; \mu; R)$, then $f \in L_1(T, \mathfrak{C}; \mu; A)$.

G16. Let $L_1{}^0(T, \mathfrak{C}; \mu; A) = L_1{}^0(\mu; A)$ denote the linear space of all integrable A-valued simple functions on T. [Here again, it is understood that the members of $L_1{}^0(\mu; A)$ are really equivalence classes of functions differing from the simple functions on no more than μ-null sets.] Given any $f \in L_1(\mu; A)$, we can choose a sequence $\{h_k\}_{k=1}^{\infty} \subset L_1{}^0(\mu; A)$ such that $h_k \to f$ in $L_1(\mu; A)$ and $h_k(t) \to f(t)$ almost everywhere on T. (Any one of the subsequences mentioned in the first paragraph of Appendix G10 will do.) Consequently, $L_1{}^0(\mu; A)$ is dense in $L_1(\mu; A)$. Moreover, $L_1(\mu; A)$ is the completion of $L_1{}^0(\mu; A)$ under the norm $\|\cdot\|_{L_1}$.

G17. Let M be a continuous linear mapping of the complex Banach space A into another such space B. If $f \in L_1(T, \mathfrak{C}; \mu; A)$, then $Mf(\cdot) \in L_1(T, \mathfrak{C}; \mu; B)$ and

$$M \int f\, d\mu = \int Mf(t)\, d\mu_t.$$

In particular, if $f \in L_1(T, \mathfrak{C}; \mu; [A; B])$, $a \in A$, and $b' \in B'$, where B' is the dual of B, then

$$b'\left[\left(\int f\, d\mu\right)a\right] = \int b'[f(t)a]\, d\mu_t.$$

G18. The following is the *theorem of dominated convergence*. If $\{f_k\}_{k=1}^{\infty} \subset L_1(\mu; A)$, if $\|f_k(t)\| \leq g(t)$ for each k, where $g \in L_1(\mu; R)$, and if $f_k(t) \to f(t)$ for almost all $t \in T$, then $f \in L_1(\mu; A)$ and, as $k \to \infty$, $\int_T \|f_k(t) - f(t)\| \, d|\mu_t| \to 0$ and $\int_T f_k \, d\mu \to \int_T f \, d\mu$.

G19. Let $p \in R$ and $p \geq 1$. Let $\{f_k\}_{k=1}^{\infty}$ be a sequence of integrable A-valued simple functions on T such that

$$\int \|f_k(\cdot) - f_j(\cdot)\|_A{}^p \, d|\mu| \to 0$$

as k and j tend to ∞ independently. Then, there exists an equivalence class F consisting of all A-valued measurable functions f on T such that any two members of F are equal to each other for almost all $t \in F$ and, for any $f \in F$, the integral

$$\int \|f_k(\cdot) - f(\cdot)\|_A{}^p \, d|\mu|$$

exists and tends to zero as $k \to \infty$. Moreover, $\|f(\cdot)\|_A{}^p \in L_1(T, \mathfrak{C}; |\mu|; R)$, and

$$\int \|f(\cdot)\|_A{}^p \, d|\mu| = \lim_{k \to \infty} \int \|f_k(\cdot)\|_A{}^p \, d|\mu|.$$

The space of all such equivalence classes F is denoted by $L_p(T, \mathfrak{C}; \mu; A) = L_p(\mu; A)$, and it is a Banach space under the norm $\|\cdot\|_{L_p}$, where

$$\|F\|_{L_p} \triangleq \left[\int \|f(\cdot)\|_A{}^p \, d|\mu| \right]^{1/p}, \qquad f \in F.$$

As before, we shall represent F by any one of its members f. Moreover, we shall speak of $L_p(\mu; A)$ as consisting of all f in all F, the partitioning of the space of all such f into equivalence classes being understood.

It is a fact that $f \in L_p(\mu; A)$ if and only if f is an A-valued measurable function on R^n and $\|f(\cdot)\|_A \in L_p(\mu; R)$.

In the special case where $\mathfrak{C}$ is the σ-algebra of Borel subsets of R^n and μ is Lebesgue measure, we denote $L_p(\mu; A)$ by $L_p(A)$ and $L_p(C)$ by L_p.

G20. The following is an extension of Holder's inequality. Let $p, q \in R$ be such that $p > 1$ and $p^{-1} + q^{-1} = 1$. Moreover, let $f \in L_p(T, \mathfrak{C}; \mu; A)$ and $g \in L_q(T, \mathfrak{C}; \mu; C)$. Then, $fg \in L_1(T, \mathfrak{C}; \mu; A)$, and

$$\left\| \int fg \, d\mu \right\|_A \leq \int \|f(\cdot)\| \, |g(\cdot)| \, d|\mu| \leq \|f\|_{L_p} \|g\|_{L_q}.$$

References

Akhiezer, N. I., and Glazman, I. M. (1963). "Theory of Linear Operators in Hilbert Space," Vol. II. Ungar, New York.

Berberian, S. K. (1961). "Introduction to Hilbert Space." Oxford Univ. Press, London and New York.

Bogdanowicz, W. (1961). A proof of Schwartz's theorem on kernels, *Studia Math.* **20**, 77–85.

Carlin, H. J., and Giordano, A. B. (1964). "Network Theory." Prentice-Hall, Englewood Cliffs, New Jersey.

Cioranescu, I. (1967). Familii compozabile de Operatori, *Stud. Cerc. Mat.* **19**, 449–454.

Copson, E. T. (1962). "Theory of Functions of a Complex Variable." Oxford Univ. Press, London and New York.

Cristescu, R. (1964). Familles composables des distributions et systems physiques lineaires, *Rev. Roumaine Math. Pures Appl.* **9**, 703–711.

Cristescu, R., and Marinescu, G. (1966). "Unele Aplicatii Ale Teoriei Distributiilor." Editura Academiei Republicii Socialiste Romania, Bucharest.

D'Amato, L. (1971). The lossless Hilbert port, Ph.D. Thesis, State Univ. of New York at Stony Brook, New York.

Dinculeanu, N. (1967). "Vector Measures." Pergamon, Oxford.

Dolezal, V. (1970). A representation of linear continuous operators on testing functions and distributions, *SIAM J. Math. Anal.* **1**, 491–506.

Dunford, N., and Schwartz, J. T. (1966). "Linear Operators," Pt. I. Wiley (Interscience), New York.

Ehrenpreis, L. (1956). On the theory of kernels of Schwartz, *Proc. Amer. Math. Soc.* **7**, 713–718.

Gask, H. (1960). A proof of Schwartz's kernel theorem, *Math. Scand.* **8**, 327–332.

Gelfand, I. M., and Vilenkin, N. Ya. (1964). "Generalized Functions," Vol. 4. Academic Press, New York.

Gross, B. (1953). "Mathematical Structure of the Theories of Viscoelasticity." Hermann, Paris.

Hackenbroch, W. (1968). Integration vektorwertiger Funktionen nach operatorwertigen Massen, *Math. Z.* **105**, 327–344.

Hackenbroch, W. (1969). Integraldarstellung einer Klasse dissipativer linearer Operatoren, *Math. Z.* **109**, 273–287.

Hackenbroch, W. (To be published.) Passivity and causality over locally compact Abelian groups.

Hille, E., and Phillips, R. S. (1957). "Functional Analysis and Semigroups." (Colloq. Publ. Vol. 31). Amer. Math. Soc., Providence, Rhode Island.

Horváth, J. (1966). "Topological Vector Spaces and Distributions," Vol. I. Addison-Wesley, Reading, Massachusetts.

Jones, D. S. (1964). "The Theory of Electromagnetism." Macmillan, New York.

Kaplan, W. (1952). "Advanced Calculus." Addison-Wesley, Reading, Massachusetts.

König, H. (1959). Zur Theorie der linearen dissipativen Transformationen, *Arch. Math.* 10, 447–451.

König, H., and Meixner, J. (1958). Lineare Systeme und lineare Transformationen, *Math. Nachr.* **19**, 265–322.

König, H., and Zemanian, A. H. (1965). Necessary and sufficient conditions for a matrix distribution to have a positive-real Laplace transform, *SIAM J. Appl. Math.* **13**, 1036–1040.

Kritt, B. (1968). Spectral decomposition of positive and positive-definite distributions of operators, *Bull. Acad. Polon. Sci. Sér. Sci. Math Astronom. Phys.* **16**, 865–870.

Kurepa, S. (1965). Quadratic and sesquilinear functionals, *Glasnik Mat.-Fiz. Astronom. Ser. II Družlvo Mat. Fiz. Hrvatske* **20**, 79–92.

Love, E. R. (1956). Linear superposition in viscoelasticity and theories of delayed effects, *Austral. J. Phys.* **9**, 1–12.

McMillan, B. (1952). Introduction to formal realizability theory, *Bell System Tech. J.* **31**, 217–279, 541–600.

Meidan, R. (1970). The mathematical foundation of translation-varying linear operators, Ph. D. Thesis, State University of New York at Stony Brook.

Meidan, R. (1972). Translation-varying linear operators, *SIAM J. Appl. Math.* **22**, 419–436.

Meixner, J. (1954). Thermodynamische Erweiterung der Nachwirkingstheorie, *Z. Phys.* **139**, 30–43.

Newcomb, R. W. (1966). "Linear Multiport Synthesis." McGraw-Hill, New York.

Pondelicek, B. (1969). A contribution to the foundation of network theory using distribution theory, *Czechoslovak Math. J.* **19** 697–710.

Robertson, A. P., and Robertson, W. (1964). "Topological Vector Spaces." Cambridge Univ. Press, London and New York.

Rudin, W. (1966). "Real and Complex Analysis." McGraw-Hill, New York.

Sabac, M. (1965). Familii compozabile de distributti si transformata Fourier, *Stud. Cerc. Mat.* **17**, 607–613.

Schaefer, H. H. (1966). "Topological Vector Spaces." Macmillan, New York.

Schwartz, L. (1957). Théorie des distributions a valeurs vectorielles, *Ann. Inst. Fourier (Grenoble)* **7**, 1–139.

Schwartz, L. (1959). Théorie des distributions a valeurs vectorielles, *Ann. Inst. Fourier (Grenoble)* **8**, 1–206.

Schwartz, L. (1966). "Théorie des Distributions." Hermann, Paris.

Schwindt, R. (1965). Lineare Transformationen vektorwertige Funktionen, Ph.D. Thesis, Univ. of Cologne, Germany, 1965.

Sebastião e Silva, J. (1960). Sur la définition et la structure des distributions vectorielles, *Portugal. Math.* **19**, 1–80; Errata: pp. 243–244.

Sz.-Nagy, B., and Foias, C. (1970). "Harmonic Analysis of Operators on Hilbert Spaces." Amer. Elsevier, New York.

Toll, J. S. (1956). Causality and the dispersion relation: Logical foundations, *Phys. Rev.* **104**, 1760–1770.

Treves, F. (1967). "Topological Vector Spaces, Distributions, and Kernels." Academic Press, New York.

Vladimirov, V. S. (1969a). On the theory of linear passive systems, *Soviet Math. Dokl.* **10**, 733–736.

Vladimirov, V. S. (1969b). Linear passive systems, *Theor. Math. Phys. (USSR)* **1**, 67–94.

Wexler, D. (1966). Solutions périodiques et presque-périodic des systemes d'equations différentielles lineáires en distributions, *J. Differential Equations* **2**, 12–32.

Willems, J. C. (1971). "The Analysis of Feedback Systems." MIT Press, Cambridge, Massachusetts.

Williamson, J. H. (1962). "Lebesque Integration." Holt, New York.

Wohlers, M. R. (1969). "Lumped and Distributed Passive Networks." Academic Press, New York.

Wohlers, M. R., and Beltrami, E. J. (1965). Distribution theory as the basis of generalized passive-network analysis, *IEEE Trans. Circuit Theory* **CT-12**, 164–170.

Wu, T. T. (1954). *Causality and the Radiation condition*, Tech. Rep. No. 211. Cruft Lab., Harvard Univ. Cambridge, Massachusetts. (Also, *J. Math. Phys.*, **3** (1962), 262.)

Youla, D. C., and Castriota, L. J. and Carlin, H. J. (1959). Bounded real scattering matrices and the foundations of linear passive network theory, *IEEE Trans. Circuit Theory* **CT-6**, 102–124.

Zaanen, A. C. (1967). "Integration." North-Holland Publ., Amsterdam.

Zemanian, A. H. (1963). An *n*-port realizability theory based on the theory of distributions, *IEEE Trans. Circuit Theory* **CT-10**, 265–274.

Zemanian, A. H. (1965). "Distribution Theory and Transform Analysis." McGraw-Hill, New York.

Zemanian, A. H. (1968a). "Generalized Integral Transformations." Wiley (Interscience), New York.

Zemanian, A. H. (1968b). The postulational foundations of linear systems, *J. Math. Anal. Appl.* **24**, 409–424.

Zemanian, A. H. (1970a). The Hilbert port, *SIAM J. Appl. Math.* **18**, 98–138.

Zemanian, A. H. (1970b). A scattering formulism for the Hilbert port, *SIAM J. Appl. Math.* **18**, 467–488.

Zemanian, A. H. (1970c). The composition of Banach-space-valued distributions, College of Eng. Tech. Rep. 177. State University of New York at Stony Brook, New York.

Zemanian, A. H. (1972a). Banach systems, Hilbert ports, and ∞-ports, *Proc. Advan. Study Inst. on Network Theory*, Knokke, Belgium, *September 1969* (R. Boite, ed.). Gordon & Breach, New York.

Zemanian, A. H. (1972b). Realizability conditions for time-varying and time-invariant Hilbert ports, *SIAM J. Appl. Math.* **22**, 612–628.

Index of Symbols

A, 4
$\mathfrak{A}$, 164
B, 4
$B(\mathscr{U}, \mathscr{V}; \mathscr{W})$, 196
$\mathscr{B}(B)$, 107
$\mathfrak{B} = \mathfrak{B}_{\mathfrak{N}}$, 153
$\hat{\mathfrak{B}} = \hat{\mathfrak{B}}_{\mathfrak{N}}$, 153
$\mathfrak{B}_P$, 43
C^n, 3
C_+, 131
$\mathscr{C}_K{}^m(A)$, 57
$\mathfrak{C}$, 23
$\mathfrak{C}_\infty$, 29
diam, 4
D^k, 14
$\mathscr{D}(A)$, 50
$\mathscr{D}_K(A)$, 50
$\mathscr{D}^m(A)$, 65
$\mathscr{D}_K{}^m(A)$, 50
$\mathscr{D}_{L_p}(A)$, 72
$\mathscr{D}^m_{R^n}(A)$, 50
$\mathscr{D}_+{}^m(A)$, 67
$\mathscr{D}_-{}^m(A)$, 67
$\mathscr{E}^m(A)$, 65
$\mathfrak{F}$, 155
$G(A)$, 24
$\mathscr{G}(A)$, 25
$\mathscr{G}(T, \mathfrak{C}; C)$, 220
$\mathscr{G}(T \times X, \mathfrak{C} \times \mathfrak{C}'; C)$, 40
$\mathscr{G}_g(H) = \mathscr{G}_g(T, \mathfrak{C}; H)$, 46
$\mathscr{G}_0(A) = \mathscr{G}_0(T, \mathfrak{C}; A)$, 24
H, 4
$\mathscr{H}$, 81
inf, 4
I, 145
$\mathscr{I}(A)$, 65
$\mathscr{I}^m(A)$, 65
$\mathscr{I}_J{}^m(A)$, 64
$\overline{\lim}$, 4
$\underline{\lim}$, 4

$L_1(\mu; A) = L_1(T, \mathfrak{C}; \mu; A)$, 219
$L_1{}^0(\mu; A) = L_1{}^0(T, \mathfrak{C}; \mu; A)$, 220
$\mathrm{L}_1^{\mathrm{loc}}(H)$, 150
$\mathrm{L}_p(\mu; A) = L_p(T, \mathfrak{C}; \mu; A)$, 221
$\mathscr{L}_{c,d}(A)$, 118
$\mathscr{L}^m_{c,d}(A)$, 66
$\mathscr{L}(w, z; A)$, 118
$\mathscr{L}^m(w, z; A)$, 66
$\mathfrak{L}$, 118
max, 4
min, 4
$\mathscr{N}[T]$, 147
$\mathfrak{N}$, 78
$\mathscr{Q}$, 23
R_+, 3
R^n, 3
$R_e{}^n$, 3
$\mathscr{R}[T]$, 147
SSVar, 24
sup, 4
supp, 4, 55
SVar, 24
$\mathscr{S}^m(A)$, 66
Var, 24
vol, 4
$\mathscr{Z}$, 156
$\gamma_{c,d,k}$, 118
γ_k, 50
$\gamma_{p,k}$, 72
γ_Φ, 53, 68
$\delta_{j,k}$, 57
∂_k, 12
$\kappa_{c,d}$, 66, 118
$\varkappa_{p,\zeta}$, 82
$\rho_{J,p}$, 64
σ_τ, 54
$\overline{\Omega}$, 2
$\mathring{\Omega}$, 2

Some Additional Special Symbols

1_+, 58, 73
$\Delta t|_k$, 12
$\triangleq$, 4
$\diamondsuit$, 4
$\bullet$, 90
$*$, 97
$\sim$, 155
$\odot$, 37, 61, 69
$\hat{\odot}$, 37
$\otimes$, 214
$\hat{\otimes}$, 215
$|\cdot|$, 10, 14, 24
$[\cdot]$, 4
$\|\cdot\| = \|\cdot\|_A$, 204
$\|\cdot\|_1$, 37
$(\cdot, \cdot)$, 4, 205
$(\cdot, \cdot]$, 4
$[\cdot, \cdot)$, 4
$[\cdot, \cdot]$, 4
$[\cdot; \cdot]$, 3, 208
$[\cdot; \cdot]^s$, 68
$[\cdot; \cdot]^\sigma$, 3, 68, 208
$\langle \cdot, \cdot \rangle$, 3
$\cdot \backslash \cdot$, 2
$\cdot \curvearrowright \cdot$, 2
$\cdot \mapsto \cdot$, 2

Index

A

[A; B]-valued distribution, 52
Absolutely convergent series, 204
Absolutely convex set, 196
Absorbent set, 196
Additive set function, 24
Adjoint operator, 210
Admittance operator, 80
Akhiezer, N. I., 178
Algebraic dual, 196
Analytic function, 17–19
  strongly, 18
  weakly, 17–18
Antilinear mapping, 197
Automorphism, 207

B

B-valued distribution, 52
Baire's category theorem, 204
Balanced set, 196
Banach space, 204
Banach system, 79
Base of continuous seminorms, 203
Base of neighborhoods, 199
Beltrami, E. J., 130, 131
Berberian, S. K., 147, 190, 193
Bijection, 3
Bilinear form, 196
Bilinear mapping, 196
Binomial coefficient, 15
Black box approach, xi
Bochner integral, 218
Bochner–Schwartz theorem, 156
Bogdanowicz, W., 81
Borel set, 216
Boundary of a set, 198
Bounded function, 5, 208
Bounded* function, 131
Bounded*-real function, 138
Bounded set, 203
Bounded topology, 53, 208

C

Canonical bilinear mapping, 214
Canonical injection, 200
Carlin, H. J., 80, 137, 144, 146, 147, 151
Cartesian product, 2, 200
Castriota, L. J., 137, 151
Cauchy filter, 204
Cauchy sequence, 204
Cauchy's integral formula, 21
Cauchy's theorem, 20
Causality, 93
Change-of-variable formula, 111–112
Characteristic function, 218
Cioranescu, I., 92
Closed graph theorem, 209
Closed set, 198
Closure, 198
Closure property, 212
Compact set, 4
Complete orthonormal set, 205
Complete space, 204
Completion of a space, 207
Complex conjugate of an operator, 138
Complexification, 137
Complex measure, 217
Composition operator, 90
Composition product, 90
Conditions E, 97
Constant distribution, 108
Continuous mapping, 199
Contour integration, 20–22
Contractive operator, 129
Convergent filter, 204
Convergent sequence, 199–200
Convex hull, 196
Convex set, 196
Convolution operator, 98, 112–115
  commutativity with shifting and differentiation, 104
Convolution product, 97
Copson, E. T., 20
Counting measure, 40
Cristescu, R., 92

D

D'Amato, L., 144, 192
Delta function, 55
Dense set, 198
Dependent variable, 2
Diameter of a set, 4
Differentiation, 12–17
  generalized, 54, 68
Dinculeanu, N., 24
Direct product, 110
Distribution, 52
  generated by a measure, 56
  independent of certain coordinates, 110–111
  L_p-type, 75
  regular, 56
Dolezal, V., 92
Domain, 2
Dominated convergence, theorem of, 220–221
Dual space, 208
Dunford, N., 33, 194

E

Ehrenpreis, L., 81
Equicontinuous set of mappings, 208
Euclidean space, 3
Exchange formula, 119

F

Filter, 204
Finite-dimensionally-ranging function, 37
Finite-dimensional subspace, 195
Finite measure, 217
Foias, C., 141
Form, 208
Fourier transform, 155
Fréchet space, 204
Function, 3
Functional, 208

G

Gask, H., 81
Gelfand, I. M., 81, 156
Generalized differentiation, 54, 68
Generalized function, 67
  [A; B]-valued, 67
  B-valued, 67

Generating family of seminorms, 203
Giordano, A. B., 80, 144, 146, 147
Glazman, I. M., 178
Graph, 2
Gross, B., xi

H

Hackenbroch, W., 23, 29, 150, 153, 157, 164, 174, 175
Hausdorff space, 199
Hilbert n-port, 143
Hilbert port, 78, 80
Hilbert space, 205
Hille, E., 194
Holder's inequality, 221
Horvath, J., 194

I

Imaginary part of an operator, 138
Impedance operator, 81
Improper integral, 11
Independent variable, 2
Induced topology, 200
Inductive-limit space, 211
∞-port, 79
Injection, 3
Inner product, 205
Integrable function, 30
Integration by parts, 17
Interior point, 198
Interval in R^n, 4
Isometric operator, 140
Isomorphism, 207

J

Jones, D. S., 78

K

Kaplan, W., 112
Kernel of $\mathfrak{N}$, 91
Kernel operator, 90
Kernel representation, 91
Kernel theorem, 85
König, H., 150, 153, 187
Kritt, B., 174
Kurepa, S., 180

L

L_p-type distribution, 75, 125–128
L_p-type testing function, 72
Laplace transform, 118
  strip of definition for, 118
Laplace-transformable distribution, 118
Lebesgue measure, 217
Leibniz's rule, 15
Linear combination, 195
Linear form, 196
Linearly independent set, 196
Linear mapping, 196
Linear space, 194
Locally convex space, 202
Locally essentially bounded function, 56
Local mapping, 157
Lossless Hilbert port, 139–143
Lossless operator, 140
Lossless scattering transform, 145
Love, E. R., xi

M

McMillan, B., xi
Mapping, 3
Marinescu, G., 92
Matched port, 146
Measurable function, 218
Measurable set, 216
Measure, 217
Meidan, R., 80, 92
Meixner, J., xi, 150
Metric space, 200
Metrizable space, 200
Multinorm, 203

N

Neighborhood, 199
Nested closed cover, 64
Newcomb, R. W., xi, 137

Norm, 202
Normal space, 65
Normed linear space, 204
Nowhere dense set, 204
Null set of a distribution, 55

O

One-to-one mapping, 3
Open set, 198
Operator, 3
Operator-valued measure, 26
Order of a derivative, 14
Orthonormal set, 205

P

Parseval's equation, 136, 155, 205
Partition, 217
Passive operator, 150
Phillips, R. S., 194
π-topology, 214
Pointwise topology, 53, 68, 208
Polarization identity, 197
Pondelicek, B., 92
Positive-definite distribution, 156, 174
Positive* mapping, 178
Positive measure 217
Positive operator, 210
PO measure, 26
Positive*-real mapping, 181
Positive sesquilinear form, 160–168, 197, 210
Primitive, 108
Principle of uniform boundedness, 209
Product σ-algebra, 216–217
Product topology, 200
Projection, 58

R

Range, 2
Real convolution operator, 138
Real distribution, 138
Real linear space, 197
Real part of an operator, 138
Real signal, 138
Realizability theory, xi
Reciprocal scattering transform, 145
Rectangular partition, 6
Refinement of a partition, 6
Regular distribution, 56
Regularization, 106–108
Relation, 2
ρ-type testing function space, 65
  degenerate, 65
Riemann integration, 6–11
Riemann sum, 6
Riesz-Fischer theorem, 205
Robertson, A. P., 194
Robertson, W., 194
Rudin, W., 33, 40, 179, 194

S

$\mathfrak{S}$-set, 65
$\mathfrak{S}$-topology, 53, 68
Sabac, M., 92
Scalar semivariation, 24
Scattering operator, 80
Scatter-passive operator, 129
Scatter-semipassive operator, 129
Schaeffer, H. H., 194
Schwarz inequality, 197, 205
Schwartz, J. T., 33, 194
Schwartz, L., 49, 81, 92, 110, 128, 156, 219
Schwindt, R., 178
Sebastiao e Silva, J., 57
Second dual space, 209
Self-adjoint operator, 210
Seminorm, 202
Semipassive operator, 150
Semivariation, 24
Separable space, 198
Separated space, 198–199
Separately continuous bilinear mapping, 213
Sequentially complete space, 204
Sequentially continuous mapping, 200
Sequentially dense set, 200
Sesquilinear form, 197
Sesquilinear mapping, 197
Shifting operator, 54
σ-additive measure, 217
σ-additive set function, 26

σ-algebra, 216
σ-finite complex measure, 219
σ-finite PO measure, 30
Simple function, 24, 218
Skew-adjoint operator, 210
Smooth function, 14
Span, 195
Strict inductive-limit space, 212
Strongly measurable function, 56
Strong operator topology, 209
Subspace, 195
Support, 4
  of a distribution, 55
Surjection, 3
System function, 114
Sz.-Nagy, B., 141

T

Tempered complex measure, 157
Tempered distribution, generated by a PO measure, 163
Tempered positive measure, 156
Tempered PO measure, 162
Tensor product, 214
Testing functions of rapid descent, 66
Time invariance, 112
Time-varying operator, 112
Toll, J. S., xi
Topological linear space, 201
Topological space, 198
Topology, 198
Topology of uniform convergence on $\mathfrak{S}$-sets, 68
Total subset, 203
Total variation, 24, 217
Transformation, 2
Translation-invariance, 112
Translation-invariant sesquilinear form, 161
Translation operator, 54
Translation-varying operator, 112
Treves, F., 194

U

Uniform continuity, 5, 97
Uniform convergence, 9
Uniform operator topology, 209
Unitary operator, 142
Unit impulse, 114
Unit-impulse response, 114

V

Vilenkin, N. Ya., 81, 156
Vladimirov, V .S., 186
Volume, 4

W

Weak operator topology, 209
Weak topology, 208
Wexler, D., 92
Willems, J. C., 157
Williamson, J. H., 136
Wohlers, M. R., xi, 130, 131
Wu, T. T., xi

Y

Youla, D. C., 137, 151

Z

Zaanen, A. C., 178, 194
Zemanian, A. H., 55, 62, 66, 79, 83, 92, 108, 114, 117, 119, 121, 131, 137, 150, 186

A CATALOG OF SELECTED

DOVER BOOKS

IN SCIENCE AND MATHEMATICS

A CATALOG OF SELECTED
DOVER BOOKS
IN SCIENCE AND MATHEMATICS

GEOMETRY OF COMPLEX NUMBERS, Hans Schwerdtfeger. Illuminating, widely praised book on analytic geometry of circles, the Moebius transformation, and two-dimensional non-Euclidean geometries. 200pp. 5⅝ × 8¼.
63830-8 Pa. $8.95

MECHANICS, J.P. Den Hartog. A classic introductory text or refresher. Hundreds of applications and design problems illuminate fundamentals of trusses, loaded beams and cables, etc. 334 answered problems. 462pp. 5⅜ × 8½. 60754-2 Pa. $9.95

TOPOLOGY, John G. Hocking and Gail S. Young. Superb one-year course in classical topology. Topological spaces and functions, point-set topology, much more. Examples and problems. Bibliography. Index. 384pp. 5⅝ × 8¼.
65676-4 Pa. $9.95

STRENGTH OF MATERIALS, J.P. Den Hartog. Full, clear treatment of basic material (tension, torsion, bending, etc.) plus advanced material on engineering methods, applications. 350 answered problems. 323pp. 5⅜ × 8½. 60755-0 Pa. $8.95

ELEMENTARY CONCEPTS OF TOPOLOGY, Paul Alexandroff. Elegant, intuitive approach to topology from set-theoretic topology to Betti groups; how concepts of topology are useful in math and physics. 25 figures. 57pp. 5⅜ × 8½.
60747-X Pa. $3.50

ADVANCED STRENGTH OF MATERIALS, J.P. Den Hartog. Superbly written advanced text covers torsion, rotating disks, membrane stresses in shells, much more. Many problems and answers. 388pp. 5⅜ × 8½. 65407-9 Pa. $9.95

COMPUTABILITY AND UNSOLVABILITY, Martin Davis. Classic graduate-level introduction to theory of computability, usually referred to as theory of recurrent functions. New preface and appendix. 288pp. 5⅜ × 8½. 61471-9 Pa. $7.95

GENERAL CHEMISTRY, Linus Pauling. Revised 3rd edition of classic first-year text by Nobel laureate. Atomic and molecular structure, quantum mechanics, statistical mechanics, thermodynamics correlated with descriptive chemistry. Problems. 992pp. 5⅜ × 8½. 65622-5 Pa. $19.95

AN INTRODUCTION TO MATRICES, SETS AND GROUPS FOR SCIENCE STUDENTS, G. Stephenson. Concise, readable text introduces sets, groups, and most importantly, matrices to undergraduate students of physics, chemistry, and engineering. Problems. 164pp. 5⅜ × 8½. 65077-4 Pa. $6.95

THE HISTORICAL BACKGROUND OF CHEMISTRY, Henry M. Leicester. Evolution of ideas, not individual biography. Concentrates on formulation of a coherent set of chemical laws. 260pp. 5⅜ × 8½. 61053-5 Pa. $6.95

THE PHILOSOPHY OF MATHEMATICS: An Introductory Essay, Stephan Körner. Surveys the views of Plato, Aristotle, Leibniz & Kant concerning propositions and theories of applied and pure mathematics. Introduction. Two appendices. Index. 198pp. 5⅜ × 8½. 25048-2 Pa. $7.95

THE DEVELOPMENT OF MODERN CHEMISTRY, Aaron J. Ihde. Authoritative history of chemistry from ancient Greek theory to 20th-century innovation. Covers major chemists and their discoveries. 209 illustrations. 14 tables. Bibliographies. Indices. Appendices. 851pp. 5⅜ × 8½. 64235-6 Pa. $18.95